AF323490

AUTOMATED INEQUALITY PROVING AND DISCOVERING

AUTOMATED INEQUALITY PROVING AND DISCOVERING

Bican Xia
Peking University, China

Lu Yang
Chinese Academy of Sciences, China

World Scientific

NEW JERSEY · LONDON · SINGAPORE · BEIJING · SHANGHAI · HONG KONG · TAIPEI · CHENNAI · TOKYO

Published by

World Scientific Publishing Co. Pte. Ltd.

5 Toh Tuck Link, Singapore 596224

USA office: 27 Warren Street, Suite 401-402, Hackensack, NJ 07601

UK office: 57 Shelton Street, Covent Garden, London WC2H 9HE

Library of Congress Cataloging-in-Publication Data
Names: Xia, Bican. | Yang, Lu, 1936–
Title: Automated inequality proving and discovering / by Bican Xia (Peking University, China),
    Lu Yang (Chinese Academy of Sciences, China).
Description: New Jersey : World Scientific, 2017. | Includes bibliographical references and index.
Identifiers: LCCN 2016018215 | ISBN 9789814759113 (hardcover : alk. paper)
Subjects: LCSH: Inequalities (Mathematics)--Data processing. | Numerical analysis--
    Data processing. | Mathematical statistics--Data processing. | Algorithms.
Classification: LCC QA295 .X53 2017 | DDC 515/.26--dc23
LC record available at https://lccn.loc.gov/2016018215

British Library Cataloguing-in-Publication Data
A catalogue record for this book is available from the British Library.

Printed in Singapore

Preface

Inequalities, originated from human beings' daily activities, are significant tools in diverse disciplines of science, technology and engineering nowadays.

Although the development of the theory of inequalities could date back to very early time and a lot of mathematicians had made contributions before the 20th century, among whom we may find many great names — Cauchy, Gauss and Hilbert to name a few, it is generally admitted that the systematic research on inequalities started from the classic work *"Inequalities"* by Hardy, Littlewood and Pólya, which was first published in 1934 and whose second edition was published in 1952 with three appendices (ten pages) added. In 1961, another book with the same title *"Inequalities"* by Beckenbach and Bellman was published, which contained many new methods and applications developed between 1934 and 1961. *"Analytic Inequalities"* by Mitrinović, which was published in 1970 and covered many topics not included in the above two, was a good dictionary for analysts.

Classic books on inequalities like the three mentioned above are collections of important inequalities and various methods to prove those inequalities. Many methods presented there are beautiful but usually special so that they can hardly be used for proving other inequalities. Thanks to Tarski's work of 1951, which pointed out that elementary algebra and geometry are decidable, one knows that it is possible to design one algorithm to prove a wide class of polynomial inequalities. Since then, benefited from numerous researchers' work, there have been lots of algorithms and implementations for polynomial inequality proving and discovering.

Our book *"Automated Inequality Proving and Discovering"*, on one hand, is different from the classic books. It focuses on practical algorithms for automated inequality proving and discovering. On the other hand, it is different from those famous books on the theory and algorithms of real algebraic geometry — "*Real Algebraic Geometry*" by Bochnak, Coste and Roy, "*Algorithms in Real Algebraic Geometry*" by Basu, Pollack and Roy, and "*Positive Polynomials and Sums of Squares*" by Marshall to name but a few. Although some classical results and related work are briefly introduced in corresponding chapters, this book is far from a systematic or comprehensive introduction to all well-known theories and algorithms in the field. It is indeed a collection of practical algorithms for polynomial inequality proving and discovering developed by the authors and their collaborators in recent years.

Resultant, subresultant and pseudo-division, which are basic operations of elimination methods and will be frequently used in the first eight chapters, are introduced in Chapter 1. Chapters 2 and 3 are devoted to the description of algorithms for hierarchically triangularizing parametric polynomial systems and semi-algebraic systems into pairwise disjoint square-free regular systems, which act as the base of hierarchical strategy for real root classification. Some classical methods and the complete discrimination system for counting real roots of polynomials are contained in Chapter 4. Real root isolation for constant semi-algebraic systems is the main content of Chapter 5. Real root classification is a special kind of quantifier elimination problems. An algorithm for solving the problem based on hierarchical strategy and various applications of the algorithm are given in Chapter 6. Two improvements on the projection operators of cylindrical algebraic decomposition for detecting nonnegativity of polynomials and computing global infimum of polynomials are clarified in Chapter 7. Dimension-decreasing algorithm can handle radicals without introducing new variables and thus prove many inequalities on triangles efficiently. Chapter 8 describes the algorithm and many examples. Chapter 9 defines two classes of polynomials of which the problem of SOS (sum of squares of polynomials) decomposition can be split into smaller sub-problems. Successive difference substitution (SDS), which can work out large examples not tractable by other tools, has been developed into a complete method for determining the nonnegativity of forms. A complete description of SDS is presented in Chapter 10. Finally, Chapter 11 introduces algorithms for proving some kinds of inequalities beyond Tarski's model.

All the algorithms have been implemented by the authors or their collaborators and illustrated by lots of examples. Some of the implementations have been integrated in Maple and thus can be available easily.

We are especially grateful to all our collaborators for the great experience of discussing and working with them. We hope that all the cooperative research achievements in this book have been interpreted correctly.

Thanks a lot to our families for their love and encouragement which is the indispensable support to finish the writing on time.

The work of the authors have been supported continuously by the "973 program" and National Science Foundation of China. We gratefully acknowledge the support provided by NSFC Grants 11290141, 11271034 and 61532019, and the help by Mr. Don Mak and Ms. E. H. Chionh at World Scientific.

The authors

Contents

Chapter 1

Basics of Elimination Method

Pseudo-division and resultant are two basic tools used by many elimination methods and are also frequently used in many algorithms in this book. So, we begin with an introduction to some related concepts and results.

If not specified in this chapter, $\mathcal{R}$ is a domain and univariate polynomials are in $\mathcal{R}[x]$. The degree of $f \in \mathcal{R}[x]$ is denoted by $\deg(f, x)$ or $\deg(f)$.

1.1 Pseudo-division

If $\mathcal{K}$ is a field, the Euclidean division in the Euclidean domain $\mathcal{K}[x]$ is well-known. For two polynomials f and $g \neq 0$ in $\mathcal{K}[x]$, there exist $q, r \in \mathcal{K}[x]$ such that $f = qg + r$ and $\deg(r, x) < \deg(g, x)$. The polynomials q and r are called respectively the *quotient* and *remainder* of f divided by g and denoted by $\mathrm{quo}(f, g)$ and $\mathrm{rem}(f, g)$, respectively.

If $\mathcal{R}$ is a domain, the concept of division in $\mathcal{R}[x]$ is generalized to the so-called *pseudo-division*, for the element in $\mathcal{R}$ is not invertible in general.

Suppose

$$f = \sum_{i=0}^{m} a_i x^i \quad \text{and} \quad g = \sum_{i=0}^{l} b_i x^i$$

are polynomials in $\mathcal{R}[x]$ with $m \geq l$. Construct a matrix as follows.

$$M = \begin{pmatrix} b_l & \cdots & b_1 & b_0 & & & \\ & b_l & \cdots & b_1 & b_0 & & \\ & & \ddots & & & \ddots & \\ & & & b_l & \cdots & b_1 & b_0 \\ a_m & a_{m-1} & \cdots & \cdots & \cdots & a_1 & a_0 \end{pmatrix}, \tag{1.1}$$

1

where all the other entries are zero except those of the coefficients of f and g. The ith column of M can be viewed as indexed by x^{m-i+1}. That is to say,

$$
M \cdot \begin{pmatrix} x^m \\ x^{m-1} \\ \vdots \\ x \\ 1 \end{pmatrix} = \begin{pmatrix} x^{m-l}g \\ x^{m-l-1}g \\ \vdots \\ g \\ f \end{pmatrix}.
$$

If $\mathcal{R}$ is a field or, at least, b_l is invertible, perform *Gaussian elimination* on M to get the following matrix

$$
\begin{pmatrix}
b_l & \cdots & b_1 & b_0 & & & \\
& b_l & \cdots & b_1 & b_0 & & \\
& & \ddots & & \ddots & & \\
& & & b_l & \cdots & b_1 & b_0 \\
0 & \cdots & \cdots & 0 & r_{l-1} & \cdots & r_0
\end{pmatrix}.
\tag{1.2}
$$

Then, the last row of the matrix is the remainder, *i.e.* $r = \sum_{i=0}^{l-1} r_i x^i$ is the *remainder* of f divided by g, denoted as $r = \mathrm{rem}(f, g)$.

If b_l is not invertible, we apply *fraction-free Gaussian elimination* on M as follows: First, multiply the last row by b_l and minus the product of the first row by a_m. Suppose the ith step ($1 \le i < m - l + 1$) is completed and the ith entry of the last row is c_i, multiply the last row by b_l and minus the product of the ith row by c_i. After $m - l + 1$ such transformations, M is transformed into a matrix of the form (1.2).

Definition 1.1. Let notations be as above. If M is transformed into a matrix of the form (1.2) by the above procedure via fraction-free Gaussian elimination, we then define $r = \sum_{i=0}^{l-1} r_i x^i$ as the *pseudo-remainder* of f *pseudo-divided* by g and denote it by $\mathrm{prem}(f, g, x)$ or $\mathrm{prem}(f, g)$.

Obviously, there exists a polynomial $q \in \mathcal{R}[x]$ such that

$$
b_l^{m-l+1} f = qg + r.
\tag{1.3}
$$

We define q as the *pseudo-quotient* of f *pseudo-divided* by g and denote it by $\mathrm{pquo}(f, g, x)$ or $\mathrm{pquo}(f, g)$. The formula (1.3) is called the *pseudo-remainder formula*.

If $c_i = 0$ for some i in the above procedure, b_l is a common divisor of the two sides of the pseudo-remainder formula. Obviously, if we modify the above procedure and do not multiply the last row by b_l when $c_i = 0$, the exponent of b_l in formula (1.3) can be decreased.

Example 1.1. Suppose

$$f = 2x^3 - x^2 + 1, \ g = 3x^2 + x - 1.$$

We compute the (pseudo-)remainder of f (pseudo-)divided by g. Construct the matrix

$$M = \begin{pmatrix} 3 & 1 & -1 & 0 \\ 0 & 3 & 1 & -1 \\ 2 & -1 & 0 & 1 \end{pmatrix}.$$

Apply Gaussian elimination and fraction-free Gaussian elimination on M to transform it respectively into

$$M_1 = \begin{pmatrix} 3 & 1 & -1 & 0 \\ 0 & 3 & 1 & -1 \\ 0 & 0 & 11/9 & 4/9 \end{pmatrix} \text{ and } M_2 = \begin{pmatrix} 3 & 1 & -1 & 0 \\ 0 & 3 & 1 & -1 \\ 0 & 0 & 11 & 4 \end{pmatrix}.$$

Therefore

$$\operatorname{rem}(f, g, x) = \frac{11}{9}x + \frac{4}{9}, \ \operatorname{prem}(f, g, x) = 11x + 4.$$

Pseudo-remainders can also be expressed explicitly. Suppose $M = (m_{ij})_{t \times s}$ is a matrix over $\mathcal{R}$ where $t \leq s$. Define the *determinant polynomial* of M to be

$$\operatorname{detpol}(M) = \det(M^{(t)})x^{s-t} + \det(M^{(t+1)})x^{s-t-1} + \cdots + \det(M^{(s)}),$$

where $M^{(j)}$ is a $t \times t$ submatrix formed by the first $t-1$ and the jth columns of M and $\det(M^{(j)})$ is the determinant of $M^{(j)}$.

Suppose

$$A_i = \sum_{j=0}^{n_i} a_{ij} x^{n_i - j} \quad (1 \leq i \leq k)$$

is a finite list of polynomials and $t = 1 + \max(n_1, \ldots, n_k)$. We denote by $\operatorname{mat}(A_1, \ldots, A_k)$ the matrix $(m_{ij})_{k \times t}$, where m_{ij} is the coefficient of A_i in x^{t-j}.

Definition 1.2. The *determinant polynomial* of a list of polynomials $A_1, \ldots, A_k$ is defined to be

$$\operatorname{detpol}(A_1, \ldots, A_k) = \operatorname{detpol}(\operatorname{mat}(A_1, \ldots, A_k)).$$

Example 1.2. Suppose

$$A_1 = x^3 + 2x + 5, \quad A_2 = 3x^2 - x - 6, \quad A_3 = -x^4 + x^3.$$

Then $t = 5, k = 3$ and

$$\mathrm{mat}(A_1, A_2, A_3) = \begin{pmatrix} 0 & 1 & 0 & 2 & 5 \\ 0 & 0 & 3 & -1 & -6 \\ -1 & 1 & 0 & 0 & 0 \end{pmatrix}.$$

Thus

$$\mathrm{detpol}(A_1, A_2, A_3)$$

$$= \begin{vmatrix} 0 & 1 & 0 \\ 0 & 0 & 3 \\ -1 & 1 & 0 \end{vmatrix} x^2 + \begin{vmatrix} 0 & 1 & 2 \\ 0 & 0 & -1 \\ -1 & 1 & 0 \end{vmatrix} x + \begin{vmatrix} 0 & 1 & 5 \\ 0 & 0 & -6 \\ -1 & 1 & 0 \end{vmatrix}$$

$$= -3x^2 + x + 6.$$

The following proposition is important, which gives an explicit expression of pseudo-remainder.

Proposition 1.1. *Suppose $f, g \in \mathcal{R}[x]$ and $\deg(f) = m \geq l = \deg(g) > 0$, then*

$$\mathrm{detpol}(x^{m-l}g, x^{m-l-1}g, \ldots, xg, g, f) = \mathrm{prem}(f, g, x).$$

Proof. Assume $f = \sum_{i=0}^{m} a_i x^i$ and $g = \sum_{i=0}^{l} b_i x^i$. It is easy to see that $\mathrm{mat}(x^{m-l}g, x^{m-l-1}g, \ldots, xg, g, f)$ is just the matrix M in Eq. (1.1). Suppose $\mathrm{prem}(f, g, x) = \sum_{i=0}^{l-1} r_i x^i$, we need to prove that $r_j = M^{(m+1-j)}$ for $j = 0, 1, \ldots, l-1$. By the procedure of pseudo-division, $M^{(m+1-j)}$ is transformed into

$$\overline{M}^{(m+1-j)} = \begin{pmatrix} b_l & \cdots & b_1 & b_0 & & \\ & b_l & \cdots & b_1 & & \\ & & \ddots & & \ddots & \\ & & & b_l & \cdots & \\ 0 & \cdots & \cdots & 0 & r_j & \end{pmatrix}.$$

Obviously, $b_l^{m-l+1} \det(M^{(m+1-j)}) = \det(\overline{M}^{(m+1-j)}) = b_l^{m-l+1} r_j$. That completes the proof. $\qquad\square$

1.2 Resultants

Generally speaking, a *resultant* of two univariate polynomials $f, g \in \mathcal{R}[x]$ is a polynomial in the coefficients of f and g such that the polynomial being zero is a necessary (and sufficient) condition for the two polynomials to have *common zeros*. Herein, a *common zero* $\bar{x}$ of f and g is an element of a certain extension field of the quotient field of $\mathcal{R}$ such that $f(\bar{x}) = g(\bar{x}) = 0$.

The theory of resultant is a classical theory of elimination methods. There exist many well-known kinds of resultants, such as the Sylvester resultant, the Bézout resultant, the Dixon resultant and the Macaulay resultant. For our purpose in this book, we only introduce the Sylvester resultant, which not only has elegant properties but also fully demonstrate the idea and techniques of classical elimination methods. Moreover, the theory of subresultants introduced in the next subsection is based on the Sylvester matrix. So, in this book, if not specified, *resultant* always means the Sylvester resultant.

Suppose

$$f = a_0 x^m + a_1 x^{m-1} + \cdots + a_{m-1} x + a_m,$$
$$g = b_0 x^l + b_1 x^{l-1} + \cdots + b_{l-1} x + b_l, \tag{1.4}$$

are two polynomials in $\mathcal{R}[x]$ with positive degrees. Construct an $(m + l) \times (m + l)$ square matrix

$$\boldsymbol{S} = \left.\left(\begin{array}{ccccccc} a_0 & a_1 & \cdots & a_m & & & \\ & a_0 & a_1 & \cdots & a_m & & \\ & & \ddots & \ddots & & \ddots & \\ & & & a_0 & a_1 & \cdots & a_m \\ b_0 & b_1 & \cdots & b_l & & & \\ & b_0 & b_1 & \cdots & b_l & & \\ & & \ddots & \ddots & & \ddots & \\ & & & b_0 & b_1 & \cdots & b_l \end{array}\right)\begin{array}{l} \left.\vphantom{\begin{array}{c} a \\ a \\ a \\ a \end{array}}\right\} l \\ \left.\vphantom{\begin{array}{c} b \\ b \\ b \\ b \end{array}}\right\} m \end{array}\right., \tag{1.5}$$

where all the other entries except those of the coefficients of f and g are zero. We call this matrix the *Sylvester matrix* of f and g with respect to x. It is clear that

$$\boldsymbol{S} \cdot \begin{pmatrix} x^{m+l-1} \\ x^{m+l-2} \\ \vdots \\ x \\ 1 \end{pmatrix} = (x^{l-1}f, x^{l-2}f, \ldots, xf, f, x^{m-1}g, x^{m-2}g, \ldots, xg, g)^{\mathrm{T}}.$$

Definition 1.3. The determinant of Sylvester matrix $\boldsymbol{S}$ is called the *Sylvester resulatnt* of f and g with respect to x and denoted by $\operatorname{res}(f,g,x)$.

The resultant $\operatorname{res}(f,g,x) = \det(\boldsymbol{S})$ is homogeneous, which is of degree l in a_i and of degree m in b_i.

We extend the definition to cover some special cases. Define

$$\operatorname{res}(f,g,x) = \begin{cases} a_0^l, & \text{if } f \in \mathcal{R}, g \notin \mathcal{R}, \\ b_0^m, & \text{if } g \in \mathcal{R}, f \notin \mathcal{R}, \\ 0, & \text{if } f = g = 0, \\ 1, & \text{if } f, g \in \mathcal{R}, \ f \text{ and } g \text{ are not both zero.} \end{cases} \tag{1.6}$$

Lemma 1.1. *Suppose f and g are polynomials of positive degrees in the form of (1.4), there exist nonzero polynomials $p, q \in \mathcal{R}[x]$ such that*

$$pf + qg = \operatorname{res}(f,g,x), \tag{1.7}$$

and $\deg(p,x) < \deg(g,x)$, $\deg(q,x) < \deg(f,x)$.

Proof.   Denote the Sylvester matrix of f and g by $\boldsymbol{S}$. Multiply the ith column by x^{m+l-i} and add to the last column of $\boldsymbol{S}$ for all $i(1 \leq i \leq m+l-1)$. That gives

$$\boldsymbol{S}' = \begin{pmatrix} a_0 & a_1 & \cdots & \cdots & a_m & & & & x^{l-1}f \\ & a_0 & a_1 & \cdots & \cdots & a_m & & & x^{l-2}f \\ & & \ddots & \ddots & & & \ddots & & \vdots \\ & & & a_0 & a_1 & \cdots & a_{m-1} & & f \\ b_0 & b_1 & \cdots & \cdots & b_l & & & & x^{m-1}g \\ & b_0 & b_1 & \cdots & \cdots & b_l & & & x^{m-2}g \\ & & \ddots & \ddots & & & \ddots & & \vdots \\ & & & b_0 & b_1 & \cdots & b_{l-1} & & g \end{pmatrix}.$$

Then

$$\operatorname{res}(f,g,x) = \det(\boldsymbol{S}) = \det(\boldsymbol{S}').$$

Expanding $\det(\boldsymbol{S}')$ by minors along the last column and collecting the terms containing f and g, respectively, will give the formula (1.7).

We prove that p and q are nonzero. If $\operatorname{res}(f,g,x) \neq 0$, the conclusion is clear. If $\operatorname{res}(f,g,x) = 0$, assume

$$p(x) = u_{l-1}x^{l-1} + \cdots + u_0, \ q(x) = v_{m-1}x^{m-1} + \cdots + v_0.$$

Then

$$\boldsymbol{S^T} \cdot \begin{pmatrix} u_{l-1} \\ \vdots \\ u_0 \\ v_{m-1} \\ \vdots \\ v_0 \end{pmatrix} = \begin{pmatrix} 0 \\ \vdots \\ 0 \\ 0 \\ \vdots \\ 0 \end{pmatrix}.$$

As $\det(\boldsymbol{S^T}) = \operatorname{res}(f,g,x) = 0$, the equations have nonzero solutions. That completes the proof. □

Theorem 1.1. *Suppose f and g are polynomials of the form (1.4) with positive degrees. Assume a_0 and b_0 are not both zero, then f and g have non-trivial common divisors if and only if $\operatorname{res}(f,g,x) = 0$.*

Proof. By Lemma 1.1, a common divisor of f and g must be a divisor of $\operatorname{res}(f,g,x)$. If $\operatorname{res}(f,g,x) \neq 0$, $\operatorname{res}(f,g,x)$ must be a nonzero element of $\mathcal{R}$. So, f and g do not have non-trivial common divisors.

On the contrary, assume $\operatorname{res}(f,g,x) = 0$. By Lemma 1.1, there exist p and q such that $pf = -qg$. Without loss of generality (w.l.o.g.), assume $a_0 \neq 0$. If f and g do not have non-trivial common divisors, then f divides q. This is a contradiction since $\deg(q,x) < \deg(f,x)$. □

It is not difficult to verify the following proposition.

Proposition 1.2. *Let f and g be as in (1.4). The resultant of f and g satisfies the following three properties.*

(1) If $f \in \mathcal{R}$ or/and $g \in \mathcal{R}$, $\operatorname{res}(f,g,x)$ satisfies Eq. (1.6);
(2) $\operatorname{res}(f,g,x) = (-1)^{ml}\operatorname{res}(g,f,x)$;
(3) Assume $l \geq m > 0$ and $r = \operatorname{rem}(g,f)$ is the remainder of g divided by f in the quotient field of $\mathcal{R}$. Let $t = \deg(r,x)$, then

$$\operatorname{res}(f,g,x) = a_0^{l-t}\operatorname{res}(f,r,x).$$

The three items of Proposition 1.2 give naturally an algorithm for computing resultants. In fact, the algebraic form satisfying the three properties is unique.

Lemma 1.2. *Assume f, g are polynomials in $\mathcal{R}[x]$. If $R(f,g,x)$ is a polynomial in the coefficients of f and g over a certain extension field of the quotient field of $\mathcal{R}$ and $R(f,g,x)$ satisfies the three properties of Proposition 1.2, then $R(f,g,x) = \operatorname{res}(f,g,x)$.*

Proof. By induction on $\min(\deg(f), \deg(g))$, the conclusion is almost obvious. $\qquad\square$

Theorem 1.2. *Suppose f and g are as in (1.4) and α_i $(i = 1, \ldots, m)$ and β_j $(j = 1, \ldots, l)$ are respectively the zeros of f and g in a certain extension field of the quotient field of $\mathcal{R}$. Then*

$$\mathrm{res}(f, g, x) = a_0^l \prod_{i=1}^{m} g(\alpha_i) = (-1)^{ml} b_0^m \prod_{j=1}^{l} f(\beta_j)$$

$$= a_0^l b_0^m \prod_{i=1}^{m} \prod_{j=1}^{l} (\alpha_i - \beta_j).$$

Herein, the products are defined to be 0 if $f = g = 0$ and to be 1 if $f \in \mathcal{R}$ and $g \in \mathcal{R}$ are not both zero.

Proof. We only need to prove the first equality. It is easy to verify that the first product satisfies the three properties in Proposition 1.2. Then the conclusion follows by Lemma 1.2. $\qquad\square$

Corollary 1.1. *Let notations be as in Theorem 1.2. The discriminant of f is defined to be*

$$\mathrm{dis}(f, x) = a_0^{2m-2} \prod_{1 \leq j < i \leq m} (\alpha_i - \alpha_j)^2.$$

Then $\mathrm{res}(f, f', x) = (-1)^{\frac{m(m-1)}{2}} a_0 \, \mathrm{dis}(f, x)$.

1.3   Subresultants

The theory of subresultants is one of the classical constructible theory, which is the basis of many well-known algorithms in algorithmic algebra and computational algebraic geometry. We introduce some basic concepts and results of subresultants in this section.

Suppose f and g are polynomials in (1.4). Without loss of generality, we assume that $m \geq l > 0$ in this section.

Definition 1.4. For each $j(0 \leq j < l)$, set

$$\boldsymbol{S}_j = \mathrm{mat}(x^{l-j-1} f, \ldots, xf, f, x^{m-j-1} g, \ldots, xg, g),$$

which is an $(m + l - 2j) \times (m + l - j)$ matrix. We call

$$S_j = \mathrm{subres}_j(f, g) = \mathrm{detpol}(\boldsymbol{S}_j) = \sum_{i=0}^{j} \det(\boldsymbol{S}_j^{(m+l-2j+i)}) x^{j-i}$$

the jth *subresultant* of f and g with respect to x and $R_j = \det(\boldsymbol{S}_j^{m+l-2j})$ the jth *principal subresultant coefficient* (PSC) of f and g with respect to x.

If $m > l + 1$, the definition of the jth subresultant S_j and the PSC R_j of f and g with respect to x is extended as follows:
$$S_l = b_0^{m-l-1}g, \quad R_l = b_0^{m-l}; \quad S_j = R_j = 0, \quad l < j < m - 1.$$
Obviously, $\deg(S_j, x) \le j$. S_j is said to be *defective* of degree t if $\deg(S_j, x) = t < j$, and *regular* otherwise.

It is easy to see that $S_0 = R_0$ is the resultant of f and g with respect to x.

Theorem 1.3. *Let $f, g \in \mathcal{R}[x]$ be as in (1.4) and $m = \deg(f, x) \ge \deg(g, x) = l > 0$. For the jth subresultant S_j ($0 \le j < m - 1$) of f and g with respect to x, there exist polynomials $p_j, q_j \in \mathcal{R}[x]$ such that*
$$p_j f + q_j g = S_j,$$
where $\deg(p_j, x) < l - j$ and $\deg(q_j, x) < m - j$.

Proof. If $l \le j < m - 1$, the conclusion is clear. We consider the case that $j < l$. It is easy to verify that the determinant of the following matrix

$$
\left(
\begin{array}{ccccccc}
a_0 & a_1 & \cdots & \cdots & a_m & & x^{l-j-1}f \\
 & a_0 & a_1 & \cdots & \cdots & a_m & x^{l-j-2}f \\
 & & \ddots & \ddots & & \ddots & \vdots \\
 & & a_0 & a_1 & \cdots & a_{m-j-1} & f \\
b_0 & b_1 & \cdots & \cdots & b_l & & x^{m-j-1}g \\
 & b_0 & b_1 & \cdots & \cdots & b_l & x^{m-j-2}g \\
 & & \ddots & \ddots & & \ddots & \vdots \\
 & & b_0 & b_1 & \cdots & b_{l-j-1} & g
\end{array}
\right)
\begin{array}{l}
\left.\rule{0pt}{3.2em}\right\} l - j \text{ rows} \\[1.5em]
\left.\rule{0pt}{3.2em}\right\} m - j \text{ rows}
\end{array}
$$

is S_j. Expanding the matrix by minors along the last column gives the result. $\qquad\square$

Proposition 1.3. *Let $f, g \in \mathcal{R}[x]$ be as in (1.4) and $m = \deg(f, x) \ge \deg(g, x) = l > 0$. Then*
$$S_{l-1} = (-1)^{m-l+1}\mathrm{prem}(f, g, x).$$

In the following subsections, we introduce some main results on subresultants. The structure of the subsections and the idea of proofs are mainly taken from [Mishra (1993)]. However, the main content is from [Xia (2003)] where the *subresultant chain theorem* is corrected and thus some lemmas are modified. See also [Wang and Xia (2004)].

1.3.1   *The Habicht Theorem*

Definition 1.5. Let $f, g \in \mathcal{R}[x]$ be as in (1.4) and $m = \deg(f, x) \geq \deg(g, x) = l > 0$. Set

$$\mu = \begin{cases} m - 1, & \text{if } m > l, \\ l, & \text{otherwise.} \end{cases}$$

Let $S_{\mu+1} = f$, $S_\mu = g$, and $S_j (0 \leq j < \mu)$ be the jth subresultant of f and g with respect to x. The sequence of polynomials

$$S_{\mu+1}, S_\mu, S_{\mu-1}, \ldots, S_0$$

in $\mathcal{R}[x]$ is called the *subresultant chain* of f and g with respect to x. The chain is said to be *regular* if every S_j is regular, and *defective* otherwise.

Let

$$R_{\mu+1} = 1, \quad \text{and} \quad R_j = \begin{cases} \mathrm{lc}(S_j, x), & \text{if } S_j \text{ is regular,} \\ 0, & \text{otherwise,} \end{cases} \quad 0 \leq j \leq \mu.$$

The sequence

$$R_{\mu+1}, R_\mu, \ldots, R_0$$

is called the *PSC chain* of f and g with respect to x.

We first study the subresultant chain of two polynomials in symbolic coefficients of degrees $n + 1$ and $n > 0$, respectively. Let

$$\begin{aligned} A &= a_0 x^{n+1} + a_1 x^n + \cdots + a_{n+1}, \\ B &= b_0 x^n + b_1 x^{n-1} + \cdots + b_n, \end{aligned} \tag{1.8}$$

which are viewed as polynomials in $\mathbb{Z}[a_0, \ldots, a_{n+1}, b_0, \ldots, b_n][x]$.

Lemma 1.3. *Let A and B be as above. Then*

(a) $S_{n-1}(A, B) = \mathrm{prem}(A, B, x)$;
(b) For $i < n - 1$, $b_0^{2(n-i-1)} \mathrm{subres}_i(A, B) = \mathrm{subres}_i(B, \mathrm{prem}(A, B, x))$.

Proof.   Firstly,

$$\begin{aligned} S_{n-1}(A, B) &= \mathrm{detpol}(A, xB, B) = (-1)^2 \mathrm{detpol}(xB, B, A) \\ &= \mathrm{prem}(A, B, x). \end{aligned}$$

Secondly, let $R = \mathrm{prem}(A, B, x)$, then

$$\begin{aligned} \mathrm{subres}_i(A, B) &= \mathrm{detpol}(x^{n-i-1} A, \ldots, A, x^{n-i} B, \ldots, B) \\ &= b_0^{-2(n-i)} \mathrm{detpol}(x^{n-i-1} b_0^2 A, \ldots, b_0^2 A, x^{n-i} B, \ldots, B) \\ &= b_0^{-2(n-i)} \mathrm{detpol}(x^{n-i-1} R, \ldots, R, x^{n-i} B, \ldots, B) \\ &= b_0^{-2(n-i)} \mathrm{detpol}(x^{n-i} B, \ldots, B, x^{n-i-1} R, \ldots, R) \\ &= b_0^{-2(n-i)+2} \mathrm{detpol}(x^{n-i-2} B, \ldots, B, x^{n-i-1} R, \ldots, R). \end{aligned}$$

Therefore

$$b_0^{2(n-i-1)} \mathrm{subres}_i(A, B) = \mathrm{subres}_i(B, \mathrm{prem}(A, B, x)), \quad 0 \le i < n - 1.$$

$\square$

Remark 1.1. In the integral domain $\mathcal{R}$, b_0 may not be invertible. We use such notations as $b_0^{-2(n-i)}$ in the above proof only for the sake of concision. This does not do harm to the correctness of the proof. For example,

$$\mathrm{subres}_i(A, B) = b_0^{-2(n-i)} \mathrm{detpol}(x^{n-i}B, \ldots, B, x^{n-i-1}R, \ldots, R)$$

can be viewed as another notation of

$$b_0^{2(n-i)} \mathrm{subres}_i(A, B) = \mathrm{detpol}(x^{n-i}B, \ldots, B, x^{n-i-1}R, \ldots, R).$$

So, we use this kind of notations freely in this section.

Theorem 1.4 (Habicht's Theorem). *Let A and B be as above,*

$$S_{n+1} = A, \ S_n = B, \ S_{n-1}, \ldots, S_1, S_0$$

be the subresultant chain of A and B, and $R_{n+1}, \ldots, R_0$ the PSC chain. Then for each $j = 1, \ldots, n$,

(a) $R_{j+1}^2 S_{j-1} = \mathrm{prem}(S_{j+1}, S_j, x)$;
(b) $R_{j+1}^{2(j-i)} S_i = \mathrm{subres}_i(S_{j+1}, S_j)$, $0 \le i < j$.

Proof.   We use induction on j.

If $j = n$, $R_{n+1} = 1$. Therefore, (a) follows by Lemma 1.3, and (b) is just the definition of subresultant. Now, assume the results hold for $n, n-1, \ldots, j$. Then for any i ($0 \le i < j - 1$),

$$
\begin{aligned}
S_i &= R_{j+1}^{-2(j-i)} \mathrm{subres}_i(S_{j+1}, S_j) && \text{(induction assumption)} \\
&= R_{j+1}^{-2(j-i)} R_j^{-2(j-1-i)} \mathrm{subres}_i(S_j, \mathrm{prem}(S_{j+1}, S_j, x)) && \text{(Lemma 1.3)} \\
&= R_{j+1}^{-2(j-i)} R_j^{-2(j-1-i)} \mathrm{subres}_i(S_j, R_{j+1}^2 S_{j-1}) && \text{(induction assumption)} \\
&= R_j^{-2(j-1-i)} \mathrm{subres}_i(S_j, S_{j-1}).
\end{aligned}
$$

That means (b) holds for $j - 1$. Especially, if $i = j - 2$,

$$R_j^2 S_{j-2} = \mathrm{subres}_{j-2}(S_j, S_{j-1}) = \mathrm{prem}(S_j, S_{j-1}, x).$$

That completes the proof.

$\square$

We then study the relation between the two subresultant chains of two polynomials and their images under ring homomorphism. Suppose ϕ is a ring homomorphism mapping $\mathcal{R}$ into $\tilde{\mathcal{R}}$. Denote also by ϕ its induced ring homomorphism mapping $\mathcal{R}[x]$ into $\tilde{\mathcal{R}}[x]$ such that

$$\phi\left(\sum_{i=0}^{k} a_i x^i\right) = \sum_{i=0}^{k} \phi(a_i) x^i.$$

Proposition 1.4. *Let $\phi : \mathcal{R} \to \tilde{\mathcal{R}}$ be a ring homomorphism and $\phi : \mathcal{R}[x] \to \tilde{\mathcal{R}}[x]$ its induced ring homomorphism. Assume f and g are polynomials in (1.4) and*

$$\tilde{a}_0 = \phi(a_0), \quad \tilde{b}_0 = \phi(b_0); \quad \tilde{m} = \deg(\phi(f), x) \geq \tilde{l} = \deg(\phi(g), x).$$

Define $\tilde{\mu} = \tilde{m} - 1$ if $\tilde{m} > \tilde{l}$; and $\tilde{\mu} = \tilde{m}$ otherwise. Then the image of the jth subresultant S_j of f and g under ϕ is equal to the jth subresultant $\tilde{S}_j$ of $\phi(f)$ and $\phi(g)$ multiplied by δ, i.e. $\phi(S_j) = \delta \tilde{S}_j$, $0 \leq j < \tilde{\mu}$, where

$$\delta = \begin{cases} 1, & \text{if } \tilde{a}_0 \tilde{b}_0 \neq 0, \\ \tilde{a}_0^{l-\tilde{l}}, & \text{if } \tilde{a}_0 \neq 0 \text{ and } \tilde{b}_0 = 0, \\ (-1)^{(m-\tilde{m})(l-j)} \tilde{b}_0^{m-\tilde{m}}, & \text{if } \tilde{a}_0 = 0 \text{ and } \tilde{b}_0 \neq 0, \\ 0, & \text{if } \tilde{a}_0 = \tilde{b}_0 = 0. \end{cases}$$

Proof. We only prove the third case ($\tilde{l} = l$) and the other cases are left to the readers. If $l < j < \tilde{\mu}$, it is obvious that $S_j = \tilde{S}_j = 0$. If $j \leq l$, we have first that

$$S_j(A, B) = (-1)^{(m-j)(l-j)} S_j(B, A).$$

Second,

$$\begin{aligned}
\phi(S_j(B, A)) &= \mathrm{detpol}(x^{m-j-1}\phi(B), \ldots, \phi(B), x^{l-j-1}\phi(A), \ldots, \phi(A)) \\
&= \tilde{b}_0^{m-\tilde{m}} \mathrm{detpol}(x^{\tilde{m}-j-1}\phi(B), \ldots, \phi(B), x^{l-j-1}\phi(A), \ldots, \phi(A)) \\
&= \tilde{b}_0^{m-\tilde{m}}(-1)^{(\tilde{m}-j)(l-j)}. \\
&\quad\quad \mathrm{detpol}(x^{l-j-1}\phi(A), \ldots, \phi(A), x^{\tilde{m}-j-1}\phi(B), \ldots, \phi(B)) \\
&= \tilde{b}_0^{m-\tilde{m}}(-1)^{(\tilde{m}-j)(l-j)} \mathrm{subres}_j(\phi(A), \phi(B)) \\
&= (-1)^{(\tilde{m}-j)(l-j)} \tilde{b}_0^{m-\tilde{m}} \tilde{S}_j.
\end{aligned}$$

Hence

$$\phi(S_j) = (-1)^{(m+\tilde{m}-2j)(l-j)} \tilde{b}_0^{m-\tilde{m}} \tilde{S}_j = (-1)^{(m-\tilde{m})(l-j)} \tilde{b}_0^{m-\tilde{m}} \tilde{S}_j. \qquad \square$$

Remark 1.2. Although we do not define the subresultant of f and g when $m < l$, the notation $S_j(B, A)$ in the above proof can be easily understood as a short notation of $\mathrm{detpol}(x^{m-j-1}B, \ldots, B, x^{l-j-1}A, \ldots, A)$.

Proposition 1.5. *Let notations be as above. Assume*
$$\phi : \mathbb{Z}[a_0, \ldots, a_{n+1}, b_0, \ldots, b_n] \to \tilde{\mathcal{R}}$$
is a ring homomorphism and $\phi : \mathbb{Z}[a_0, \ldots, a_{n+1}, b_0, \ldots, b_n][x] \to \tilde{\mathcal{R}}[x]$ is its induced ring homomorphism.

If $\phi(S_{j+1})$ is regular and $\phi(S_j)$ is defective of degree t, then

(a) $\phi(S_{j-1}) = \phi(S_{j-2}) = \cdots = \phi(S_{t+1}) = 0$;

(b) if $j = n$,
$$\phi(S_t) = [\mathrm{lc}(\phi(S_{n+1}), x)\,\mathrm{lc}(\phi(S_n), x)]^{n-t}\,\phi(S_n);$$

if $j < n$,
$$\phi(R_{j+1})^{j-t}\phi(S_t) = \mathrm{lc}(\phi(S_j), x)^{j-t}\phi(S_j);$$

(c) if $j = n$,
$$\phi(S_{t-1}) = [-\mathrm{lc}(\phi(S_{n+1}), x)]^{n-t}\,\mathrm{prem}(\phi(S_{n+1}), \phi(S_n), x);$$

if $j < n$,
$$\phi(-R_{j+1})^{j-t+2}\phi(S_{t-1}) = \mathrm{prem}(\phi(S_{j+1}), \phi(S_j), x).$$

Proof. By the Habicht Theorem, for each i $(0 \le i < j)$,
$$R_{j+1}^{2\,(j-i)}S_i = \mathrm{subres}_i(S_{j+1}, S_j).$$
Besides, $\deg(\phi(S_{j+1}), x) = j + 1$, $\deg(\phi(S_j), x) = t$. Therefore
$$\phi(R_{j+1})^{2\,(j-i)}\phi(S_i)$$
$$= \phi(\mathrm{subres}_i(S_{j+1}, S_j))$$
$$= \phi(\mathrm{lc}(S_{j+1}, x))^{j-t}\mathrm{subres}_i(\phi(S_{j+1}), \phi(S_j)) \quad (\text{Proposition 1.4})$$
$$= \mathrm{lc}(\phi(S_{j+1}), x)^{j-t}\mathrm{subres}_i(\phi(S_{j+1}), \phi(S_j)).$$
On the other hand,
$$\mathrm{subres}_i(\phi(S_{j+1}), \phi(S_j))$$
$$= \begin{cases} 0, & \text{if } t < i < j, \\ \mathrm{lc}(\phi(S_j), x)^{j-t}\phi(S_j), & \text{if } i = t, \\ (-1)^{j-t+2}\mathrm{prem}(\phi(S_{j+1}), \phi(S_j), x), & \text{if } i = t - 1. \end{cases}$$

Notice that
$$\phi(R_{j+1}) = \begin{cases} \mathrm{lc}(\phi(S_{j+1}), x), & \text{if } j < n, \\ 1, & \text{if } j = n. \end{cases}$$

That completes the proof. $\qquad\square$

1.3.2 *Subresultant Chain Theorem*

The Subresultant Chain Theorem under the case $m > l$ was redeveloped by Loos in [Loos (1983)] based on Habicht's work [Habicht (1948)]. Ho and Yap [Ho and Yap (1996)] gave a complete treatment of Loos' approach by introducing the concept of pseudo-subresultants for the case when the degrees of the two polynomials are not equal. B. Mishra [Mishra (1993)] claimed that the theorem is valid also for the case when the two degrees are equal. In this section, we point out that the theorem should be modified when the degrees of the two polynomials are the same.

Theorem 1.5 (Subresultant Chain $m > l$). *Let*

$$S_{\mu+1}, S_\mu, \ldots, S_0$$

be the subresultant chain of f $(= S_{\mu+1})$ and g $(= S_\mu)$ in (1.4) with $m > l$ and

$$R_{\mu+1}, R_\mu, \ldots, R_0,$$

be its PSC coefficients. If for some j $(1 \le j \le \mu)$, S_{j+1} and S_j are both regular, then

$$R_{j+1}^2 S_{j-1} = \mathrm{prem}(S_{j+1}, S_j, x).$$

If S_{j+1} is regular but S_j is defective of degree t $(< j)$, then

$$S_{j-1} = S_{j-2} = \cdots = S_{t+1} = 0,$$
$$R_{j+1}^{j-t} S_t = \mathrm{lc}(S_j, x)^{j-t} S_j,$$
$$(-1)^{j-t} R_{j+1}^{j-t+2} S_{t-1} = \mathrm{prem}(S_{j+1}, S_j, x).$$

Theorem 1.6 (Subresultant Chain $m = l$). *Let notations be as above but $m = l$. If S_μ and $S_{\mu-1}$ are both regular, then*

$$R_\mu S_{\mu-2} = \mathrm{prem}(S_\mu, S_{\mu-1}, x).$$

If S_μ is regular but $S_{\mu-1}$ is defective of degree t, then

$$S_{\mu-2} = S_{\mu-3} = \cdots = S_{t+1} = 0,$$
$$S_t = \mathrm{lc}(S_{\mu-1}, x)^{\mu-t-1} S_{\mu-1},$$
$$(-1)^{\mu-t+1} R_\mu S_{t-1} = \mathrm{prem}(S_\mu, S_{\mu-1}, x).$$

If for some j $(1 \le j < \mu - 1)$, S_{j+1} and S_j are both regular, then

$$R_{j+1}^2 S_{j-1} = \mathrm{prem}(S_{j+1}, S_j, x), \quad j < \mu - 1.$$

If S_{j+1} is regular but S_j is defective of degree t, then
$$S_{j-1} = S_{j-2} = \cdots = S_{t+1} = 0,$$
$$R_{j+1}^{j-t} S_t = \mathrm{lc}(S_j, x)^{j-t} S_j,$$
$$(-1)^{j-t} R_{j+1}^{j-t+2} S_{t-1} = \mathrm{prem}(S_{j+1}, S_j, x), \quad j < \mu - 1.$$

It is easy to see that the relations under the case $j < \mu - 1$ in Theorem 1.6 do not hold under the case $j = \mu - 1$.

Consider two polynomials in symbolic coefficients
$$A = c_0 x^{\mu+1} + \cdots + c_{\mu+1}, \quad B = d_0 x^\mu + \cdots + d_\mu.$$
Let
$$S_{\mu+1}^*, S_\mu^*, \ldots, S_0^* \quad \text{and} \quad R_{\mu+1}^*, R_\mu^*, \ldots, R_0^*$$
be the subresultant chain and the PSC chain of A and B with respect to x, respectively.

Polynomials f, g can be viewed as the images of A, B under the ring homomorphism ϕ :
$$\phi(c_i) = \mathrm{coef}(f, x^{\mu+1-i}), \quad 0 \le i \le \mu + 1,$$
$$\phi(d_j) = \mathrm{coef}(g, x^{\mu-j}), \quad 0 \le j \le \mu.$$

Proof of Theorem 1.5 We first prove the theorem under the case $m > l$, where $\mu = m - 1$ and $S_{\mu+1}$ is regular. By Proposition 1.4, we have the following results:

- For any i $(0 \le i < \mu)$,
$$\phi(S_i^*) = \phi(\mathrm{subres}_i(A, B)) = a_0^{\mu-l} \mathrm{subres}_i(\phi(A), \phi(B)) = a_0^{\mu-l} S_i.$$
- $\phi(S_i^*)$ is regular if and only if S_i is regular; $\phi(S_i^*)$ is defective of degree t if and only if S_i is defective of degree t.
- If S_i $(0 \le i < \mu)$ is regular, then
$$\phi(R_i^*) = \phi(\mathrm{lc}(S_i^*, x)) = \mathrm{lc}(\phi(S_i^*), x) = a_0^{\mu-l} R_i.$$

(1) If S_{j+1} and S_j are both regular, by the Habicht Theorem,
$$(R_{j+1}^*)^2 S_{j-1}^* = \mathrm{prem}(S_{j+1}^*, S_j^*, x) = \mathrm{detpol}(x S_j^*, S_j^*, S_{j+1}^*).$$
If $j = \mu$ or $j = \mu - 1$, then S_μ is regular. By Proposition 1.4, $\phi(S_i^*) = S_i$ $(0 \le i \le \mu + 1)$, and thus the conclusion follows. If $j < \mu - 1$, then
$$\begin{aligned}
\phi(R_{j+1}^*)^2 \phi(S_{j-1}^*) &= a_0^{2(\mu-l)} R_{j+1}^2 a_0^{\mu-l} S_{j-1} \\
&= \mathrm{detpol}(x \phi(S_j^*), \phi(S_j^*), \phi(S_{j+1}^*)) \\
&= a_0^{3(\mu-l)} \mathrm{detpol}(x S_j, S_j, S_{j+1}) \\
&= a_0^{3(\mu-l)} \mathrm{prem}(S_{j+1}, S_j, x).
\end{aligned}$$

That is

$$R_{j+1}^2 S_{j-1} = \mathrm{prem}(S_{j+1}, S_j, x).$$

(2) If S_{j+1} is regular but S_j is defective of degree t.

(a) If $j = \mu$, i.e. $S_{\mu+1}$ is regular but S_μ is defective of degree $t\ (= l)$, then by the definition of subresultant,

$$S_{\mu-1} = S_{\mu-2} = \cdots = S_{t+1} = 0, \text{ and } R_{\mu+1} = 1.$$

Therefore

$$R_{\mu+1}^{\mu-t} S_t = S_t = \mathrm{lc}(S_\mu, x)^{\mu+1-t-1} S_\mu = \mathrm{lc}(S_\mu, x)^{\mu-t} S_\mu.$$

Finally,

$$\begin{aligned}
(-R_{\mu+1})^{\mu-t+2} S_{t-1} &= (-1)^{\mu-t+2} S_{t-1} \\
&= (-1)^{\mu-t+2}(-1)^{\mu+1-t+1}\mathrm{prem}(S_{\mu+1}, S_\mu, x) \\
&= \mathrm{prem}(S_{\mu+1}, S_\mu, x).
\end{aligned}$$

(b) If $j = \mu - 1$, then S_μ is regular. By Proposition 1.4, $\phi(S_i^*) = S_i\ (0 \le i \le \mu + 1)$. Then the conclusion is clear.

(c) If $j < \mu - 1$, i.e. $\phi(S_{j+1}^*)$ is regular but $\phi(S_j^*)$ is defective of degree t, then by Proposition 1.5,

$$\begin{aligned}
\phi(S_{j-1}^*) = \phi(S_{j-2}^*) &= \cdots = \phi(S_{t+1}^*) = 0, \\
\phi(R_{j+1}^*)^{j-t}\phi(S_t^*) &= \mathrm{lc}(\phi(S_j^*), x)^{j-t}\phi(S_j^*), \\
\phi(-R_{j+1}^*)^{j-t+2}\phi(S_{t-1}^*) &= \mathrm{prem}(\phi(S_{j+1}^*), \phi(S_j^*), x).
\end{aligned}$$

By the above discussion, the above relations give

$$\begin{aligned}
a_0^{\mu-l} S_{j-1} &= \cdots = a_0^{\mu-l} S_{t+1} = 0, \\
a_0^{(j-t)(\mu-l)} R_{j+1}^{j-t} a_0^{\mu-l} S_t &= [a_0^{\mu-l}\mathrm{lc}(S_j, x)]^{j-t} a_0^{\mu-l} S_j, \\
(-a_0^{\mu-l} R_{j+1})^{j-t+2} a_0^{\mu-l} S_{t-1} &= \mathrm{prem}(a_0^{\mu-l} S_{j+1}, a_0^{\mu-l} S_j, x) \\
&= a_0^{\mu-l} a_0^{(\mu-l)(j-t+2)}\mathrm{prem}(S_{j+1}, S_j, x).
\end{aligned}$$

Therefore

$$\begin{aligned}
S_{j-1} &= \cdots = S_{t+1} = 0, \\
R_{j+1}^{j-t} S_t &= \mathrm{lc}(S_j, x)^{j-t} S_j, \\
(-R_{j+1})^{j-t+2} S_{t-1} &= \mathrm{prem}(S_{j+1}, S_j, x).
\end{aligned}$$

That completes the proof of Theorem 1.5.

Proof of Theorem 1.6 We now prove the theorem under the case $m = l$, where $\mu = m = l$, $S_{\mu+1}$ is defective of degree m and S_μ is regular. By Proposition 1.4, we have the following results:

- For any i $(0 \le i < \mu)$,
$$\phi(S_i^*) = \phi(\text{subres}_i(A, B)) = (-1)^{\mu-i} b_0 \text{subres}_i(\phi(A), \phi(B))$$
$$= (-1)^{\mu-i} b_0 S_i.$$

- $\phi(S_i^*)$ is regular if and only if S_i is regular; $\phi(S_i^*)$ is defective of degree t if and only if S_i is defective of degree t.

- If S_i $(0 \le i < \mu)$ is regular, then
$$\phi(R_i^*) = (-1)^{\mu-i} b_0 R_i.$$

(1) Assume S_{j+1} and S_j are both regular. By the Habicht Theorem,
$$(R_{j+1}^*)^2 S_{j-1}^* = \text{prem}(S_{j+1}^*, S_j^*, x) = \text{detpol}(x S_j^*, S_j^*, S_{j+1}^*).$$

(a) If $j = \mu - 1$,
$$(R_\mu^*)^2 S_{\mu-2}^* = \text{prem}(S_\mu^*, S_{\mu-1}^*, x) = \text{detpol}(x S_{\mu-1}^*, S_{\mu-1}^*, S_\mu^*).$$

Hence
$$\phi((R_\mu^*)^2 S_{\mu-2}^*) = R_\mu^2 \cdot (-1)^2 \cdot b_0 \cdot S_{\mu-2}$$
$$= \text{detpol}(x \cdot (-1) \cdot b_0 \cdot S_{\mu-1}, (-1) \cdot b_0 \cdot S_{\mu-1}, S_\mu)$$
$$= b_0^2 \, \text{prem}(S_\mu, S_{\mu-1}, x).$$

Because $R_\mu = b_0$,
$$R_\mu S_{\mu-2} = \text{prem}(S_\mu, S_{\mu-1}, x).$$

(b) If $j < \mu - 1$,
$$(-1)^{2\,(\mu-j-1)} b_0^2 \cdot R_{j+1}^2 \cdot (-1)^{\mu-j+1} b_0 \cdot S_{j-1}$$
$$= \text{detpol}(x \cdot (-1)^{\mu-j} \cdot b_0 \cdot S_j, (-1)^{\mu-j} \cdot b_0 \cdot S_j, (-1)^{\mu-j-1} \cdot b_0 \cdot S_{j+1})$$
$$= (-1)^{3\,(\mu-1)-1} b_0^3 \, \text{detpol}(x S_j, S_j, S_{j+1})$$
$$= (-1)^{3\,(\mu-1)-1} b_0^3 \, \text{prem}(S_{j+1}, S_j, x).$$

Hence
$$R_{j+1}^2 S_{j-1} = \text{prem}(S_{j+1}, S_j, x).$$

(2) Assume S_{j+1} is regular but S_j is defective of degree t, i.e. $\phi(S_{j+1}^*)$ is regular but $\phi(S_j^*)$ is defective of degree t. By Proposition 1.5,
$$\phi(S_{j-1}^*) = \phi(S_{j-2}^*) = \cdots = \phi(S_{t+1}^*) = 0, \tag{1.9}$$
$$\phi(R_{j+1}^*)^{j-t} \phi(S_t^*) = \text{lc}(\phi(S_j^*), x)^{j-t} \phi(S_j^*), \tag{1.10}$$
$$\phi(-R_{j+1}^*)^{j-t+2} \phi(S_{t-1}^*) = \text{prem}(\phi(S_{j+1}^*), \phi(S_j^*), x). \tag{1.11}$$

Obviously, formula (1.9) implies $S_{j-1} = S_{j-2} = \cdots = S_{t+1} = 0$.

(a) If $j = \mu - 1$, formula (1.10) becomes

$$R_\mu^{\mu-t-1} \cdot (-1)^{\mu-t} b_0 \cdot S_t = (-b_0)^{\mu-t-1} \cdot \mathrm{lc}(S_{\mu-1}, x)^{\mu-t-1} \cdot (-b_0) \cdot S_{\mu-1}.$$

Note that $R_\mu = b_0$, which implies

$$S_t = \mathrm{lc}(S_{\mu-1}, x)^{\mu-t-1} S_{\mu-1}.$$

Besides, formula (1.11) becomes

$$(-1)^{\mu-t+1} R_\mu^{\mu-t+1} \cdot (-1)^{\mu-t+1} b_0 \cdot S_{t-1} = \mathrm{prem}(S_\mu, -b_0 S_{\mu-1}, x)$$
$$= (-b_0)^{\mu-t+1} \mathrm{prem}(S_\mu, S_{\mu-1}, x).$$

Therefore $(-1)^{\mu-t+1} R_\mu S_{t-1} = \mathrm{prem}(S_\mu, S_{\mu-1}, x)$.

(b) If $j < \mu - 1$, it can be verified similarly that

$$R_{j+1}^{j-t} S_t = \mathrm{lc}(S_j, x)^{j-t} S_j,$$
$$(-R_{j+1})^{j-t+2} S_{t-1} = \mathrm{prem}(S_{j+1}, S_j, x).$$

That completes the proof of Theorem 1.6.

Theorems 1.5 and 1.6 provide an effective algorithm for computing subresultant chain via pseudo-division. If $\deg(S_{\mu+1}, x) = \deg(S_\mu, x)$, $S_{\mu+1}$ is defective and thus the theorems do not give a way for computing $S_{\mu-1}$. However, by Proposition 1.3,

$$S_{\mu-1} = -\mathrm{prem}(S_{\mu+1}, S_\mu, x).$$

By Theorems 1.5 and 1.6 and the discussion above, we may give the following algorithm (Algorithm 1.1) for computing the subresultant chain of two polynomials.

Notation 1.1. For a list (or set) $L = [a_1, ..., a_n]$, let $\mathsf{op}(L)$ denote $a_1, ..., a_n$. Then, when we describe an algorithm, $L \leftarrow [\mathsf{op}(L), b]$ means appending b to the list L.

1.3.3   *Subresultant Polynomial Remainder Sequence*

Definition 1.6. Let p_1 and p_2 be two polynomials in $\mathcal{R}[x]$ with $\deg(p_1) \geq \deg(p_2)$. A sequence of nonzero polynomials $p_1, p_2, \ldots, p_t$ in $\mathcal{R}[x]$ is called the *subresultant polynomial remainder sequence* of p_1 and p_2 with respect to x, if

$$p_{i+2} = \mathrm{prem}(p_i, p_{i+1}, x)/\beta_{i+2}, \quad 1 \leq i \leq t-2,$$
$$\mathrm{prem}(p_{t-1}, p_t, x) = 0,$$

Algorithm 1.1 `SubRes`

Input: Two polynomials $f, g \in \mathcal{R}[x]$ with $\deg(f) \geq \deg(g) > 0$

Output: The subresultant chain of f and g with respect to x

1: $m \leftarrow \deg(f); l \leftarrow \deg(g);$
2: **if** $m > l$ **then** $j \leftarrow m - 1$ **else** $j \leftarrow l$ **endif**;
3: $S_{j+1} \leftarrow f;\ S_j \leftarrow g;\ R_{j+1} \leftarrow 1;\ \mu \leftarrow j;\ L \leftarrow [f, g];$
4: **if** $m = l$ **then**
5: $\quad S_{j-1} \leftarrow -\operatorname{prem}(S_{j+1}, S_j, x);$
6: $\quad L \leftarrow [\operatorname{op}(L), S_{j-1}];\ j \leftarrow j - 1;\ R_{j+1} \leftarrow \operatorname{lc}(S_{j+1}, x);$
7: **end if**
8: **do**
9: $\quad$ **if** $S_j = 0$ **then** $r \leftarrow -1$ **else** $r \leftarrow \deg(S_j)$ **endif**;
10: $\quad$ **for** i **from** $r + 1$ **to** $j - 1$ **do** $S_i \leftarrow 0;\ L \leftarrow [\operatorname{op}(L), 0]$ **enddo**;
11: $\quad$ **if** $0 <= r$ **and** $r < j$ **then**
12: $\quad\quad$ **if** $m = l$ **and** $j = \mu - 1$ **then**
13: $\quad\quad\quad S_r \leftarrow \operatorname{lc}(S_j, x)^{j-r} S_j;$
14: $\quad\quad\quad S_{r-1} \leftarrow \operatorname{prem}(S_{j+1}, S_j, x)/((-1)^{j-r} R_{j+1});$
15: $\quad\quad$ **else**
16: $\quad\quad\quad S_r \leftarrow \operatorname{lc}(S_j, x)^{j-r} S_j/(R_{j+1}^{j-r});$
17: $\quad\quad\quad S_{r-1} \leftarrow \operatorname{prem}(S_{j+1}, S_j, x)/((-R_{j+1})^{j-r+2});$
18: $\quad\quad$ **endif**;
19: $\quad\quad$ **if** $r > 0$ **then** $L \leftarrow [\operatorname{op}(L), S_r, S_{r-1}]$ **else** $L \leftarrow [\operatorname{op}(L), S_r]$ **endif**;
20: $\quad$ **elseif** $0 < r$ **and** $r = j$ **then**
21: $\quad\quad$ **if** $m = l$ **and** $j = \mu - 1$ **then**
22: $\quad\quad\quad S_{r-1} \leftarrow \operatorname{prem}(S_{j+1}, S_j, x)/R_{j+1}$
23: $\quad\quad$ **else**
24: $\quad\quad\quad S_{r-1} \leftarrow \operatorname{prem}(S_{j+1}, S_j, x)/R_{j+1}^2$
25: $\quad\quad$ **endif**;
26: $\quad\quad L \leftarrow [\operatorname{op}(L), S_{r-1}]$
27: $\quad$ **endif**;
28: $\quad$ **if** $r <= 0$ **then return** L
29: $\quad$ **else** $\quad j \leftarrow r - 1;\ R_{j+1} \leftarrow \operatorname{lc}(S_{j+1}, x);$
30: $\quad$ **endif**;
31: **enddo**

where

$$
\begin{aligned}
&\beta_3 = (-1)^{\delta_2},\ \beta_{i+1} = (-1)^{\delta_i} \psi_{i-1}^{\delta_i - 1} I_{i-1}, && i = 3, \ldots, t - 1, \\
&I_1 = 1,\ I_i = \operatorname{lc}(p_i, x), && i = 2, \ldots, t, \\
&\delta_i = d_{i-1} - d_i + 1, && i = 2, \ldots, t, \\
&d_i = \deg(p_i, x), && i = 1, \ldots, t, \\
&\psi_1 = 1,\ \psi_2 = I_2^{\delta_2 - 1},\ \psi_i = \psi_{i-1} \left(\frac{I_i}{\psi_{i-1}} \right)^{\delta_i - 1}, && i = 3, \ldots, t.
\end{aligned}
$$

We establish the relation between the subresultant polynomial remainder sequence and the subresultant chain, which guarantees that subresultant polynomial remainder sequence is well-defined. That is to say, $p_i \in \mathcal{R}[x]$ for $i \geq 3$ whenever $p_1, p_2 \in \mathcal{R}[x]$.

The subresultant chain theorem shows that the subrestltant chains have the so-called "block structure".

Definition 1.7. Let f and g be as above with $m = \deg(f, x) \geq \deg(g, x) = l > 0$ and let

$$\mathcal{S} : S_{\mu+1}, S_\mu, \ldots, S_0$$

be the subresultant chain of f and g with respect to x. A sequence

$$d_1, d_2, \ldots, d_t$$

of steadily decreasing nonnegative integers is called the *block indices* of $\mathcal{S}$, if $d_1 = \mu + 1$, for each $2 \leq i \leq t$, S_{d_i} is regular, and for $0 \leq j \leq \mu$, $j \notin \{d_2, \ldots, d_t\}$, S_j is defective.

The sequence of regular subresultants $S_{d_2}, \ldots, S_{d_t}$ is called the *subresultant regular subchain* of f and g with respect to x.

The interesting block structure of $\mathcal{S}$ can be described as follows. The first block consists of the single term $S_{\mu+1}$. For any $i(2 \leq i \leq t)$,

$$S_{d_i} \neq 0, \ S_{d_i} \sim S_{d_{i-1}-1}, \quad \text{and} \quad S_{d_{i-1}-2} = \cdots = S_{d_i+1} = 0.$$

So, we say that the *i*th *nonzero block* of $\mathcal{S}$ has the form

$$S_{d_{i-1}-1}, 0, \ldots, 0, S_{d_i},$$

where $S_{d_{i-1}-1} \sim S_{d_i}$, and $d_{i-1} - 1 \geq d_i$. If $d_t > 0$, then

$$S_{d_t-1} = \cdots = S_0 = 0;$$

This is the last block, called the *zero block* of $\mathcal{S}$.

Proposition 1.6. *Let f and g be as in (1.4) and $m \geq l > 0$. Suppose $\mathcal{S} : S_{\mu+1}, S_\mu, \ldots, S_0$ is the subresultant chain of f and g with respect to x and $d_1, d_2, \ldots, d_t$ is the block indices of $\mathcal{S}$.*

(a) If $m > l$, for $i = 1, \ldots, t - 1$,

$$R_{d_i}^{\delta_{i+1}-2} S_{d_{i+1}} = \mathrm{lc}(S_{d_i-1}, x)^{\delta_{i+1}-2} S_{d_i-1}, \tag{1.12}$$

$$(-R_{d_i})^{\delta_{i+1}} S_{d_{i+1}-1} = \mathrm{prem}(S_{d_i}, S_{d_i-1}, x), \tag{1.13}$$

where $\delta_{i+1} = d_i - d_{i+1} + 1$.

(b) If $m = l$, setting

$$\delta_2 = 1, \quad \delta_{i+1} = d_i - d_{i+1} + 1, \quad i \geq 2,$$

we have that Eq. (1.12) and Eq. (1.13) hold for $i = 3, \ldots, t-1$ and

$$S_{d_2} = S_{d_1-1}, \quad S_{d_3} = \mathrm{lc}(S_{d_2-1}, x)^{\delta_3 - 2} S_{d_2-1},$$
$$-S_{d_2-1} = \mathrm{prem}(S_{d_1}, S_{d_1-1}, x), \tag{1.14}$$
$$(-1)^{\delta_3} R_{d_2} S_{d_3-1} = \mathrm{prem}(S_{d_2}, S_{d_2-1}, x).$$

Proof. (a) $m > l$. If $d_i - 1 = d_{i+1}$, then $\delta_{i+1} - 2 = 0$. So Eq. (1.12) is an identity. Equation (1.13) can be obtained by taking $j + 1 = d_i$ in Theorem 1.5.

If $d_i - 1 > d_{i+1}$, taking

$$S_{j+1} = S_{d_i} \text{ and } S_t = S_{d_{i+1}}$$

in Theorem 1.5, we have

$$S_j = S_{d_i-1}, \quad S_{t-1} = S_{d_{i+1}-1}.$$

On the other hand,

$$R_{j+1} = R_{d_i}, \quad j - t = d_i - d_{i+1} - 1 = \delta_{i+1} - 2.$$

The conclusion follows.

(b) $m = l$. We only need to notice that $d_1 = \mu + 1$, $d_2 = \mu$. Then the conclusion (1.14) is either obvious or can be obtained directly from Theorem 1.6 by letting $j = \mu - 1$. The proof for the case $i > 2$ is similar to that for $m > l$. $\qquad \square$

The above proposition shows that

$$S_{d_i+1} \sim S_{d_i-1}, \quad S_{d_i+1-1} \sim \mathrm{prem}(S_{d_i}, S_{d_i-1}, x).$$

Hence

$$S_{d_i+2} \sim \mathrm{prem}(S_{d_i}, S_{d_i-1}, x) \sim \mathrm{prem}(S_{d_i}, S_{d_i+1}, x).$$

Under the assumption of Proposition 1.6, if $p_1 = f, p_2 = g, \ldots, p_k$ is a polynomial remainder sequence of f and g, then $k = t$ and $S_{d_i} \sim p_i$, $i = 1, \ldots, t$.

Theorem 1.7. *Let $f, g \in \mathcal{R}[x]$ be as in (1.4) and $m \geq l > 0$. Suppose $\mathcal{S} : S_{\mu+1}, S_\mu, \ldots, S_0$ is the subresultant chain of f and g with respect to x, $d_1, d_2, \ldots, d_t$ is the block indices of $\mathcal{S}$ and $p_1 = f, p_2 = g, p_3, \ldots, p_t$ is the subresultant polynomial remainder sequence of f and g with respect to x. Then we have*

(a) $p_i = S_{d_{i-1}-1}$ *for* $i = 1, \ldots, t$, *where* $d_0 - 1 = \mu + 1$. *In other words,*

$$S_{d_0-1}, S_{d_1-1}, \ldots, S_{d_{t-1}-1}$$

is the subresultant polynomial remainder sequence of f and g with respect to x.

(b) $\psi_i = R_{d_i}$ *for* $i = 1, \ldots, t$ *with only one exception that* $\psi_2 = 1 \neq R_{d_2}$ *when* $m = l$.

Proof. Because $S_{d_i} \sim p_i$ $(1 \leq i \leq t)$, the definition of $\delta_i = d_{i-1} - d_i + 1$ in Definition 1.6 is consistent with the above notation of δ_i.

We use induction on i. If $i = 1$, by the definition,

$$p_1 = f = S_{\mu+1} = S_{d_1} = S_{d_0-1}, \quad \psi_1 = 1 = R_{\mu+1} = R_{d_1}.$$

If $i = 2$, $p_2 = S_\mu = S_{d_1-1}$. When $m > l$, taking $i = 1$ in formula (1.12), we have

$$R_{d_2} = I_2^{\delta_2-1} = \psi_2.$$

When $m = l$, we have $\delta_2 = 1$ and thus $\psi_2 = 1$.

If $i = 3$, by Proposition 1.6,

$$\mathrm{prem}(p_1, p_2, x) = (-1)^{\delta_2} S_{d_2-1}.$$

The above relation holds for either $m > l$ or $m = l$. So, $p_3 = S_{d_2-1}$. When $m > l$, setting $i = 2$ in formula (1.12), we have $\psi_3 = R_{d_3}$. When $m = l$, since $\psi_2 = 1$ and

$$S_{d_3} = \mathrm{lc}(S_{d_2-1}, x)^{\delta_3-2} S_{d_2-1},$$

we still have $\psi_3 = R_{d_3}$.

If $i = 4$, $\beta_4 = (-1)^{\delta_3} \cdot \psi_2^{\delta_3-1} \cdot \mathrm{lc}(S_{d_1-1}, x)$. When $m > l$, we have

$$\beta_4 = (-1)^{\delta_3} \cdot R_{d_2}^{\delta_3-1} \cdot \mathrm{lc}(S_{d_1-1}, x).$$

When $m = l$, we have

$$\beta_4 = (-1)^{\delta_3} \cdot R_{d_2}.$$

Hence, we have $p_4 = S_{d_3-1}$ when $m \geq l$.

Now we assume the two conclusions of the theorem hold respectively when $j \leq i$ and $j \leq i - 1$ $(i \geq 4)$. Then by formula (1.12),

$$R_{d_i} = \left(\frac{I_i}{R_{d_{i-1}}}\right)^{\delta_i-1} R_{d_{i-1}}.$$

First, by the induction assumption, it is clear that $\psi_i = R_{d_i}$. Second,

$$\operatorname{prem}(p_{i-1}, p_i, x) = \operatorname{prem}(S_{d_{i-2}-1}, S_{d_{i-1}-1}, x)$$

$$= \operatorname{prem}\left(\left(\frac{R_{d_{i-2}}}{I_{i-1}}\right)^{\delta_{i-1}-2} \cdot S_{d_{i-1}}, S_{d_{i-1}-1}, x\right)$$

$$= \left(\frac{R_{d_{i-2}}}{I_{i-1}}\right)^{\delta_{i-1}-2} \cdot \operatorname{prem}(S_{d_{i-1}}, S_{d_{i-1}-1}, x)$$

$$= \left(\frac{R_{d_{i-2}}}{I_{i-1}}\right)^{\delta_{i-1}-2} \cdot (-R_{d_{i-1}})^{\delta_i} \cdot S_{d_i-1}$$

$$= (-1)^{\delta_i} R_{d_{i-1}}^{\delta_i-1} I_{i-1} S_{d_i-1}$$

$$= (-1)^{\delta_i} \psi_{i-1}^{\delta_i-1} I_{i-1} S_{d_i-1}$$

$$= \beta_{i+1} S_{d_i-1}.$$

That completes the proof. $\qquad\square$

Chapter 2

Zero Decomposition of Polynomial System

2.1 Notations

Suppose $\{u_1, \ldots, u_d, x_1, \ldots, x_n\}$ is a set of indeterminates with a given order

$$u_1 \prec \ldots \prec u_d \prec x_1 \prec \ldots \prec x_n$$

where $\{u_1, \ldots, u_d\}$ $(d \geq 0)$ and $\{x_1, \ldots, x_n\}$ $(n \geq 1)$ are the sets of parameters and variables, respectively. Let $\boldsymbol{u} = \{u_1, \ldots, u_d\}$ and $\boldsymbol{x} = \{x_1, \ldots, x_n\}$. Sometimes, with a little abuse of notations, $\boldsymbol{u}$ and $\boldsymbol{x}$ also denote $(u_1, \ldots, u_d)$ and $(x_1, \ldots, x_n)$, respectively. Suppose $\mathcal{K}$ is a field with characteristic 0 and $\overline{\mathcal{K}}$ is its algebraic closure. Let $\mathcal{K}[\boldsymbol{u}]$ be the ring of polynomials in $\boldsymbol{u}$ with coefficients in $\mathcal{K}$ and $\mathcal{K}(\boldsymbol{u})$ be the rational function field.

A non-empty finite subset P of $\mathcal{K}[\boldsymbol{u}][\boldsymbol{x}](= \mathcal{K}[\boldsymbol{u}, \boldsymbol{x}])$ is said to be a *polynomial set* or a *system*. If $\mathrm{P} \subset \mathcal{K}[\boldsymbol{u}, \boldsymbol{x}] \backslash \mathcal{K}[\boldsymbol{x}]$, it is called a *parametric system*. If $\mathrm{P} \subset \mathcal{K}[\boldsymbol{x}]$, it is called a *constant system*. For a system $\mathrm{P} \subset \mathcal{K}[\boldsymbol{u}, \boldsymbol{x}]$ $(\mathcal{K}[\boldsymbol{x}])$, $\langle \mathrm{P} \rangle_{\mathcal{K}[\boldsymbol{u}, \boldsymbol{x}]}$ $(\langle \mathrm{P} \rangle_{\mathcal{K}[\boldsymbol{x}]})$ denotes the ideal generated by P in $\mathcal{K}[\boldsymbol{u}, \boldsymbol{x}]$ $(\mathcal{K}[\boldsymbol{x}])$.

For any f in $\mathcal{K}[\boldsymbol{u}, \boldsymbol{x}] \backslash \{0\}$ $(\mathcal{K}[\boldsymbol{x}] \backslash \{0\})$ and for any $x \in \boldsymbol{x}$, if x appears in f, f can be regarded as a univariate polynomial in x, namely

$$f = c_0 x^m + c_1 x^{m-1} + \ldots + c_m$$

where $c_0, c_1, \ldots, c_m$ are polynomials in $\mathcal{K}[\boldsymbol{u}][\boldsymbol{x} \backslash \{x\}]$ $(\mathcal{K}[\boldsymbol{x} \backslash \{x\}])$ and $c_0 \neq 0$.

Then m is the *leading degree* of f w.r.t. x and is denoted by $\deg(f, x)$. Note that if x does not appear in f, $\deg(f, x) = 0$. The *class* of f is the biggest index k such that $\deg(f, x_k) > 0$. If $\deg(f, x_i) = 0$ for every i $(1 \leq i \leq n)$, then the *class* of f is 0. The class of f in $\mathcal{K}[\boldsymbol{u}, \boldsymbol{x}] \backslash \{0\}$ $(\mathcal{K}[\boldsymbol{x}] \backslash \{0\})$ is denoted by $\mathrm{cls}(f)$. If $k = \mathrm{cls}(f) > 0$, x_k is the *main variable* of f and is denoted by $\mathrm{mvar}(f)$ or $\mathrm{lv}(f)$.

25

Assume that $f = c_0 x_k^m + c_1 x_k^{m-1} + \ldots + c_m$ where $k = \mathrm{cls}(f) > 0$ and $c_0 \neq 0$, then c_0, denoted by $\mathrm{lc}(f)$ or lc_f, is called the *leading coefficient* or the *initial* of f and x_k^m, denoted by $\mathrm{rank}(f)$, the *rank* of f. For a polynomial set P in $\mathcal{K}[\boldsymbol{u}, \boldsymbol{x}]$, lc_P (or $\mathrm{lc}(\mathrm{P})$), $\mathrm{mvar}(\mathrm{P})$ and $\mathrm{rank}(\mathrm{P})$ denote $\Pi_{p \in \mathrm{P}} \mathrm{lc}_p$, $\{\mathrm{mvar}(p) | p \in \mathrm{P}\}$ and $\{\mathrm{rank}(p) | p \in \mathrm{P}\}$, respectively.

Let $\tilde{\mathcal{K}}$ be an arbitrary extension field of $\mathcal{K}$. The common solutions in $\tilde{\mathcal{K}}$ of a polynomial set $\mathrm{P} \subset \mathcal{K}[\boldsymbol{u}, \boldsymbol{x}]$ is called the *zero set* of P in $\tilde{\mathcal{K}}$ or the *variety* generated by P in $\tilde{\mathcal{K}}$ and is denoted by $\mathrm{V}(\mathrm{P})$, that is,

$$\mathrm{V}(\mathrm{P}) = \{(v_1, \ldots, v_d, a_1, \ldots, a_n) \in \tilde{\mathcal{K}}^{d+n} | f(v_1, \ldots, v_d, , a_1, \ldots, a_n) = 0, \forall f \in \mathrm{P}\}.$$

Particularly, if $\tilde{\mathcal{K}}$ is an extension field of $\mathcal{K}(\boldsymbol{u})$,

$$\mathrm{V}(\mathrm{P}) = \{(a_1, \ldots, a_n) \in \tilde{\mathcal{K}}^n | f(\boldsymbol{u}, a_1, \ldots, a_n) = 0, \forall f \in \mathrm{P}\}.$$

If it is necessary to indicate the field $\mathcal{K}'$, we use $\mathrm{V}_{\mathcal{K}'}(\mathrm{P})$. The cases where $\mathcal{K}'$ is $\overline{\mathcal{K}}$, $\overline{\mathcal{K}(\boldsymbol{u})}$ or $\overline{\mathcal{K}(\boldsymbol{u}')}$ (see formula (2.2)) are often discussed in this chapter.

Let Q be another polynomial set, $\mathrm{V}(\mathrm{P} \backslash \mathrm{Q})$ means $\mathrm{V}(\mathrm{P}) \setminus \mathrm{V}(\mathrm{Q})$. If a polynomial set contains only a single polynomial, say f, we use $\mathrm{V}(f)$ instead of $\mathrm{V}(\{f\})$, $\mathrm{V}(\mathrm{P}, f)$ instead of $\mathrm{V}(\mathrm{P} \cup \{f\})$, and $\mathrm{V}(\mathrm{P} \backslash f)$ instead of $\mathrm{V}(\mathrm{P} \backslash \{f\})$.

The so-called *zero decomposition* is to decompose the zero set defined by a polynomial set into a union of finitely many *constructible sets*, e.g., $\mathrm{V}(\mathrm{P}) = \bigcup_i \mathrm{V}(\mathrm{T}_i \backslash \mathrm{Q}_i)$. Usually, these constructible sets are defined by some *triangular sets* (see Definition 2.1), *i.e.* every T_i is of *triangular form*, e.g., $f_1(x_1), f_2(x_1, x_2), \ldots, f_n(x_1, ..., x_n)$. So, *triangular decomposition* is usually another name for zero decomposition.

2.2   Wu's Zero Decomposition

Definition 2.1. Let $\mathrm{T} = [f_1, f_2, \ldots, f_t]$ be a finite nonempty list of nonzero polynomials in $\mathcal{K}[\boldsymbol{u}, \boldsymbol{x}]$. T is said to be a *triangular set* if

$$0 < \mathrm{cls}(f_1) < \mathrm{cls}(f_2) < \cdots < \mathrm{cls}(f_t).$$

Remark 2.1. Note that our definition of triangular set is slightly different from others. By Definition 2.1, if a polynomial list contains elements in $\mathcal{K}[\boldsymbol{u}]$, it is not a triangular set.

In general, a triangular set can be written as

$$T = \begin{bmatrix} f_1(\boldsymbol{u}, x_1, \ldots, x_{k_1}), \\ f_2(\boldsymbol{u}, x_1, \ldots, x_{k_1}, \ldots, x_{k_2}), \\ \cdots \\ f_t(\boldsymbol{u}, x_1, \ldots, x_{k_1}, \ldots, x_{k_2}, \ldots, x_{k_t}) \end{bmatrix}, \tag{2.1}$$

where

$$0 < k_1 < k_2 < \cdots < k_t \le n,$$
$$k_i = \mathrm{cls}(f_i), \quad x_{k_i} = \mathrm{lv}(f_i), \quad i = 1, \ldots, t.$$

Remark 2.2. Sometimes, for a given triangular set in the form of (2.1) in $\mathcal{K}[\boldsymbol{u}, \boldsymbol{x}]$ or $\mathcal{K}[\boldsymbol{x}]$, we may re-name the main variables, *e.g.*, set $y_i = x_{k_i}$, and view the other variables as parameters. Then the triangular set can be written as

$$T = [f_1(\boldsymbol{u}', y_1), f_2(\boldsymbol{u}', y_1, y_2), \ldots, f_t(\boldsymbol{u}', y_1, \ldots, y_t)], \tag{2.2}$$

where $\boldsymbol{u}'$ is the union set of $\boldsymbol{u}$ and those non-main variables.

Definition 2.2. Assume f is a polynomial in $\mathcal{K}[\boldsymbol{u}, \boldsymbol{x}]$ and T is a triangular set of the form (2.1). Set

$$R_t = f,$$
$$R_{t-1} = \mathrm{prem}(R_t, f_t, x_{k_t}),$$
$$R_{t-2} = \mathrm{prem}(R_{t-1}, f_{t-1}, x_{k_{t-1}}),$$
$$\cdots,$$
$$R_0 = \mathrm{prem}(R_1, f_1, x_{k_1}).$$

The last pseudo-remainder R_0 is called the *successive pseudo-remainder* (or *pseudo-remainder*) of f with respect to T, denoted by $\mathrm{prem}(f; T)$ or $\mathrm{prem}(f; f_t, \ldots, f_1)$. The above process of computation is called *successive pseudo-division*. If P is a polynomial set,

$$\mathrm{prem}(P; T) = \{\mathrm{prem}(p; T) \mid p \in P\}.$$

Using pseudo-remainder formula (1.3), one can easily obtain the so-called *pseudo-remainder formula* for successive pseudo-division as follows: there exist polynomials q_i $(1 \le i \le t)$ and integers d_i such that

$$I_1^{d_1} \cdots I_t^{d_t} \cdot f = \sum_{i=1}^{t} q_i f_i + R_0, \tag{2.3}$$

where I_i is the initial of f_i.

Assume f is a polynomial with $\mathrm{cls}(f) = k$. A polynomial g is said to be *reduced* with respect to f if $\deg(g, x_k) < \deg(f, x_k)$. If T is a triangular set and g is reduced with respect to every polynomial of T, g is said to be *reduced* with respect to T.

Successive pseudo-division does not guarantee the elimination of any main variables. However, it is true that

$$\deg(R_0, x_{k_i}) < \deg(f_i, x_{k_i}) \quad (1 \le i \le t).$$

That is, R_0 is reduced with respect to T.

Definition 2.3. T is called a *contradictory ascending set* if T is a single nonzero element of $\mathcal{K}[\boldsymbol{u}]$. A triangular set $\mathrm{T} = [f_1, f_2, \ldots, f_t]$ is called a *non-contradictory ascending set* (or *non-contradictory ascending chain*) if each f_j is reduced with respect to f_i $(1 \le i < j \le t)$. An *ascending set* is either a non-contradictory ascending set or a contradictory ascending set.

Remark 2.3. Note that a contradictory ascending set is not a triangular set by Definition 2.1.

Definition 2.4. A non-contradictory ascending set C in $\mathcal{K}[\boldsymbol{u}, \boldsymbol{x}]$ is called a *characteristic set* of a polynomial set $\mathrm{P} \subseteq \mathcal{K}[\boldsymbol{u}, \boldsymbol{x}]$ if

$$\mathrm{C} \subseteq \langle \mathrm{P} \rangle_{\mathcal{K}[\boldsymbol{u}, \boldsymbol{x}]} \quad \text{and} \quad \mathrm{prem}(\mathrm{P}; \mathrm{C}) = \{0\}.$$

Theorem 2.1 (Wu's well-ordering principle). *[Wu (1978, 1994a)] There exists an algorithm which, for an input polynomial set P in $\mathcal{K}[\boldsymbol{u}, \boldsymbol{x}]$, outputs either a contradictory ascending chain meaning that $\mathrm{V}(\mathrm{P}) = \emptyset$, or a (non-contradictory) characteristic set $\mathrm{C} : [f_1, \ldots, f_t]$ of P such that*

$$\mathrm{V}(\mathrm{P}) \subseteq \mathrm{V}(\mathrm{C}), \quad \mathrm{V}(\mathrm{C} \backslash \mathrm{lc}(\mathrm{C})) \subseteq \mathrm{V}(\mathrm{P}).$$

By the well-ordering principle, it is obvious that

$$\mathrm{V}(\mathrm{P}) = \mathrm{V}(\mathrm{C} \backslash \mathrm{lc}(\mathrm{C})) \cup \bigcup_{j=1}^{t} \mathrm{V}(\mathrm{P}, I_j), \tag{2.4}$$

where I_j is the initial of f_j. So, we can call the algorithm repeatedly to decompose all $\mathrm{V}(\mathrm{P}, I_i)$ and obtain new characteristic sets with new initials. The process must terminate, *i.e.* each recursive call returns a contradictory ascending chain (to guarantee the termination, one may make use of $\mathrm{V}(\mathrm{P}, I_i) = \mathrm{V}(\mathrm{P}, I_i, \mathrm{C})$). Finally, one can obtain a series of ascending sets $\mathrm{C}_1, \ldots, \mathrm{C}_m$ such that

(1) C_i is either a contradictory ascending chain or a (non-contradictory) characteristic set in $\mathcal{K}[\boldsymbol{u}, \boldsymbol{x}]$;

(2) If $m = 1$, $V(P) = \emptyset$. Otherwise, suppose $\mathbb{S} = \{C_i | 1 \le i \le m$ and C_i is a (non-contradictory) characteristic set$\}$, then

$$V(P) = \bigcup_{C \in \mathbb{S}} V(C \backslash \mathrm{lc}(C)). \tag{2.5}$$

The set of ascending chains $C_1, \ldots, C_m$ or the formula (2.5) is called *Wu's zero decomposition* of P in $\mathcal{K}[\boldsymbol{u}, \boldsymbol{x}]$ and the algorithm for computing Wu's zero decomposition, omitted here, is denoted as `WuCharSet` in this book.

2.3 Relatively Simplicial Decomposition (RSD)

Let P be a polynomial set and f a polynomial in $\mathcal{K}[\boldsymbol{u}, \boldsymbol{x}]$. We are interested in the following relations between $V(P)$ and $V(f)$:

(1) $V(P) \bigcap V(f) = \emptyset$;
(2) $V(P) \subseteq V(f)$.

In the first case, P and f are called *coprime*; in the second case, P is called *integrally dependent* with respect to f; P and f are called *dependent* otherwise. In Case (1) or (2), P is said to be *relatively simplicial* with respect to f.

Particularly, when P is a *regular chain* (see Definition 2.6), we will discuss how to determine the above relations of P and f. Another key concern is how to deal with the *dependent* case by reducing to the *relatively simplicial* case.

2.3.1 *Regular Chain*

The concept of regular chain was first introduced in [Yang and Zhang (1991)], where it was called *proper chain*. In [Yang *et al.* (1992)], it was renamed as *normal chain*. See also [Yang and Zhang (1994)]. An equivalent concept, *regular chain*, was introduced independently in [Kalkbrener (1993)]. In [Wang (2000, 2002)], the concept was generalized to the triangular system case.

For two polynomials f and p in $\mathcal{K}[\boldsymbol{u}, \boldsymbol{x}]$, $\mathrm{res}(f, p, \mathrm{mvar}(p))$ is denoted simply by $\mathrm{res}(f, p)$.

Definition 2.5. Assume f is a polynomial in $\mathcal{K}[\boldsymbol{u}, \boldsymbol{x}]$ and T is a triangular set of the form (2.1). Set

$$
\begin{aligned}
R_t &= f, \\
R_{t-1} &= \mathrm{res}(R_t, f_t), \\
R_{t-2} &= \mathrm{res}(R_{t-1}, f_{t-1}), \\
&\quad\vdots \\
R_0 &= \mathrm{res}(R_1, f_1).
\end{aligned}
$$

The last resultant R_0 is called the *successive resultant* (or *resultant*) of f with respect to T, denoted by $\mathrm{res}(f; \mathrm{T})$ or $\mathrm{res}(f; f_t, \ldots, f_1)$.

By Lemma 1.1, it is easy to know that: there exist nonzero polynomials g_i $(0 \le i \le t)$ such that

$$
R_0 = \mathrm{res}(f; f_t, \ldots, f_1) = g_0 f + \sum_{i=1}^{t} g_i f_i. \tag{2.6}
$$

The main variables of T are guaranteed no show in R_0. If $t = n$, $R_0 \in \mathcal{K}[\boldsymbol{u}]$.

Definition 2.6. A triangular set $\mathrm{T} = \{f_1, \ldots, f_r\}$ in $\mathcal{K}[\boldsymbol{u}, \boldsymbol{x}]$ is said to be a *regular chain* in $\mathcal{K}[\boldsymbol{u}, \boldsymbol{x}]$, if $\mathrm{lc}_{f_1} \ne 0$ and for each i $(1 < i \le r)$, $\mathrm{res}(\mathrm{lc}_{f_i}; f_{i-1}, \ldots, f_1) \ne 0$.
If $\mathrm{mvar}(\mathrm{T}) = \boldsymbol{x}$, T is called a *zero-dimensional regular chain*.

Definition 2.7. A triangular set $\mathrm{T} = \{f_1, \ldots, f_r\}$ in $\mathcal{K}[\boldsymbol{u}, \boldsymbol{x}]$ is said to be *squarefree* in $\mathcal{K}[\boldsymbol{u}, \boldsymbol{x}]$, if $\mathrm{discrim}(f_1) \ne 0$ and for each i $(1 < i \le r)$, $\mathrm{res}(\mathrm{discrim}(f_i); f_{i-1}, \ldots, f_1) \ne 0$.

It is easy to prove by Theorem 1.1 that

Proposition 2.1. *If* T *is a regular chain in* $\mathcal{K}[\boldsymbol{u}, \boldsymbol{x}]$ *and* $\tilde{\mathcal{K}}$ *is an algebraically closed extension field of* $\mathcal{K}$, *then* $\mathrm{V}_{\tilde{\mathcal{K}}}(\mathrm{T} \backslash \mathrm{lc_T}) \ne \emptyset$.

So, if we compute a regular chain decomposition of a given system, the decomposition does not suffer from the so-called "redundant components problem" because a regular chain always has solutions.

Theorem 2.2. *Suppose* f *is a polynomial in* $\mathcal{K}[\boldsymbol{u}, \boldsymbol{x}]$ *and* T *is a regular chain in the form of (2.2), i.e.*

$$
\mathrm{T} = [f_1(\boldsymbol{u}', y_1), f_2(\boldsymbol{u}', y_1, y_2), \ldots, f_t(\boldsymbol{u}', y_1, \ldots, y_t)],
$$

where $\boldsymbol{u}'$ is the union set of $\boldsymbol{u}$ and those non-main variables. Let $\tilde{\mathcal{K}}$ be an algebraically closed extension field of $\mathcal{K}(\boldsymbol{u}')$, then

$$V_{\tilde{\mathcal{K}}}(\mathrm{T}) \cap V_{\tilde{\mathcal{K}}}(f) = \emptyset \Longleftrightarrow \mathrm{res}(f;\mathrm{T}) \neq 0.$$

Proof. If $V_{\tilde{\mathcal{K}}}(\mathrm{T}) \cap V_{\tilde{\mathcal{K}}}(f) \neq \emptyset$, assume $(y_1^*,\ldots,y_t^*) \in V_{\tilde{\mathcal{K}}}(\mathrm{T}) \cap V_{\tilde{\mathcal{K}}}(f)$. Then substituting $(y_1^*,\ldots,y_t^*)$ in (2.6) will give $\mathrm{res}(f;\mathrm{T}) = 0$.

Now, assume $V_{\tilde{\mathcal{K}}}(\mathrm{T}) \cap V_{\tilde{\mathcal{K}}}(f) = \emptyset$. Then $V_{\tilde{\mathcal{K}}}(\mathrm{T}\backslash\mathrm{lc_T}) \cap V_{\tilde{\mathcal{K}}}(f) = \emptyset$. By Proposition 2.1, suppose $\mathbf{y}_{t-1} = (y_1^*,\ldots,y_{t-1}^*)$ is an arbitrary point of $V_{\tilde{\mathcal{K}}}(\{f_1,\ldots,f_{t-1}\}\backslash\mathrm{lc_T})$. So, $\mathrm{lc}_{f_t}(\mathbf{y}_{t-1}) \neq 0$. Because $f(\mathbf{y}_{t-1},y_t)$ and $f_t(\mathbf{y}_{t-1},y_t)$ have no common zeros, by Theorem 1.1 and Proposition 1.4,

$$R_{t-1}(\mathbf{y}_{t-1}) = \mathrm{lc}_{f_t}(\mathbf{y}_{t-1}) \cdot \mathrm{res}(f(\mathbf{y}_{t-1},y_t), f_t(\mathbf{y}_{t-1},y_t)) \neq 0.$$

Therefore $V_{\tilde{\mathcal{K}}}(\{f_1,\ldots,f_{t-1}\}\backslash\mathrm{lc_T}) \cap V_{\tilde{\mathcal{K}}}(R_{t-1}) = \emptyset$. Then we can use similar deduction to discuss $R_{t-2},\ldots,R_0$ one by one. Finally, we get that $R_0 = \mathrm{res}(f;\mathrm{T}) \neq 0$. $\qquad\square$

Corollary 2.1. *Let T be a triangular set as in (2.2) and $\tilde{\mathcal{K}}$ an algebraically closed extension field of $\mathcal{K}(\boldsymbol{u}')$, then T is a regular chain if and only if*

$$\mathrm{lc}_{T_1} \neq 0 \text{ and for each } i(1 < i \leq t), V_{\tilde{\mathcal{K}}}(\{f_1,\ldots,f_{i-1}\}) \cap V_{\tilde{\mathcal{K}}}(\mathrm{lc}_{T_i}) = \emptyset.$$

Theorem 2.2 provides a method to determine whether a polynomial and a regular chain are coprime in $\tilde{\mathcal{K}}$. Note that if the triangular set is not regular, we do not have similar result. For example, let $f = z + y$ and $\mathrm{T} = [f_1, f_2, f_3]$ where

$$f_1 = x + 1, f_2 = (x+1)y^2 + y + 1, f_3 = (x+1)z^2 - z - 1.$$

It is easy to verify that $\mathrm{res}(f; f_3, f_2, f_1) = 0$. However, $V(\mathrm{T}) \cap V(f) = \emptyset$.

For a polynomial and a regular chain, we wish to have a simple criterion as Theorem 2.2 on whether they are integrally dependent. A natural idea is to consider the successive pseudo-remainder of the polynomial with respect to the regular chain.

Lemma 2.1. *Let T be a triangular set as in (2.2) and f any polynomial in $\mathcal{K}[\boldsymbol{u},\boldsymbol{x}]$. Then*

$$\mathrm{prem}(f;\mathrm{T}) = 0 \Longrightarrow V(\mathrm{T}\backslash\mathrm{lc_T}) \subseteq V(f).$$

Proof. Use pseudo-remainder formula (2.3). $\qquad\square$

However, $\mathrm{prem}(f; \mathrm{T}) = 0$ is not a necessary and sufficient condition for $V(\mathrm{T}\backslash \mathrm{lc_T}) \subseteq V(f)$ as indicated by the following Theorem 2.3.

Definition 2.8. Let I be an ideal in a ring $\mathcal{R}$. The *radical* of I, denoted by $\sqrt{I}$, is the set

$$\{g \in \mathcal{R} \mid g^m \in I \text{ for some integer } m \geq 1\}.$$

The *saturated ideal* of I with respect to an element $f \in \mathcal{R}$, denoted by $I : f^\infty$, is the set

$$\{g \in \mathcal{R} \mid f^s g \in I \text{ for some integer } s \geq 1\}.$$

The *saturated ideal* of a triangular set T in $\mathcal{K}[\boldsymbol{u}, \boldsymbol{x}]$ is defined as

$$\mathrm{sat}(\mathrm{T})_{\mathcal{K}[\boldsymbol{u}, \boldsymbol{x}]} = \langle \mathrm{T} \rangle_{\mathcal{K}[\boldsymbol{u}, \boldsymbol{x}]} : \mathrm{lc}_\mathrm{T}^\infty.$$

Theorem 2.3. *[Wang (2002)](pp. 72–73, pp. 180–182) Let* T *be a regular chain in the form of (2.2),* f *any polynomial in* $\mathcal{K}[\boldsymbol{u}, \boldsymbol{x}]$ *and* $\tilde{\mathcal{K}}$ *an algebraically closed extension field of* $\mathcal{K}(\boldsymbol{u}')$. *Then*

$$\mathrm{prem}(f; \mathrm{T}) = 0 \iff f \in \mathrm{sat}(\mathrm{T}),$$

$$V_{\tilde{\mathcal{K}}}(\mathrm{T}\backslash \mathrm{lc_T}) \subseteq V_{\tilde{\mathcal{K}}}(f) \iff f \in \sqrt{\mathrm{sat}(\mathrm{T})}.$$

In other words,

$$\mathrm{prem}(f; \mathrm{T}) = 0 \iff \mathrm{lc}_\mathrm{T}^s f \in \langle \mathrm{T} \rangle_{\mathcal{K}[\boldsymbol{u}, \boldsymbol{x}]} \text{ for some positive integer } s, \text{ and}$$

$$V_{\tilde{\mathcal{K}}}(\mathrm{T}\backslash \mathrm{lc_T}) \subseteq V_{\tilde{\mathcal{K}}}(f) \iff \mathrm{prem}(f^d; \mathrm{T}) = 0 \text{ for some positive integer } d.$$

2.3.2   *RSD*

As defined before, if a regular chain T and a polynomial f are integrally dependent or coprime, T is said to be relatively simplicial with respect to f. Roughly speaking, being relatively simplicial means that either f is zero at all the zeros of T or f is non-zero at all the zeros of T. If T is dependent but not relatively simplicial with respect to f, we want to compute a zero decomposition of T, *e.g.*,

$$V(\mathrm{T}) = \bigcup V(\mathrm{T}_i)$$

such that each T_i is a regular chain and relatively simplicial with respect to f.

In this subsection, for a zero-dimensional regular chain in $\mathcal{K}[\boldsymbol{x}]$, we introduce an algorithm for computing such decomposition. The algorithm was first proposed in [Yang *et al.* (1992)] and called the WR algorithm. See also [Yang *et al.* (1995, 1996b)]. In this book, we call it the RSD (Relatively Simplicial Decomposition) algorithm.

Let f be a polynomial in $\mathcal{K}[x_1, \ldots, x_k]$ and

$$\mathrm{T} = [f_1(x_1), f_2(x_1, x_2), \ldots, f_k(x_1, \ldots, x_k)]$$

a zero-dimensional regular chain in $\mathcal{K}[x_1, \ldots, x_k]$ with $\mathrm{lv}(f_i) = x_i$. Viewing f and f_k as polynomials in x_k, we compute their subresultant chain with respect to x_k: $S_{\mu+1}, S_\mu, \ldots, S_0$. Using the notations in Chapter 1, we denote the corresponding principal subresultant coefficients (PSC) by R_j $(0 \le j \le \mu + 1)$.

Theorem 2.4. *Let notations be as above. If*

$$\mathrm{prem}(R_0; f_{k-1}, \ldots, f_1) = \cdots = \mathrm{prem}(R_{i-1}; f_{k-1}, \ldots, f_1) = 0$$

but

$$\mathrm{res}(R_i; f_{k-1}, \ldots, f_1) \ne 0,$$

then S_i is a common divisor with highest degree of f and f_k in $\mathcal{K}[x_1, \ldots, x_k]/\mathrm{sat}([f_1, \ldots, f_{k-1}])$.

This theorem is a direct corollary of the theory of subresultants over $\mathcal{K}[x_1, \ldots, x_k]/\mathrm{sat}([f_1, \ldots, f_{k-1}])$. It should be pointed out that, if $k = 1$, the condition of the theorem should be understood as

$$R_0 = \cdots = R_{i-1} = 0 \quad \text{but} \quad R_i \ne 0.$$

Theorem 2.5. *The algorithm RSD terminates correctly.*

Proof. First, consider Lines 6-9. When the algorithm goes into Line 6, we have $\mathrm{prem}(f; \mathrm{T}) \ne 0$ but $\mathrm{res}(f; \mathrm{T}) = 0$. On one hand, by Theorem 2.2, f and f_t have non-trivial common divisors. On the other hand, perform pseudo-division of f by f_t and let

$$\mathrm{lc}_{f_t}^s f = Q f_t + R.$$

Because $\mathrm{prem}(f; \mathrm{T}) \ne 0$,

$$R \notin \mathrm{sat}([f_1, \ldots, f_{t-1}]).$$

Then f_t does not divide f in $\mathcal{K}[x_1, \ldots, x_t]/\mathrm{sat}([f_1, \ldots, f_{t-1}])$. That is to say the common divisor f_{t1} of f and f_t in $\mathcal{K}[x_1, \ldots, x_t]/\mathrm{sat}([f_1, \ldots, f_{t-1}])$ is a

---

Algorithm 2.1 RSD

Input: A polynomial $f \in \mathcal{K}[x_1, \ldots, x_t]$ and a zero-dimensional regular chain $\mathrm{T} : [f_1, \ldots, f_t] \subset \mathcal{K}[x_1, \ldots, x_t]$.

Output: A zero decomposition T_i of T

$$V(\mathrm{T}) = \bigcup_{i \in A} V(\mathrm{T}_i)$$

such that each T_i ($i \in A$ and A is a finite set) is a regular chain and relatively simplicial with respect to f.

1: **if** $\mathrm{prem}(f; \mathrm{T}) = 0$ **or** $\mathrm{res}(f; \mathrm{T}) \neq 0$ **then**
2:     **return** $\{\mathrm{T}\}$
3: **end if**
4: Let j be the least nonnegative integer such that

$$\mathrm{prem}(R_j(f, f_t); f_{t-1}, \ldots, f_1) \neq 0,$$

where $R_j(f, f_t)$ is the principal subresultant coefficient corresponding to the jth subresultant $S_j(f, f_t)$ of f and f_t with respect to x_t;

5: **if** $\mathrm{res}(R_j(f, f_t); f_{t-1}, \ldots, f_1) \neq 0$ **then**
6:     compute, by Theorem 2.4, a common divisor with highest degree of f and f_t in $\mathcal{K}[x_1, \ldots, x_t]/\mathrm{sat}([f_1, \ldots, f_{t-1}])$ and denote it by f_{t1};
7:     $f_{t2} \leftarrow \mathrm{pquo}(f_t, f_{t1}, x_t)$;
8:     $\mathrm{T}_1 \leftarrow [f_1, \ldots, f_{t-1}, f_{t1}]$; $\mathrm{T}_2 \leftarrow [f_1, \ldots, f_{t-1}, f_{t2}]$;
9:     **return** $\mathrm{RSD}(\mathrm{T}_1, f) \bigcup \mathrm{RSD}(\mathrm{T}_2, f)$;
10: **else**
11:     $L' \leftarrow \mathrm{RSD}([f_1, \ldots, f_{t-1}], R_j(f, f_t))$;
12:     **return** $\bigcup_{[\bar{f}_1, \ldots, \bar{f}_{t-1}] \in L'} \mathrm{RSD}([\bar{f}_1, \ldots, \bar{f}_{t-1}, f_t], f)$;
13: **end if**

---

non-trivial divisor with degree less than that of f_t. Therefore the degrees of f_{t1} and f_{t2} are both strictly less than that of f_t.

Second, consider Line 11. It is clear that when we call RSD recursively, the number of polynomials in the ascending chain decreases. If there is only one polynomial in the ascending chain at a certain step of the recursion process, by Theorem 2.4 and the discussion following it, we are actually computing the greatest common divisor of two univariate polynomials, which can be effectively obtained.

In other words, in each recursive call of Algorithm RSD, either the number of polynomials in the ascending chain decreases or the degrees of polynomials in the ascending chain decrease. Therefore, the algorithm must terminate.

Suppose the output of the algorithm is $\{T_i \mid i \in A\}$. It is obvious that, for each T_i, either $\operatorname{prem}(f; T_i) = 0$ or $\operatorname{res}(f; T_i) \neq 0$. That means T_i is relatively simplicial with respect to f.

Finally, we show that each output ascending chain is regular. Note that only at Lines 6-7 is the ascending chain split. First, f_{t1} in Line 6 is computed by Theorem 2.4 and thus

$$\operatorname{res}(\operatorname{lc}_{f_{t1}}; f_{t-1}, \ldots, f_1) \neq 0.$$

Second, f_{t2} is the pseudo-quotient of f_t pseudo-divided by f_{t1}. So, $\operatorname{lc}_{f_{t2}}$ divides lc_{f_t}. Therefore, $\operatorname{lc}_{f_{t2}}(\xi) \neq 0$ for any $\xi \in V(\{f_{t-1}, \ldots, f_1\})$ since T is regular (which means $\operatorname{lc}_{f_t}(\xi) \neq 0$ by Corollary 2.1). In other words, by Definition 2.6 and Corollary 2.1, $[f_1, \ldots, f_{t-1}, f_{t1}]$ and $[f_1, \ldots, f_{t-1}, f_{t2}]$ are both regular. $\qquad\square$

Remark 2.4.

(1) Of course, we can modify the algorithm slightly so that it outputs two sets of regular chains, say $\mathcal{H}$ and $\mathcal{G}$, such that $\mathcal{H}$ is the set of regular chains integrally dependent with respect to f and $\mathcal{G}$ is the set of regular chains coprime with respect to f. For convenience, in the following we often refer the output of RSD to the form $(\mathcal{H}, \mathcal{G})$.

(2) If T_i is a regular chain output by the Algorithm RSD and is integrally dependent with respect to f, it indeed satisfies a stronger condition, *i.e.* $\operatorname{prem}(f; T_i) = 0$.

(3) If $T = [f_1, \ldots, f_t] \subset \mathcal{K}[x_1, \ldots, x_n]$, where $t < n$, is a regular chain, then T is not zero-dimensional. We may re-name the main variables of T as $y_1, \ldots, y_t$ and the other variables as parameter $\boldsymbol{u} = (u_1, \ldots, u_d)$ $(t + d = n)$ and then T can be viewed as a zero-dimensional regular chain in $\mathcal{K}(\boldsymbol{u})[y_1, \ldots, y_t]$. Let $\tilde{\mathcal{K}}$ be an algebraically closed extension field of $\mathcal{K}(\boldsymbol{u})$. For a polynomial $f \in \mathcal{K}[x_1, \ldots, x_n]$, $\operatorname{RSD}(T, f)$ gives a relatively simplicial decomposition of T with respect to f over $\tilde{\mathcal{K}}$. To get a decomposition over $\overline{\mathcal{K}}$, we need to do some further computation. We shall discuss this in the next section and Chapters 3 and 6.

Example 2.1. [Yang *et al.* (1996b)] Compute an RSD decomposition of a regular chain $T = [f_1, \ldots, f_8]$ with respect to a polynomial g, where

$$f_1 = 4x_1^2 - 3,$$
$$f_2 = 2x_2 - 1,$$
$$f_3 = x_3 - 1,$$
$$f_4 = x_4^2 - 3,$$
$$f_5 = 4x_5^2 - 8x_5 + 1,$$
$$f_6 = 2x_6 - 4x_5 + 3,$$
$$f_7 = ((4 - 2x_1)x_4 + 2x_1 - 3)x_7 - 2x_1 + 2,$$
$$f_8 = 2(2 - x_1)x_8 + x_7 - 2,$$
$$g = x_5 x_8 - x_6 x_7.$$

As $\mathrm{prem}(g; T) \neq 0$ but $\mathrm{res}(g; T) = 0$, Algorithm RSD goes into Line 5. Let

$$R^{(8)} = R_0(g, f_8)$$

be the psc of the 0th subresultant, $S_0(g, f_8)$, of g and f_8 with respect to x_8. Since

$$\mathrm{res}(R^{(8)}; f_7, \ldots, f_1) = 0,$$

the algorithm goes into Line 11: call RSD $([f_1, \ldots, f_7], R^{(8)})$. Then it computes $R^{(7)} = R_0(R^{(8)}, f_7)$. Since

$$\mathrm{res}(R^{(7)}; f_6, \ldots, f_1) = 0,$$

the algorithm goes into Line 11 again: call RSD $([f_1, \ldots, f_6], R^{(7)})$. By several such recursive calls, the following is computed

$$R^{(6)} = R_0(R^{(7)}, f_6), \ \ R^{(5)} = R_0(R^{(6)}, f_5)$$

one by one and we have

$$\mathrm{res}(R^{(i)}; f_{i-1}, \ldots, f_1) = 0, \ \ i = 5, 6, 7, 8.$$

When calling RSD $([f_1, \ldots, f_4], R^{(5)})$, f_4 is split into f_{41} and f_{42} at Lines 6-7, where

$$f_{41} = -8625x_4 + 9896x_4 x_1 - 14844 + 17250x_1,$$
$$f_{42} = 9896x_4 x_1 - 17250x_1 - 8625x_4 + 14844.$$

In the computation branch returned with $[f_1, f_2, f_3, f_{42}, f_5]$, the algorithm always returns at Line 2 and we get a component

$$T_1 = [f_1, f_2, f_3, f_{42}, f_5, f_6, f_7, f_8],$$

which is coprime with respect to g.

Consider another computation branch returned with $[f_1, f_2, f_3, f_{41}, f_5]$:

$$\text{RSD}\,([f_1, f_2, f_3, f_{41}, f_5], R^{(6)}).$$

At Lines 6-7, f_5 is split into f_{51} and f_{52} where

$$f_{51} = 212356x_5 - 245252x_1x_5 + 457608x_1 - 396295,$$

$$f_{52} = -981008x_1x_5 + 131584x_1 + 849424x_5 - 113668.$$

All the computation following these two new branches returns at Line 2 and we get two new components:

$$\text{T}_2 = [f_1, f_2, f_3, f_{41}, f_{52}, f_6, f_7, f_8],$$

$$\text{T}_3 = [f_1, f_2, f_3, f_{41}, f_{51}, f_6, f_7, f_8],$$

where T_2 and g are coprime and T_3 and g are integrally dependent (moreover $\text{prem}(g; \text{T}_3) = 0$). The RSD decomposition was completed.

2.4 Weakly RSD

In this section, polynomials are in $\mathcal{K}[u, x]$, $\hat{\mathcal{K}}$ is an extension field of $\mathcal{K}$ and $\tilde{\mathcal{K}}$ is an extension field of $\mathcal{K}(u)$.

We introduce a new concept, *weakly relatively simplicial decomposition* (WRSD), which is a weaker concept compared to relatively simplicial decomposition (RSD) proposed in the last section. Since the WRSD is over $\tilde{\mathcal{K}}$, to prepare for the discussion on decompositions over $\hat{\mathcal{K}}$, we introduce another new concept *regular-decomposition-unstable variety* (RDU) and give an algorithm to compute WRSD and RDU simultaneously.

The main content of Section 2.4 and Section 2.5 is from [Tang *et al.* (2014); Chen *et al.* (2014, 2015)].

2.4.1 *Concepts and Definitions*

To start with a simpler case, we restrict ourselves to the so-called *generic zero-dimensional system* in this section. More general cases will be discussed in Chapter 6.

Definition 2.9. Suppose P is a system in $\mathcal{K}[u, x]$ and $\tilde{\mathcal{K}}$ is an algebraically closed extension field of $\mathcal{K}(u)$.

A finite set $\mathcal{T}$ of triangular sets in $\mathcal{K}[u, x]$ is said to be a *parametric triangular decomposition* of P in $\mathcal{K}[u, x]$ if $V_{\tilde{\mathcal{K}}}(P) = \cup_{T \in \mathcal{T}} V_{\tilde{\mathcal{K}}}(T \backslash lc_T)$.

If $\mathcal{T} = \emptyset$ or $\mathcal{V}_{\tilde{\mathcal{K}}}(\mathrm{T}\backslash \mathrm{lc}_\mathrm{T}) \neq \emptyset$ for any $\mathrm{T} \in \mathcal{T}$, the parametric triangular decomposition is said to be *non-redundant*.

If $\mathcal{T}$ is a finite set of regular chains in $\mathcal{K}[\boldsymbol{u}, \boldsymbol{x}]$, the parametric triangular decomposition is said to be a *parametric regular decomposition*.

Definition 2.10. Suppose $\mathcal{T}$ is a parametric triangular decomposition of a system P in $\mathcal{K}[\boldsymbol{u}, \boldsymbol{x}]$. If $\mathrm{mvar}(\mathrm{T}) = \boldsymbol{x}$ for each triangular set $\mathrm{T} \in \mathcal{T}$, P is said to be a *generic zero-dimensional system*. Otherwise, P is said to be a *generic positive dimensional system*.

Particularly, if $\mathrm{mvar}(\mathrm{C}_i) = \boldsymbol{x}$ for every non-contradictory ascending chain C_i in Wu's zero decomposition (2.5), P is generic zero-dimensional.

Definition 2.11. Let T be a zero-dimensional regular chain in $\mathcal{K}[\boldsymbol{u}, \boldsymbol{x}]$, $f \in \mathcal{K}[\boldsymbol{u}, \boldsymbol{x}]$ and $\tilde{\mathcal{K}}$ an algebraically closed extension field of $\mathcal{K}(\boldsymbol{u})$. Suppose $\mathcal{H}$ and $\mathcal{G}$ are two finite sets of zero-dimensional regular chains in $\mathcal{K}[\boldsymbol{u}, \boldsymbol{x}]$. If

    (1) $\mathrm{V}_{\tilde{\mathcal{K}}}(\mathrm{T}, f) = \cup_{\mathrm{H} \in \mathcal{H}} \mathrm{V}_{\tilde{\mathcal{K}}}(\mathrm{H})$ and
    (2) $\mathrm{V}_{\tilde{\mathcal{K}}}(\mathrm{T}\backslash f) = \cup_{\mathrm{G} \in \mathcal{G}} \mathrm{V}_{\tilde{\mathcal{K}}}(\mathrm{G})$,

then $(\mathcal{H}, \mathcal{G})$ is said to be a *weakly relatively simplicial decomposition* (WRSD) of T with respect to f over $\tilde{\mathcal{K}}$.

Remark 2.5. Note that an RSD is a WRSD but the converse is not true. For instance, $(\{\{x_1^2, x_2\}\}, \{\{x_1 + u, x_2\}\})$ is a WRSD but not an RSD of $\{(x_1 + u)x_1^2, x_2\}$ with respect to $x_1 + x_2$ in $\mathbb{R}[u][x_1, x_2]$ because $\mathrm{prem}(x_1 + x_2; \{x_1^2, x_2\}) = x_1 \neq 0$.

To consider decompositions over $\hat{\mathcal{K}}$, an extension field of $\mathcal{K}$, we need to consider the specialization of the parameter in the system.

For each $a = (a_1, \ldots, a_d) \in \hat{\mathcal{K}}^d$,

$$\begin{aligned}
\phi_a : \mathcal{K}[\boldsymbol{u}, \boldsymbol{x}] &\longrightarrow \hat{\mathcal{K}}[\boldsymbol{x}] \\
f(\boldsymbol{u}, \boldsymbol{x}) &\longrightarrow f(a, \boldsymbol{x})
\end{aligned}$$

is a homomorphism and $\phi_a(f)$ is denoted by $f(a)$. For a non-empty finite set $\mathrm{P} \subset \mathcal{K}[\boldsymbol{u}, \boldsymbol{x}]$, $\mathrm{P}(a)$ denotes the set $\{f(a) | f \in \mathrm{P}\}$. $\mathrm{P}(a) = \emptyset$ if $\mathrm{P} = \emptyset$.

Definition 2.12. [Chen *et al.* (2007)] Let T be a regular chain in $\mathcal{K}[\boldsymbol{u}, \boldsymbol{x}]$ and $a \in \hat{\mathcal{K}}^d$. If $\mathrm{T}(a)$ is a regular chain in $\hat{\mathcal{K}}[\boldsymbol{x}]$ and $\mathrm{rank}(\mathrm{T}(a)) = \mathrm{rank}(\mathrm{T})$, we say that the regular chain T *specializes well* at a.

For any $\mathrm{G} \subset \mathcal{K}[\boldsymbol{u}]$, $\mathrm{V}^{\boldsymbol{u}}(\mathrm{G})$ denotes the set

$$\{(a_1, \ldots, a_d) \in \hat{\mathcal{K}}^d | g(a_1, \ldots, a_d) = 0, \forall g \in \mathrm{G}\}.$$

For any $f \in \mathcal{K}[\boldsymbol{u}, \boldsymbol{x}]$, all the coefficients $c_1, \ldots, c_t$ of f in $\boldsymbol{x}$ are polynomials in $\mathcal{K}[\boldsymbol{u}]$. Then $V^{\boldsymbol{u}}(f)$ denotes $V^{\boldsymbol{u}}(\{c_1, \ldots, c_t\})$.

Lemma 2.2. *[Chen* et al. *(2007)] Let* T *be a regular chain in* $\mathcal{K}[\boldsymbol{u}, \boldsymbol{x}]$. *Then* T *specializes well at a if and only if* $a \in \hat{K}^d \backslash V^{\boldsymbol{u}}(\mathrm{res}(\mathrm{lc}_{\mathrm{T}}; \mathrm{T}))$.

Since the definition of regular chain in $\mathcal{K}[\boldsymbol{u}, \boldsymbol{x}]$ in this book is not exactly the same as that in [Chen *et al.* (2007)] as mentioned in Remark 2.1. Therefore, Lemma 2.2 here is stated differently.

Definition 2.13. Suppose $(\mathcal{H}, \mathcal{G})$ is a WRSD of a zero-dimensional regular chain T with respect to a polynomial f in $\mathcal{K}[\boldsymbol{u}, \boldsymbol{x}]$ and $\hat{\mathcal{K}}$ is an algebraically closed extension field of $\mathcal{K}$. A WRSD $(\mathcal{H}, \mathcal{G})$ is said to be *stable* at $a \in \hat{\mathcal{K}}^d$ if
    (1) T specializes well at a,
    (2) $V_{\hat{\mathcal{K}}}(\mathrm{T}(a), f(a)) = \cup_{\mathrm{H} \in \mathcal{H}} V_{\hat{\mathcal{K}}}(\mathrm{H}(a))$ and H specializes well at a for any $\mathrm{H} \in \mathcal{H}$, and
    (3) $V_{\hat{\mathcal{K}}}(\mathrm{T}(a) \backslash f(a)) = \cup_{\mathrm{G} \in \mathcal{G}} V_{\hat{\mathcal{K}}}(\mathrm{G}(a))$ and G specializes well at a for any $\mathrm{G} \in \mathcal{G}$.

Definition 2.14. Let $\mathcal{T}$ be a parametric triangular decomposition of a given generic zero-dimensional system P in $\mathcal{K}[\boldsymbol{u}, \boldsymbol{x}]$ such that $V_{\tilde{\mathcal{K}}}(\mathrm{P}) = \cup_{\mathrm{T} \in \mathcal{T}} V_{\tilde{\mathcal{K}}}(\mathrm{T} \backslash \mathrm{lc}_{\mathrm{T}})$. For a given $a \in \hat{\mathcal{K}}^d$, if

$$V_{\hat{\mathcal{K}}}(\mathrm{P}(a)) = \cup_{\mathrm{T} \in \mathcal{T}} V_{\hat{\mathcal{K}}}(\mathrm{T}(a) \backslash \mathrm{lc}_{\mathrm{T}(a)}) \quad \text{and} \quad \mathrm{rank}(\mathrm{T}) = \mathrm{rank}(\mathrm{T}(a)) \; \forall \mathrm{T} \in \mathcal{T},$$

then the decomposition $\mathcal{T}$ is said to be *stable* at a.
    Assume that $\mathcal{T}$ is a parametric *regular* decomposition of P. $\mathcal{T}$ is said to be *stable* at a if each $\mathrm{T} \in \mathcal{T}$ specializes well at a and $V_{\hat{\mathcal{K}}}(\mathrm{P}(a)) = \cup_{\mathrm{T} \in \mathcal{T}} V_{\hat{\mathcal{K}}}(\mathrm{T}(a) \backslash \mathrm{lc}_{\mathrm{T}(a)})$.

Definition 2.15. Let $\mathcal{T}$ be a parametric regular decomposition of a given generic zero-dimensional system P in $\mathcal{K}[\boldsymbol{u}, \boldsymbol{x}]$ and $\mathcal{V}$ an affine variety in $\hat{\mathcal{K}}^d$ with $\dim(\mathcal{V}) < d$, where $\dim(\mathcal{V})$ is the dimension of $\mathcal{V}$. If $\mathcal{T}$ is stable at any $a \in \hat{\mathcal{K}}^d \backslash \mathcal{V}$, then $\mathcal{T}$ is said to be a *generic regular decomposition* (GRD) of P and $\mathcal{V}$ is said to be a *regular-decomposition-unstable variety* (RDU) of $\mathcal{T}$.
    A pair $[\mathcal{T}, p]$ is called a *generic regular decomposition* of P in $\mathcal{K}[\boldsymbol{u}, \boldsymbol{x}]$, if $\mathcal{T}$ is a decomposition as above and $V^{\boldsymbol{u}}(p)$ is the RDU. That is to say,
    (1) $\mathcal{T}$ is a finite set of zero-dimensional regular chains in $\mathcal{K}[\boldsymbol{u}, \boldsymbol{x}]$ such that $V_{\tilde{\mathcal{K}}}(\mathrm{P}) = \cup_{\mathrm{T} \in \mathcal{T}} V_{\tilde{\mathcal{K}}}(\mathrm{T})$;
    (2) p is a polynomial in $\mathcal{K}[\boldsymbol{u}]$ such that for any $a \in \hat{\mathcal{K}}^d \backslash V^{\boldsymbol{u}}(p)$, $V(\mathrm{P}(a)) = \cup_{\mathrm{T} \in \mathcal{T}} V(\mathrm{T}(a))$ and T specializes well at a for any $\mathrm{T} \in \mathcal{T}$.

2.4.2 *Algorithm* WRSD

Algorithm 2.2 WRSD

Input: A zero-dimensional regular chain $T = \{T_1, \ldots, T_n\}$ in $\mathcal{K}[\boldsymbol{u}, \boldsymbol{x}]$, a polynomial $f \in \mathcal{K}[\boldsymbol{u}, \boldsymbol{x}]$, variables $\boldsymbol{x} = \{x_1, \ldots, x_n\}$

Output: $[\mathcal{H}, \mathcal{G}, p]$, where

    (1) $(\mathcal{H}, \mathcal{G})$ is a WRSD of T with respect to f;

    (2) $p \in \mathcal{K}[\boldsymbol{u}]$ such that for any $a \in \hat{\mathcal{K}}^d \backslash \mathrm{V}^{\boldsymbol{u}}(p)$, $(\mathcal{H}, \mathcal{G})$ is stable at a.

 1: $\mathcal{H} \leftarrow \emptyset$; $\mathcal{G} \leftarrow \emptyset$; $p \leftarrow \mathrm{res}(\mathrm{lc}_T; \mathrm{T})$;

 2: **if** f is not reduced *w.r.t.* T **then return** WRSD$(\mathrm{T}, \mathrm{prem}(f; \mathrm{T}), \boldsymbol{x})$; **end if**

 3: **if** $f = 0$ **then return** $[\{\mathrm{T}\}, \emptyset, p]$; **end if**

 4: **if** $\mathrm{cls}(f) = 0$ **then return** $[\emptyset, \{\mathrm{T}\}, f \cdot p]$; **end if**

 5: **if** $\mathrm{cls}(f) \neq n$ **then**

 6:     $W \leftarrow$ WRSD$(\{T_1, \ldots, T_{\mathrm{cls}(f)}\}, f, \{x_1, \ldots, x_{\mathrm{cls}(f)}\})$;

 7:     $\mathcal{H} \leftarrow \mathrm{map}(t \to t \cup \{T_{\mathrm{cls}(f)+1}, \ldots, T_n\}, W_1)$;

 8:     $\mathcal{G} \leftarrow \mathrm{map}(t \to t \cup \{T_{\mathrm{cls}(f)+1}, \ldots, T_n\}, W_2)$; **return** $[\mathcal{H}, \mathcal{G}, p \cdot W_3]$;

 9: **end if**

10: **if** $\mathrm{res}(f; \mathrm{T}) \neq 0$ **then**

11:     $p \leftarrow p \cdot \mathrm{res}(f; \mathrm{T})$; $\mathcal{G} \leftarrow \{\mathrm{T}\}$; **return** $[\mathcal{H}, \mathcal{G}, p]$;

12: **end if**

13: compute the regular subresultant chain $S_{d_v}, \ldots, S_{d_0}$ of T_n and f w.r.t. x_n;

14: **if** $n = 1$ **then**

15:     $q \leftarrow \mathrm{pquo}(T_1, S_{d_1}, x_1)$; $W \leftarrow$ WRSD$(\{q\}, f, \boldsymbol{x})$;

16:     $\mathcal{H} \leftarrow \{\{S_{d_1}\}\}$; $\mathcal{G} \leftarrow W_2$; $p \leftarrow W_3$;

17: **else**

18:     $\boldsymbol{x}_{n-1} \leftarrow \boldsymbol{x} \backslash \{x_n\}$; $W \leftarrow$ WRSD$(\{T_1, \ldots, T_{n-1}\}, S_{d_0}, \boldsymbol{x}_{n-1})$;

19:     $\mathcal{H}_0 \leftarrow W_1$; $\mathcal{G}_0 \leftarrow W_2$; $p \leftarrow W_3$; $\mathcal{G} \leftarrow \mathcal{G} \cup \mathrm{map}(t \to t \cup \{T_n\}, \mathcal{G}_0)$;

20:     $i \leftarrow 0$; $S_{d_{v+1}} \leftarrow T_n$;

21:     **while** $\mathcal{H}_i \neq \emptyset$ **do**

22:        $i \leftarrow i + 1$; $\mathcal{H}_i \leftarrow \emptyset$; $\mathcal{G}_i \leftarrow \emptyset$;

23:        Let R_{d_i} be the d_ith psc of T_n and f w.r.t. x_n;

24:        **for** $\mathrm{H} \in \mathcal{H}_{i-1}$ **do**

25:           $W \leftarrow$ WRSD$(\mathrm{H}, R_{d_i}, \boldsymbol{x}_{n-1})$; $\mathcal{H}_i \leftarrow \mathcal{H}_i \cup W_1$; $\mathcal{G}_i \leftarrow \mathcal{G}_i \cup W_2$; $p \leftarrow p \cdot W_3$;

26:        **end for**

27:        **for** $\mathrm{G} \in \mathcal{G}_i$ **do**

28:           $\mathcal{H} \leftarrow \mathcal{H} \cup \{\mathrm{G} \cup \{S_{d_i}\}\}$; $q \leftarrow \mathrm{pquo}(T_n, S_{d_i}, x_n)$;

29:           **if** $\deg(q, x_n) > 0$ **then**

30:              $W \leftarrow$ WRSD$(\mathrm{G} \cup \{q\}, f, \boldsymbol{x}_{n-1})$; $\mathcal{G} \leftarrow \mathcal{G} \cup W_2$; $p \leftarrow p \cdot W_3$;

31:           **end if**

32:        **end for**

33:     **end while**

34: **end if**

35: **return** $[\mathcal{H}, \mathcal{G}, p]$

Now we present Algorithm `WRSD` for computing weakly relatively simplicial decompositions.

We introduce some notations used in the pseudocodes. Assume that `Alg` is an algorithm and $p_1, \ldots, p_t$ is a sequence of inputs of this algorithm. If the output of $\mathtt{Alg}(p_1, \ldots, p_t)$ is a finite list $[q_1, \ldots, q_s]$, q_i is denoted by $\mathtt{Alg}(p_1, \ldots, p_t)_i$ for any i $(1 \leq i \leq s)$ and also said to be the ith output of $\mathtt{Alg}(p_1, \ldots, p_t)$. Given a finite set $S = \{s_1, \ldots, s_t\}$, $\mathtt{op}(S)$ denotes the finite sequence $s_1, \ldots, s_t$ and $\mathtt{map}(s \to \phi(s), S)$ denotes the set $\phi(S) = \{\phi(s_1), \ldots, \phi(s_t)\}$ for a mapping ϕ on S.

2.4.3 *Correctness of Algorithm* WRSD

Let us show the termination and correctness of Algorithm `WRSD`. Roughly speaking, Algorithm `WRSD` is based on Lemma 2.3 which is inspired by similar results presented in [Wang (2000, 2001)]. Note that the results shown in Lemma 2.3 are not covered in [Wang (2000, 2001)].

Lemma 2.3. *Given two polynomials f and g in $\mathcal{K}[\boldsymbol{u}, \boldsymbol{x}]$ with $0 < \deg(g, x_n) < \deg(f, x_n)$. Let $S_{d_v}, \ldots, S_{d_1}, S_{d_0}$ be the regular subresultant chain of f and g with respect to x_n. Let $S_{d_{v+1}} = f$ and R_{d_i} the d_ith PSC of f and g with respect to x_n for any i $(0 \leq i \leq v+1)$. Assume q_{d_i} is the pseudo-quotient of f pseudo-divided by S_{d_i} with respect to x_n for any i $(1 \leq i \leq v)$. Then*

(1) $\mathrm{V}_{\tilde{\mathcal{K}}}(\{f, g\}\backslash \mathrm{lc}_f) = \cup_{i=1}^{v+1} \mathrm{V}_{\tilde{\mathcal{K}}}(\{S_{d_i}, R_{d_{i-1}}, \ldots, R_{d_0}\}\backslash \mathrm{lc}_f R_{d_i});$

(2) $\mathrm{V}_{\tilde{\mathcal{K}}}(f \backslash \mathrm{glc}_f) = \mathrm{V}_{\tilde{\mathcal{K}}}(f \backslash \mathrm{glc}_f R_{d_0}) \cup \cup_{i=1}^{v} \mathrm{V}_{\tilde{\mathcal{K}}}(\{q_{d_i}, R_{d_{i-1}}, \ldots, R_{d_0}\}\backslash \mathrm{glc}_f R_{d_i}).$

Proof. Assume that $S_{\mu+1}, S_\mu, \ldots, S_1, S_0$ is the subresultant chain of f and g with respect to x_n. Remark that $S_{d_0} = S_0 = \mathrm{res}(f, g)$, $S_\mu = g$, $S_{\mu+1} = f$ and $S_{d_v} = \mathrm{lc}_g^c g$ where c is a non-negative integer.

(1) For any $(a_1, \ldots, a_n) \in \mathrm{V}_{\tilde{\mathcal{K}}}(\{f, g\}\backslash \mathrm{lc}_f)$, let $b = (a_1, \ldots, a_{n-1})$. If $g(b) = 0$, by the definition of PSC, $R_{d_i}(b) = 0$ and thus $R_{d_i}(a) = 0$ for any i $(1 \leq i \leq v)$. Hence,

$$(a_1, \ldots, a_n) \in \mathrm{V}_{\tilde{\mathcal{K}}}(\{S_{d_{v+1}}, R_{d_v}, \ldots, R_{d_0}\}\backslash \mathrm{lc}_f R_{d_{v+1}}).$$

If $\deg(g(b), x_n) > 0$, since $\deg(g(b), x_n) < \deg(f(b), x_n) = \deg(f, x_n)$, it is reasonable to assume that the subresultant chain of $f(b)$ and $g(b)$ with respect to x_n is $\widetilde{S}_{\mu+1}, \ldots, \widetilde{S}_0$ and the associated PSC's are $\widetilde{R}_{\mu+1}, \ldots, \widetilde{R}_0$. Note that $\widetilde{S}_\mu = g(b), \widetilde{S}_{\mu+1} = f(b)$ and by Proposition 1.4, we know that $S_j(b) = \mathrm{lc}_f(b)^{r_j} \widetilde{S}_j$ where r_j is a non-negative integer for any j $(1 \leq j \leq \mu + 1)$.

According to the theories of subresultant chains, there exists an integer j $(1 \leq j \leq \mu)$ such that $\widetilde{R}_j \neq 0$ and $\widetilde{R}_0 = \ldots = \widetilde{R}_{j-1} = 0$. Then $R_j(b) \neq 0$ and $R_0(b) = \ldots = R_{j-1}(b) = 0$. In addition, $\widetilde{S}_j$ is the greatest common divisor of $f(b)$ and $g(b)$ in $\tilde{\mathcal{K}}[x_n]$ and $\deg(\widetilde{S}_j, x_n) = j$. Hence $\widetilde{S}_j(a_n) = 0$ by $f(b)(a_n) = g(b)(a_n) = 0$. Note that $\deg(S_j, x_n) = \deg(S_j(b), x_n) = \deg(\widetilde{S}_j, x_n) = j$, so there exists some i $(1 \leq i \leq v)$ such that $d_i = j$. Therefore, $(a_1, \ldots, a_n) \in V_{\tilde{\mathcal{K}}}(\{S_{d_i}, R_{d_i-1}, \ldots, R_{d_0}\}\backslash \mathrm{lc}_f R_{d_i})$.

On the other hand, for any

$$(a_1, \ldots, a_n) \in V_{\tilde{\mathcal{K}}}(\{S_{d_{v+1}}, R_{d_v}, \ldots, R_{d_0}\}\backslash \mathrm{lc}_f R_{d_{v+1}}),$$

let $b = (a_1, \ldots, a_{n-1})$. As $R_{d_i}(b) = 0$ for any $i(1 \leq i \leq v)$, $g(b) = 0$ follows from Proposition 1.4. Hence, $(a_1, \ldots, a_n) \in V_{\tilde{\mathcal{K}}}(\{f, g\}\backslash \mathrm{lc}_f)$. Similarly, for any i $(1 \leq i \leq v)$ and for any $(a_1, \ldots, a_n) \in V_{\tilde{\mathcal{K}}}(\{S_{d_i}, R_{d_i-1}, \ldots, R_{d_0}\}\backslash \mathrm{lc}_f R_{d_i})$, it is not difficult to check $(a_1, \ldots, a_n) \in V_{\tilde{\mathcal{K}}}(\{f, g\}\backslash \mathrm{lc}_f)$.

(2) The proof is similar to that of (1). $\square$

Lemma 2.4. *Let $f \in \mathcal{K}[\boldsymbol{u}, \boldsymbol{x}]$ and $\mathrm{T} = \{T_1, \ldots, T_n\}$ be a zero-dimensional regular chain in $\mathcal{K}[\boldsymbol{u}, \boldsymbol{x}]$. If $S_0 = \mathrm{res}(f; \mathrm{T}) \neq 0$, then $\mathrm{res}(f(a); \mathrm{T}(a)) \neq 0$ for any $a \in \hat{\mathcal{K}}^d \backslash V^{\boldsymbol{u}}(S_0 \mathrm{res}(\mathrm{lc}_{\mathrm{T}}; \mathrm{T}))$.*

Proof. It is not difficult to prove the conclusion by induction on n. $\square$

Lemma 2.5. *Let $f \in \mathcal{K}[\boldsymbol{u}, \boldsymbol{x}]$ and $\mathrm{T} = \{T_1, \ldots, T_n\}$ be a zero-dimensional regular chain in $\mathcal{K}[\boldsymbol{u}, \boldsymbol{x}]$. Suppose $f_1 = \mathrm{prem}(f; \mathrm{T})$. Then*

$$V_{\tilde{\mathcal{K}}}(\mathrm{T}, f) = V_{\tilde{\mathcal{K}}}(\mathrm{T}, f_1) \;\; and \;\; V_{\tilde{\mathcal{K}}}(\mathrm{T}\backslash f) = V_{\tilde{\mathcal{K}}}(\mathrm{T}\backslash f_1).$$

Furthermore,

$$V_{\hat{\mathcal{K}}}(\mathrm{T}(a), f(a)) = V_{\hat{\mathcal{K}}}(\mathrm{T}(a), f_1(a)) \;\; and \;\; V_{\hat{\mathcal{K}}}(\mathrm{T}(a)\backslash f(a)) = V_{\hat{\mathcal{K}}}(\mathrm{T}(a)\backslash f_1(a))$$

for any $a \in \hat{\mathcal{K}}^d \backslash V^{\boldsymbol{u}}(\mathrm{res}(\mathrm{lc}_{\mathrm{T}}; \mathrm{T}))$.

Proof. It is easy to prove the conclusion by the definition of successive pseudo-division and Lemma 2.2. $\square$

Theorem 2.6. *Algorithm* WRSD *terminates correctly.*

Proof. The termination is similar to the termination of Algorithm RSD in the last section. For a given zero-dimensional regular chain $\mathrm{T} = \{T_1, \ldots, T_n\}$ in $\mathcal{K}[\boldsymbol{u}, \boldsymbol{x}]$ and a polynomial f in $\mathcal{K}[\boldsymbol{u}, \boldsymbol{x}]$, let $[\mathcal{H}, \mathcal{G}, p] = \mathrm{WRSD}(\mathrm{T}, f, \boldsymbol{x})$. Now we prove the correctness by induction on the recursive depth s of $\mathrm{WRSD}(\mathrm{T}, f, \boldsymbol{x})$. Note that we only need to prove that $\mathcal{H}, \mathcal{G}$ and p satisfy the conditions stated in Definitions 2.11 and 2.13.

When $s = 0$, *i.e.* the algorithm returns at Line 3, Line 4 or Line 6 without recursive calls, the conclusion follows from Lemma 2.2 and Lemma 2.4. Assume that the conclusion holds when $s < N$ ($N > 0$) and suppose $s = N$.

If the N-level recursive call occurs at Line 2 or Line 6, the conclusion follows from the induction hypothesis and Lemma 2.5. Now we prove the conclusion when the N-level recursive call occurs at Line 15 or Line 18, which means f is reduced with respect to T, $\mathrm{mvar}(f) = x_n$ and $\mathrm{res}(f;\mathrm{T}) = 0$. Suppose

$$S_{\mu+1}, S_\mu, \ldots, S_1, S_0$$

is the subresultant chain of T_n and f with respect to x_n in $\mathcal{K}[\boldsymbol{u}, \boldsymbol{x}_{n-1}][x_n]$ where $\boldsymbol{x}_{n-1} = \boldsymbol{x} \backslash \{x_n\}$ and

$$S_{d_\upsilon}, \ldots, S_{d_1}, S_{d_0}$$

is the associated regular subresultant chain. Note that $\deg(T_n, x_n) > \deg(f, x_n) > 0$ since f is reduced with respect to T.

If $n = 1$, S_{d_1} is the greatest common divisor of T_1 and f in $\mathcal{K}[\boldsymbol{u}][x_1]$ and hence $V_{\tilde{\mathcal{K}}}(\{T_1, f\}) = V_{\tilde{\mathcal{K}}}(\{S_{d_1}\})$. Then condition (1) in Definition 2.11 holds. Suppose $q = \mathrm{pquo}(T_1, S_{d_1}, x_1)$. Remark that $\deg(q, x_1) > 0$ and there exists a positive integer k such that $k \geq 2$ and $\mathrm{lc}_{S_{d_1}}^k T_1 = S_{d_1} q$. Thus $V_{\tilde{\mathcal{K}}}(\mathrm{T}) = V_{\tilde{\mathcal{K}}}(S_{d_1} q)$. Note that $V_{\tilde{\mathcal{K}}}(S_{d_1}) \subset V_{\tilde{\mathcal{K}}}(f)$. So $V_{\tilde{\mathcal{K}}}(\mathrm{T} \backslash f) = V_{\tilde{\mathcal{K}}}(S_{d_1} q \backslash f) = V_{\tilde{\mathcal{K}}}(q \backslash f)$. Therefore condition (2) in Definition 2.11 follows from the induction hypothesis. Remark that $\mathrm{lc}_{S_{d_1}}$ is a factor of lc_q and according to Algorithm WRSD, lc_q is a factor of p. Thus for any $a \in \hat{\mathcal{K}}^d \backslash V^{\boldsymbol{u}}(p)$, $\mathrm{lc}_{T_1}(a) \neq 0$ and $\deg(f(a), x_1) \geq d_1 > 0$ by the definition of subresultant. Obviously, condition (1) in Definition 2.13 holds. Besides, according to Proposition 1.4, $S_{d_1}(a)$ is the great common divisor of $f(a)$ and $T_1(a)$ in $\hat{\mathcal{K}}[x_1]$. Thus $V(\mathrm{T}(a) \cup f(a)) = V(S_{d_1}(a))$ and condition (2) in Definition 2.13 holds. Since $\mathrm{lc}_{S_{d_1}}(a)^k T_1(a) = S_{d_1}(a) q(a)$ and $V(S_{d_1}(a)) \subset V(f(a))$, $V(\mathrm{T}(a) \backslash f(a)) = V(S_{d_1}(a) q(a) \backslash f(a)) = V(q(a) \backslash f(a))$. Therefore condition (3) in Definition 2.13 follows from the induction hypothesis.

If $n > 1$, let $S_{d_{\upsilon+1}} = S_{\mu+1}$ and $\mathrm{T}_{n-1} = \{T_1, \ldots, T_{n-1}\}$. Suppose R_{d_i} is the principal subresultant coefficient of T_n and f with respect to x_n for any i ($0 \leq i \leq \upsilon + 1$) and assume that

$$\mathcal{H}_0 = \mathrm{WRSD}(\mathrm{T}_{n-1}, S_{d_0}, \boldsymbol{x}_{n-1})_1 \quad \text{and} \quad \mathcal{G}_0 = \mathrm{WRSD}(\mathrm{T}_{n-1}, S_{d_0}, \boldsymbol{x}_{n-1})_2.$$

Remark that $\mathcal{H}_0 \neq \emptyset$ because $\mathrm{res}(R_{d_0}; \mathrm{T}_{n-1}) = \mathrm{res}(f; \mathrm{T}) = 0$. For any i ($1 \leq i$), let

$$\mathcal{H}_i = \cup_{\mathrm{H} \in \mathcal{H}_{i-1}} \mathrm{WRSD}(\mathrm{H}, R_{d_i}, \boldsymbol{x}_{n-1})_1 \quad \text{and} \quad \mathcal{G}_i = \cup_{\mathrm{H} \in \mathcal{H}_{i-1}} \mathrm{WRSD}(\mathrm{H}, R_{d_i}, \boldsymbol{x}_{n-1})_2$$

until there exists an integer l $(1 \le l \le v+1)$ such that $\mathcal{H}_l = \emptyset$. That means $\mathcal{H}_l = \emptyset$ and $\mathcal{H}_j \ne \emptyset$ for any j $(0 \le j < l)$. We can always get this integer l owing to the fact that $S_{d_{v+1}} = T_n$.

Then we have two sequences $\mathcal{H}_0, \mathcal{H}_1, \ldots, \mathcal{H}_l$ and $\mathcal{G}_0, \mathcal{G}_1, \ldots, \mathcal{G}_l$. Let

$$L_1 = \{i \mid 1 \le i \le l, \mathcal{G}_i \ne \emptyset\}.$$

According to Algorithm WRSD, the first output of $\mathtt{WRSD}(T, f, \boldsymbol{x})$ is $\mathcal{H} = \cup_{i \in L_1} \cup_{G \in \mathcal{G}_i} (G \cup \{S_{d_i}\})$. It is not difficult to see that $\mathcal{H}$ is a finite set of zero-dimensional regular chains in $\mathcal{K}[\boldsymbol{u}, \boldsymbol{x}]$. By Lemma 2.3(1), we know that

$$V_{\tilde{\mathcal{K}}}(T \cup \{f\}) = \cup_{i=1}^{v+1}(V_{\tilde{\mathcal{K}}}(T_{n-1}) \cap V_{\tilde{\mathcal{K}}}(\{S_{d_i}, R_{d_i-1}, \ldots, R_{d_0}\} \backslash \mathrm{lc}_{T_n} R_{d_i})).$$

For any i $(1 \le i \le v+1)$, if $i \in L_1$, according to the induction hypothesis and the construction of $\mathcal{G}_i$, we get

$$V_{\tilde{\mathcal{K}}}(T_{n-1}) \cap V_{\tilde{\mathcal{K}}}(\{S_{d_i}, R_{d_i-1}, \ldots, R_{d_0}\} \backslash \mathrm{lc}_{T_n} R_{d_i}) = \cup_{G \in \mathcal{G}_i} V_{\tilde{\mathcal{K}}}(G \cup \{S_{d_i}\}).$$

If $l < i \le v+1$, according to the induction hypothesis and $\mathcal{H}_l = \emptyset$, similarly, we know that

$$V_{\tilde{\mathcal{K}}}(T_{n-1}) \cap V_{\tilde{\mathcal{K}}}(\{S_{d_i}, R_{d_i-1}, \ldots, R_{d_0}\} \backslash \mathrm{lc}_{T_n} R_{d_i}) = \emptyset.$$

If $1 \le i \le l$ and $i \notin L_1$, similarly, we get

$$V_{\tilde{\mathcal{K}}}(T_{n-1}) \cap V_{\tilde{\mathcal{K}}}(\{S_{d_i}, R_{d_i-1}, \ldots, R_{d_0}\} \backslash \mathrm{lc}_{T_n} R_{d_i}) = \emptyset.$$

So

$$V_{\tilde{\mathcal{K}}}(T \cup \{f\}) = \cup_{i \in L_1} \cup_{G \in \mathcal{G}_i} V_{\tilde{\mathcal{K}}}(G \cup \{S_{d_i}\}) = \cup_{H \in \mathcal{H}} V_{\tilde{\mathcal{K}}}(H)$$

and hence condition (1) in Definition 2.11 holds. Furthermore, as discussed above, we figure out that $V_{\tilde{\mathcal{K}}}(T \cup \{f\}) = \emptyset$ if and only if $\mathcal{H} = \emptyset$. Actually, when $\mathrm{res}(f; T) = 0$, $V_{\tilde{\mathcal{K}}}(T \cup \{f\})$ cannot be $\emptyset$ according to Theorem 2.2 and thus $\mathcal{H} \ne \emptyset$. Similarly, we can prove that condition (2) in Definition 2.11 holds on the basis of Lemma 2.3(2). Besides, it can also be shown that $V_{\tilde{\mathcal{K}}}(T \backslash f) = \emptyset$ if and only if $\mathcal{G} = \emptyset$.

For any $a \in \hat{\mathcal{K}}^d \backslash V^{\boldsymbol{u}}(p)$, T specializes well at a by Line 1 and Lemma 2.2 and thus condition (1) in Definition 2.13 holds. It is also easy to check that H specializes well at a for any $H \in \mathcal{H}$ by the induction hypothesis and we only need to prove that $V(T(a) \cup \{f(a)\}) = \cup_{H \in \mathcal{H}} V(H(a))$.

If $\deg(f(a), x_n) = 0$, it is easy to see that the equality holds.

If $\deg(f(a), x_n) > 0$, let the subresultant chain of $T_n(a)$ and $f(a)$ with respect to x_n be $\widetilde{S}_{\mu+1}, \widetilde{S}_\mu, \ldots, \widetilde{S}_0$. By Proposition 1.4, we know that $S_i(a) = \mathrm{lc}_{T_n}(a)^{r_i} \widetilde{S}_i$ where r_i is a non-negative integer for any i $(0 \le i \le \mu+1)$. Set

$$L_2 = \{i \mid 1 \le i \le v+1, R_{d_i}(a) \ne 0\}.$$

It is not difficult to check that $L_1 \subset L_2$ by the induction hypothesis and we may assume that $L_2 = \{j_1, \ldots, j_k, j_{k+1}\}$ $(k \geq 1)$ such that $0 < d_{j_1} < \ldots < d_{j_k} < d_{j_{k+1}} = d_{v+1}$. Then $\widetilde{S}_0, \widetilde{S}_{d_{j_1}}, \ldots, \widetilde{S}_{d_{j_k}}$ is the regular subresultant chain of $T_n(a)$ and $f(a)$ with respect to x_n. By Lemma 2.3(1),

$$V(T(a) \cup \{f(a)\})$$
$$= \cup_{j_t \in L_2}(V(T_{n-1}(a)) \cap V(\{\widetilde{S}_{d_{j_t}}, \widetilde{R}_{d_{j_{t-1}}}, \ldots, \widetilde{R}_{d_0}\}\backslash \mathrm{lc}_{T_n(a)}\widetilde{R}_{d_{j_t}})).$$

For any $j_t \in L_2$, if $j_t \in L_1$, then by the induction hypothesis,

$$V(T_{n-1}(a)) \cap V(\{\widetilde{S}_{d_{j_t}}, \widetilde{R}_{d_{j_{t-1}}}, \ldots, \widetilde{R}_{d_{j_1}}, \widetilde{R}_{d_0}\}\backslash \mathrm{lc}_{T_n(a)}\widetilde{R}_{d_{j_t}})$$
$$= \cup_{G \in \mathcal{G}_{j_t}} V(G(a) \cup \{S_{d_{j_t}}(a)\}).$$

For any $j_t \in L_2\backslash L_1$, if $j_t \leq l$, $\cup_{H \in \mathcal{H}_{j_t-1}} V(H(a)\backslash R_{d_{j_t}}(a)) = \emptyset$ by the induction hypothesis and $\mathcal{G}_{j_t} = \emptyset$. Then

$$V(T_{n-1}(a)) \cap V(\{\widetilde{S}_{d_{j_t}}, \widetilde{R}_{d_{j_{t-1}}}, \ldots, \widetilde{R}_{d_0}\}\backslash \mathrm{lc}_{T_n(a)}\widetilde{R}_{d_{j_t}}) = \emptyset.$$

If $j_t > l$, by the induction hypothesis, $\cup_{H \in \mathcal{H}_{l-1}} V(H(a)\cup\{R_{d_l}(a)\}) = \emptyset$ since $\mathcal{H}_l = \emptyset$. Then

$$V(T_{n-1}(a)) \cap V(\{\widetilde{S}_{d_{j_t}}, \widetilde{R}_{d_{j_{t-1}}}, \ldots, \widetilde{R}_{d_0}\}\backslash \mathrm{lc}_{T_n(a)}\widetilde{R}_{d_{j_t}}) = \emptyset.$$

So

$$V(T(a) \cup \{f(a)\}) = \cup_{j_t \in L_1} \cup_{G \in \mathcal{G}_{j_t}} V(G(a) \cup \{S_{j_t}(a)\}) = \cup_{H \in \mathcal{H}} V(H(a))$$

and hence condition (2) in Definition 2.13 holds. Similarly, we can check that condition (3) in Definition 2.13 holds by Lemma 2.3(2). $\qquad\square$

We use the following simple example to illustrate the main steps of Algorithm WRSD.

Example 2.2. Consider a polynomial $f = ux_1(x_1-x_2)(x_1+x_2)$ and a zero-dimensional regular chain $T = \{T_1, T_2\}$ in $\mathbb{R}[u, x_1, x_2]$ where $T_1 = x_1^4 - ux_1^3$, $T_2 = x_2^2 + (u + x_1)x_2$, $x_1 \prec x_2$ are variables and u is a parameter.

Since f is not reduced *w.r.t.* T, by Line 2 of Algorithm WRSD, we compute WRSD$(T, f_1, \{x_1, x_2\})$ where $f_1 = \mathrm{prem}(f; T) = ux_1^3 + ux_1^2x_2 + u^2x_1x_2$.

Step 1: Note that $f_1 \neq 0$, $\mathrm{cls}(f_1) = 2$ and $\mathrm{res}(f_1; T) = 0$. Thus at Line 13, we compute the regular subresultant chain $S_{d_2}, S_{d_1}, S_{d_0}$ of T_2 and f with respect to x_2 where

$$S_{d_2} = x_2^2 + (x_1 + u)x_2, S_{d_1} = (ux_1^2 + u^2x_1)x_2 + ux_1^3, S_{d_0} = -2u^3x_1^5 - u^4x_1^4.$$

Step 2: Since $n = 2$, we recursively compute WRSD$(\{T_1\}, S_{d_0}, \{x_1\})$ at Line 18.

Step 2.1: Note that $S_{d_0} \neq 0$, $\mathrm{cls}(S_{d_0}) = 1$ and $\mathrm{res}(S_{d_0}; \{T_1\}) = 0$. Thus at Line 13, we compute the regular subresultant chain $S_{1d_3}, S_{1d_2}, S_{1d_1}, S_{1d_0}$ of T_1 and S_{d_0} with respect to x_1 where

$$S_{1d_3} = -2u^3 x_1^5 - u^4 x_1^4, \; S_{1d_2} = x_1^4 - u x_1^3, \; S_{1d_1} = -3u^5 x_1^3, \; S_{1d_0} = 0.$$

Step 2.2: Note that the first output of $\mathrm{WRSD}(\{T_1\}, S_{d_0}, \{x_1\})$ is $\{\{S_{1d_1}\}\}$, by Line 16.

Step 2.3: We compute $q_1 = \mathrm{pquo}(T_1, S_{1d_1}, x_1) = -3u^5 x_1 + 3u^6$ and recursively compute $\mathrm{WRSD}(\{q_1\}, S_{d_0}, \{x_1\})$. However, note that $\mathrm{lc}_{q_1} = -3u^5$ and $\mathrm{res}(q_1; S_{d_0}, x_1) = 729 u^{33} \neq 0$. Thus at Line 11,

$$\mathrm{WRSD}(\{q_1\}, S_{d_0}, \{x_1\}) = [\emptyset, \{\{q_1\}\}, -2187 u^{36}].$$

Step 3: From Steps 2.1-2.3, we finally get

$$\mathrm{WRSD}(\{T_1\}, S_{d_0}, \{x_1\}) = [\{\{S_{1d_1}\}\}, \{\{q_1\}\}, -2187 u^{36}].$$

Now at Lines 18-19, $\mathcal{H}_0 = \{\{S_{1d_1}\}\}$, $\mathcal{G}_0 = \{\{q_1\}\}$, $\mathcal{G} = \{\{q_1, T_2\}\}$ and $p = -2187 u^{36}$.

Step 4: Since $\mathcal{H}_0 \neq \emptyset$, we should enter the "while" loop.

Step 4.1: $R_{d_1} = u x_1^2 + u^2 x_1$.

Step 4.2: At Line 25, we compute $\mathrm{WRSD}(\{S_{1d_1}\}, R_{d_1}, \{x_1\})$. Similarly as in **Step 2**, we get

$$\mathrm{WRSD}(\{S_{1d_1}\}, R_{d_1}, \{x_1\}) = [\{\{-3u^5 x_1^2\}\}, \emptyset, -3u^5]$$

and then $\mathcal{H}_1 = \{\{-3u^5 x_1^2\}\}$, $\mathcal{G}_1 = \emptyset$ and $p = -2187 u^{36} \cdot (-3u^5) = 6561 u^{41}$.

Step 4.3: Now note that $\mathcal{G}_1 = \emptyset$, thus we skip Lines 27-32. Since $\mathcal{H}_1 \neq \emptyset$, we should enter the "while" loop again. Then $R_{d_2} = 1$.

Step 4.4: According to Line 4,

$$\mathrm{WRSD}(\{-3u^5 x_1^2\}, R_{d_2}, \{x_1\}) = [\emptyset, \{\{-3u^5 x_1^2\}\}, -3u^5].$$

By Lines 24-26, $\mathcal{H}_2 = \emptyset$, $\mathcal{G}_2 = \{\{-3u^5 x_1^2\}\}$ and $p = 6561 u^{41} \cdot (-3u^5) = -19683 u^{46}$.

Step 4.5: By Line 28, $\mathcal{H} = \{\{-3u^5 x_1^2, x_2^2 + (x_1 + u) x_2\}\}$.

Step 4.6: Compute $\mathrm{pquo}(T_2, S_{d_2}, x_2) = 1$. Because $\mathcal{H}_2 = \emptyset$, we step out the "while" loop.

Finally, we complete the computation and get

$$\begin{aligned}
&\mathrm{WRSD}(\mathrm{T}, f, \{x_1, x_2\}) \\
&= [\{\{-3u^5 x_1^2, x_2^2 + (x_1 + u) x_2\}\}, \{\{-3u^5 x_1 + 3u^6, x_2^2 + (x_1 + u) x_2\}\}, \\
&\quad -19683 u^{46}].
\end{aligned}$$

2.5 Generic Regular Decomposition

For a given polynomial system, there exist some well-known algorithms for computing various kinds of regular decompositions of the system. See, for example, [Yang *et al.* (1992); Kalkbrener (1993); Yang *et al.* (1995, 1996b); Wang (2000, 2001, 2002); Chen *et al.* (2007)]. Since the main purpose of this book is not the zero decompositions of polynomial systems, we do not introduce those algorithms. On the other hand, for the purpose of real root classification of semi-algebraic systems in Chapter 6 of this book, we need a special kind of regular decomposition, called *generic regular decomposition,* which is based on WRSD.

In this section, all the systems are assumed to be generic zero-dimensional. More general cases are to be discussed in Chapter 6.

We first give an algorithm which computes a regular decomposition for a given generic zero-dimensional triangular set.

Algorithm 2.3 ZDtoRC

Input: A triangular set $T = \{T_1, \ldots, T_n\}$ in $\mathcal{K}[\boldsymbol{u}, \boldsymbol{x}]$ satisfying $\mathrm{mvar}(T) = \boldsymbol{x}$, variables $\boldsymbol{x} = \{x_1, \ldots, x_n\}$

Output: $[\mathcal{G}, p]$, where

(1) $\mathcal{G}$ is a finite set of zero-dimensional regular chains in $\mathcal{K}[\boldsymbol{u}, \boldsymbol{x}]$ such that $V_{\tilde{\mathcal{K}}}(T \backslash \mathrm{lc}_T) = \cup_{G \in \mathcal{G}} V_{\tilde{\mathcal{K}}}(G)$;

(2) p is a polynomial in $\mathcal{K}[\boldsymbol{u}]$ such that for any $a \in \hat{\mathcal{K}}^d \backslash V^{\boldsymbol{u}}(p)$, $V(T(a) \backslash \mathrm{lc}_T(a)) = \cup_{G \in \mathcal{G}} V(G(a))$ and G specializes well at a for any $G \in \mathcal{G}$.

1: **if** T is a regular chain **then return** $[\{T\}, \mathrm{res}(\mathrm{lc}_T, T)]$ **end if**;

2: Find the minimal integer k $(1 \leq k < n)$ such that $T_k = \{T_1, \ldots, T_k\}$ is a regular chain but $T_{k+1} = \{T_1, \ldots, T_k, T_{k+1}\}$ is not a regular chain;

3: $W \leftarrow \mathrm{WRSD}(T_k, \mathrm{lc}_{T_{k+1}}, \{x_1, \ldots, x_k\})$;

4: **if** $W_2 = \emptyset$ **then return** $[\emptyset, W_3]$ **end if**;

5: $p \leftarrow W_3$; $\mathcal{G} \leftarrow \emptyset$;

6: **for** R in W_2 **do**

7:    $R \leftarrow \{\mathrm{op}(R), T_{k+1}, \ldots, T_n\}$; $Z \leftarrow \mathrm{ZDtoRC}(R, \boldsymbol{x})$;

8:    $\mathcal{G} \leftarrow \mathcal{G} \cup Z_1$; $p \leftarrow p \cdot Z_2$;

9: **end for**

10: **return** $[\mathcal{G}, p]$

Theorem 2.7. *Algorithm* ZDtoRC *terminates correctly.*

Proof. If the input T is a regular chain, then the termination holds obviously and the correctness follows from Lemma 2.2. Now we assume that T is not a regular chain and let k be the minimal integer k ($1 \leq k < n$) such that $T_k = \{T_1, \ldots, T_k\}$ is a regular chain and $\{T_1, \ldots, T_{k+1}\}$ is not a regular chain. Note that this assumption is reasonable owing to the fact that at least $\{T_1\}$ is a regular chain in $\mathcal{K}[\boldsymbol{u}, \boldsymbol{x}]$. Let $T_{>k} = \{T_{k+1}, \ldots, T_n\}$.

Assume that $\mathtt{ZDtoRC}(T, \boldsymbol{x})$ does not terminate. Then we can get at least one regular chain $R \in \mathtt{WRSD}(T_k, \mathrm{lc}_{T_{k+1}}, \{x_1, \ldots, x_k\})_2$ such that $\mathtt{ZDtoRC}(R \cup T_{>k}, \boldsymbol{x})$ cannot terminate. According to Algorithm $\mathtt{ZDtoRC}$, there exists k_2 ($1 \leq k < k_2 < n$) such that k_2 is the minimal integer such that $R \cup \{T_{k+1}, \ldots, T_{k_2}\}$ is a regular chain but $R \cup \{T_{k+1}, \ldots, T_{k_2+1}\}$ is not a regular chain. Since $\mathtt{ZDtoRC}(R \cup T_{>k}, \boldsymbol{x})$ does not terminate, in the same manner we can get an infinite sequence of positive integers $k = k_1 < k_2 < \ldots < k_t < \ldots$. However, all positive integers in this infinite sequence must be no more than n. A contradiction. Therefore, Algorithm $\mathtt{ZDtoRC}$ terminates. Then it is not difficult to prove the correctness by induction on the recursive depth. $\qquad\square$

Now, by combining Wu's method and Algorithm $\mathtt{ZDtoRC}$, we can get an algorithm for computing the so-called generic regular decomposition (see Definition 2.15).

Algorithm 2.4 $\mathtt{GRDforZD}$

Input: A generic zero-dimensional system P in $\mathcal{K}[\boldsymbol{u}][\boldsymbol{x}]$, variables $\boldsymbol{x}$
Output: $[\mathcal{T}, p]$, a generic regular decomposition of P
 1: Compute a Wu's decomposition $\{C_1, \ldots, C_m\}$ of P in $\mathcal{K}[\boldsymbol{u}][\boldsymbol{x}]$;
 2: $\mathcal{T} \leftarrow \emptyset$; $p \leftarrow 1$;
 3: **for** $i = 1$ **to** m **do**
 4:     **if** C_i is a contradictory ascending chain **then**
 5:       $p \leftarrow p \cdot \mathrm{op}(C_i)$;
 6:     **else**
 7:       $W \leftarrow \mathtt{ZDtoRC}(C_i, \boldsymbol{x})$;
 8:       $\mathcal{T} \leftarrow \mathcal{T} \cup W_1$; $p \leftarrow p \cdot W_2$;
 9:     **end if**
10: **end for**
11: **return** $[\mathcal{T}, p]$

Lemma 2.6. *Given a zero-dimensional regular chain* T *in* $\mathcal{K}[\boldsymbol{u}][\boldsymbol{x}]$ *and a finite set of polynomials* $P \subset \mathcal{K}[\boldsymbol{u}][\boldsymbol{x}]$, *suppose* $\mathrm{V}_{\bar{\mathcal{K}}}(T) \subset \mathrm{V}_{\bar{\mathcal{K}}}(P)$. *Then*

$V(T(a)) \subset V(P(a))$ *for any* $a \in \hat{\mathcal{K}}^d \backslash V^{\boldsymbol{u}}(\mathrm{res}(\mathrm{lc}_T, T))$.

Proof. By Theorem 2.2, $V_{\tilde{\mathcal{K}}}(T \backslash \mathrm{lc}_T) = V_{\tilde{\mathcal{K}}}(T)$. Then $V_{\tilde{\mathcal{K}}}(T \backslash \mathrm{lc}_T) \subset V_{\tilde{\mathcal{K}}}(P)$. By Theorem 2.3, $P \subset \sqrt{\mathrm{sat}(T)}_{\mathcal{K}[\boldsymbol{u}][\boldsymbol{x}]}$ and hence for any $f \in P$, there exist a positive integer k such that $f^k \in \mathrm{sat}(T)_{\mathcal{K}[\boldsymbol{u}][\boldsymbol{x}]}$. By Theorem 2.3, $\mathrm{prem}(f^k, T) = 0$. Remark that $f^k \in \mathcal{K}[\boldsymbol{u}][\boldsymbol{x}]$ and $T \subset \mathcal{K}[\boldsymbol{u}][\boldsymbol{x}]$, so

$$V_{\hat{\mathcal{K}}}(T \backslash \mathrm{lc}_T) \subset V_{\hat{\mathcal{K}}}(f^k) = V_{\hat{\mathcal{K}}}(f).$$

For any $a = (a_1, \ldots, a_d) \in \hat{\mathcal{K}}^d \backslash V^{\boldsymbol{u}}(\mathrm{res}(\mathrm{lc}_T, T))$, $T(a)$ is a zero-dimensional regular chain in $\hat{\mathcal{K}}[\boldsymbol{x}]$ and $\mathrm{lc}_T(a) \neq 0$ by Lemma 2.2. So

$$\mathrm{res}(\mathrm{lc}_T(a), T(a)) = \mathrm{res}(\mathrm{lc}_{T(a)}, T(a)) \neq 0$$

and thus for any $b = (b_1, \ldots, b_n) \in V(T(a))$, $b \notin V(\mathrm{lc}_T(a))$. That implies

$$(a_1, \ldots, a_d, b_1, \ldots, b_n) \in V_{\hat{\mathcal{K}}}(T \backslash \mathrm{lc}_T) \subset V_{\hat{\mathcal{K}}}(P).$$

Therefore, $b \in V(P(a))$ and then $V(T(a)) \subset V(P(a))$. $\qquad \square$

Theorem 2.8. *Algorithm* GRDforZD *terminates correctly.*

Proof. Since the termination follows from the termination of Algorithm WRSD and Algorithm ZDtoRC, we only need to prove the correctness. Assume that P is a generic zero-dimensional system in $\mathcal{K}[\boldsymbol{u}][\boldsymbol{x}]$ and $\{C_1, \ldots, C_m\}$ is a Wu's decomposition of P computed by Wu's method. According to Wu's method and Algorithm ZDtoRC, we know that the claim (1) in the specification of Algorithm GRDforZD holds.

Now we prove the claim (2) in the specification of Algorithm GRDforZD by induction on m. If $m = 1$, C_1 is a characteristic set of P in $\mathcal{K}[\boldsymbol{u}][\boldsymbol{x}]$ and $C_1 = \{C_1\} \subset \mathcal{K}[\boldsymbol{u}]$ by Wu's method. That means $V_{\tilde{\mathcal{K}}}(P) = \emptyset$. According to Algorithm GRDforZD the first output of GRDforZD$(P, \boldsymbol{x})$ is $\emptyset$ and the second output is exactly C_1. In fact, for any $a \notin V^{\boldsymbol{u}}(C_1)$, $C_1(a) \in \hat{\mathcal{K}}$. Note that $C_1(a) \in \langle P(a) \rangle_{\hat{\mathcal{K}}[\boldsymbol{x}]}$ by $C_1 \in \langle P \rangle_{\mathcal{K}[\boldsymbol{u}][\boldsymbol{x}]}$. Thus $V(P(a)) \subset V(C_1(a)) = \emptyset$ and the claim (2) in the specification of Algorithm GRDforZD holds.

Assume that the conclusion holds for $m < N$ $(N > 1)$. If $m = N$, suppose $C_1 = \{C_{11}, \ldots, C_{1t}\}$ is the characteristic set of P computed by Wu's method. Since $m > 1$, we know that $C_1 \not\subset \mathcal{K}[\boldsymbol{u}]$ and

$$V_{\tilde{\mathcal{K}}}(P) = V_{\tilde{\mathcal{K}}}(C_1 \backslash \mathrm{lc}_{C_1}) \cup \cup_{i=1}^t V_{\tilde{\mathcal{K}}}(P \cup C_1 \cup \{\mathrm{lc}_{C_{1i}}\}).$$

Let GRDforZD$(P, \boldsymbol{x}) = [\mathcal{T}, p]$ and ZDToRC$(C_1, \boldsymbol{x}) = [\mathcal{T}_1, p_1]$. Then $\mathcal{T}_1 \subset \mathcal{T}$ and p_1 is a factor of p.

Let GRDforZD$(P \cup C_1 \cup \{\mathrm{lc}_{C_{1i}}\}, \boldsymbol{x}) = [\mathcal{T}_{2i}, p_{2i}]$ for every i $(1 \leq i \leq t)$, then $\mathcal{T}_1 \cup \cup_{i=1}^t \mathcal{T}_{2i} = \mathcal{T}$ and $p = p_1 \cdot \Pi_{i=1}^t p_{2i}$ by Wu's method and Algorithm

`GRDforZD`. We only prove the conclusion when $\mathcal{T}_1 \neq \emptyset$ and $\mathcal{T}_{2i} \neq \emptyset$ for every i ($1 \leq i \leq t$). In fact, if $\mathcal{T}_1 = \emptyset$ or there exists i ($1 \leq i \leq t$) such that $\mathcal{T}_{2i} = \emptyset$, the proof is similar.

For any $a \notin V^{\boldsymbol{u}}(p)$, we know that $a \notin V^{\boldsymbol{u}}(p_1)$ and $a \notin V^{\boldsymbol{u}}(p_{2i})$ for every i ($1 \leq i \leq t$). Hence for every i ($1 \leq i \leq t$),

$$V((P \cup C_1 \cup \{\mathrm{lc}_{C_{1i}}\})(a)) = \cup_{T \in \mathcal{T}_{2i}} V(T(a))$$

by the induction hypothesis. Therefore, in order to prove $V(P(a)) = \cup_{T \in \mathcal{T}} V(T(a))$, we only need to show

$$V(P(a)) = \cup_{T \in \mathcal{T}_1} V(T(a)) \cup \cup_{i=1}^{t} V((P \cup C_1 \cup \{\mathrm{lc}_{C_{1i}}\})(a)).$$

Note that $\cup_{i=1}^{t} V((P \cup C_1 \cup \{\mathrm{lc}_{C_{1i}}\})(a)) = V((P \cup C_1 \cup \{\mathrm{lc}_{C_1}\})(a))$. So we only need to prove

$$V(P(a)) = \cup_{T \in \mathcal{T}_1} V(T(a)) \cup V((P \cup C_1 \cup \{\mathrm{lc}_{C_1}\})(a)).$$

As a matter of fact, $\cup_{T \in \mathcal{T}_1} V(T(a)) = V(C_1(a) \backslash \mathrm{lc}_{C_1}(a))$ by Algorithm `ZDtoRC`. Note that $C_1(a) \subset \langle P(a) \rangle_{\hat{\mathcal{K}}[\boldsymbol{x}]}$, so $V(P(a)) \subset V(C_1(a))$. Then

$$V(P(a)) \subset V(C_1(a) \backslash \mathrm{lc}_{C_1}(a)) \cup V((P \cup C_1 \cup \{\mathrm{lc}_{C_1}\})(a))$$
$$= \cup_{T \in \mathcal{T}_1} V(T(a)) \cup V((P \cup C_1 \cup \{\mathrm{lc}_{C_1}\})(a)).$$

On the other hand, by claim (1), $V_{\tilde{\mathcal{K}}}(P) = \cup_{T \in \mathcal{T}} V_{\tilde{\mathcal{K}}}(T)$ and thus for any $T \in \mathcal{T}_1 \subset \mathcal{T}$, $V_{\tilde{\mathcal{K}}}(T) \subset V_{\tilde{\mathcal{K}}}(P)$. According to Algorithm `ZDtoRC`, we know that $\mathrm{res}(\mathrm{lc}_T, T)(a) \neq 0$. By Lemma 2.6,

$$\cup_{T \in \mathcal{T}_1} V(T(a)) \cup V((P \cup C_1 \cup \{\mathrm{lc}_{C_1}\})(a)) \subset V(P(a))$$

and we are done. $\qquad\square$

2.6 Zero Decomposition Keeping Multiplicity

All the existing methods for zero decomposition mainly focus on the decomposition of the set of zeros and do not care about the multiplicities of (isolated) zeros. A popular viewpoint is that zero decompositions destroy the structure of the multiplicities of zeros. That is, by zero decomposition (triangular decomposition), the zero set of a given zero-dimensional polynomial system can be decomposed into several constructible sets defined by triangular sets of polynomials while the multiplicities of the points in those constructible sets do not match those of the zeros of the original polynomial system. In this section, we have a close look at this problem and introduce the main result of [Li *et al.* (2010)].

Can we have a zero decomposition which keeps multiplicity? More precisely, for a zero-dimensional polynomial system S, can we compute finitely many pairs $[T_i, g_i]$ such that

$$\mathrm{MZero}(\mathrm{S}) = \bigcup_i \mathrm{MZero}(\mathrm{T}_i \backslash g_i)?$$

Herein $\mathrm{MZero}(\cdot)$ stands for the set of zeros (counted with multiplicity), T_i is a (triangular) polynomial set, g_i is a polynomial, $\mathrm{MZero}(\mathrm{T}_i \backslash g_i) = \{\xi \mid \forall f \in \mathrm{T}_i, f(\xi) = 0, g_i(\xi) \neq 0\}$ (counted with multiplicity), and

$$\mathrm{MZero}(\mathrm{T}_i \backslash g_i) \cap \mathrm{MZero}(\mathrm{T}_j \backslash g_j) = \varnothing \text{ for all } i \neq j.$$

That is to say, for any zero ξ of S, there exists only one pair $[T_j, g_j]$ such that ξ is a zero of T_j and $g_j(\xi) \neq 0$ and the multiplicity of ξ as a zero of S is the same as that of it as a zero of T_j. Such decomposition is said to *keep multiplicity*.

For a known zero ξ of a zero-dimensional system S, one can compute the multiplicity of ξ through Gröbner bases computation. Furthermore, one can even compute the multiplicity structure of ξ by the method in [Dayton and Zeng (2005); Dayton *et al.* (2011)] based on dual space. We believe that, with the help of Gröbner bases computation, one can get an algorithm for computing zero decomposition keeping multiplicity. Obviously, our interest here is different from those in [Cox *et al.* (1998)] and [Dayton and Zeng (2005)]. We intend to know whether existing/modified triangularization methods can keep multiplicity. Our interest is also different from that in [Li (2003)], where a new definition of multiplicity based on non-standard analysis was given and was proven to be equivalent to the classical definition. Zero decomposition is not a topic in [Li (2003)].

We intend to compute a zero decomposition which keeps multiplicity for a zero-dimensional system by modifying Wu's method. So, we study the relation between Wu's method and the multiplicity of zero in this section. Although, as everybody may have known, the characteristic sets computed by Wu's zero decomposition algorithm do not keep multiplicity in general, we prove the following result: let C be the first characteristic set computed by Wu's method for a given polynomial set P, then C keeps the multiplicities of zeros, *i.e.* for any isolated zero ξ of $V(C) \backslash V(J)$ where J is the product of the initials of C, the multiplicity of ξ related to C is equal to that of ξ related to P.

We hope that a zero decomposition keeping multiplicity is also triangular. If so, the multiplicities of zeros of each component can be read out directly [Zhang *et al.* (2011)]. In this section, we also give a zero decomposition theorem and an algorithm based on Wu's method, which computes a zero decomposition keeping multiplicity for a given zero-dimensional polynomial system. As pointed out by [Li *et al.* (2010)], if the system satisfies some additional constraints, the zero decomposition is of triangular form.

2.6.1 *Multiplicity*

First, let us recall the definition of *local (intersection) multiplicity*. We follow the notations in [Cox *et al.* (1998)]. Although some notations and definitions can be described in a more general way, we restrict ourselves to the ring $\mathbb{C}[\boldsymbol{x}] = \mathbb{C}[x_1, \ldots, x_n]$ since we are interested in the complex zeros of zero-dimensional polynomial systems.

For $\xi = (\xi_1, \ldots, \xi_n) \in \mathbb{C}^n$, we denote by M_ξ the maximal ideal generated by $\{x_1 - \xi_1, \ldots, x_n - \xi_n\}$ in $\mathbb{C}[\boldsymbol{x}]$, and write

$$\mathbb{C}[\boldsymbol{x}]_{M_\xi} = \left\{ \frac{f}{g} \mid f, g \in \mathbb{C}[\boldsymbol{x}], g(\xi_1, \ldots, \xi_n) \neq 0 \right\}.$$

It is well-known that $\mathbb{C}[\boldsymbol{x}]_{M_\xi}$ is the so-called *local ring*.

Definition 2.16. Let I be a zero-dimensional ideal in $\mathbb{C}[\boldsymbol{x}]$, and assume that $\xi \in V(I)$, the zero set of I in $\mathbb{C}$. Then the *multiplicity* of ξ as a point in $V(I)$ is defined to be

$$\dim_{\mathbb{C}} \mathbb{C}[\boldsymbol{x}]_{M_\xi} / I\mathbb{C}[\boldsymbol{x}]_{M_\xi},$$

the dimension of the quotient space $\mathbb{C}[\boldsymbol{x}]_{M_\xi} / I\mathbb{C}[\boldsymbol{x}]_{M_\xi}$ as a vector space over $\mathbb{C}$.

There is another equivalent definition of local multiplicity based on dual space.

For every index array $j = [j_1, \ldots, j_n] \in \mathbb{N}^n$, define the differential operator

$$\partial_j \equiv \partial_{j_1 \cdots j_n} \equiv \partial_{x_1^{j_1} \cdots x_n^{j_n}} \equiv \frac{1}{j_1! \cdots j_n!} \frac{\partial^{j_1 + \cdots + j_n}}{\partial x_1^{j_1} \cdots \partial x_n^{j_n}}$$

and the order of ∂_j is $|\partial_j| = \sum_{l=1}^{n} j_l$.

Consider a system of m polynomials, $\{f_1(\boldsymbol{x}), \ldots, f_m(\boldsymbol{x})\}$, in $n(n \leq m)$ variables with an isolated zero $\xi \in \mathbb{C}^n$. Let $I = \langle f_1(\boldsymbol{x}), \ldots, f_m(\boldsymbol{x}) \rangle$. A functional at $\xi \in \mathbb{C}^n$ is defined as follows:

$$\partial_j[\xi] : \mathbb{C}[\boldsymbol{x}] \to \mathbb{C},$$

where

$$\partial_j[\xi](f) = (\partial_j f)(\xi), \ \forall f \in \mathbb{C}[x].$$

All functionals at ξ that vanish on I form a vector space $D_\xi(\mathrm{I})$,

$$D_\xi(\mathrm{I}) \equiv \{v = \sum_{j \in \mathbb{N}^n} v_j \partial_j[\xi] \mid v_j \in \mathbb{C}, \ \forall f \in \mathrm{I}, v(f) = 0\}.$$

The vector space $D_\xi(\mathrm{I})$ is called the *dual space* of I at ξ. For $\alpha = 0, 1, \ldots,$ $D_\xi^\alpha(\mathrm{I})$ consists of functionals in $D_\xi(\mathrm{I})$ with order bounded by α.

Definition 2.17. [Dayton and Zeng (2005)] The local *multiplicity* of zero ξ of a zero-dimensional ideal $\mathrm{I} \subseteq \mathbb{C}[x]$ is m if the dual space $D_\xi(\mathrm{I})$ is of dimension m.

Remark 2.6. According to [Dayton and Zeng (2005)], Definition 2.17 is equivalent to Definition 2.16.

Proposition 2.2. [Dayton and Zeng (2005)] *Let σ be the smallest α such that* $\dim(D_\xi^\alpha(\mathrm{I})) = \dim(D_\xi^{\alpha+1}(\mathrm{I}))$, *then* $D_\xi(\mathrm{I}) = D_\xi^\sigma(\mathrm{I})$. *Furthermore we have* $\sigma < m = \dim(D_\xi(\mathrm{I}))$.

Lemma 2.7. *Assume ξ is an isolated zero of I and the local multiplicity of ξ is m. If $\xi \in \mathrm{V}(g(x))$ where $g(x) \in \mathbb{C}[x]$, then for any $v \in D_\xi(\mathrm{I})$, any $h(x) \in \mathbb{C}[x]$ and $l \geq m$, we have $v(h(x)g(x)^l) = 0$.*

Proof. If $\partial_j \equiv \partial_{j_1 \cdots j_n} \in D_\xi(\mathrm{I}) = D_\xi^\sigma(\mathrm{I}), |\partial_j| \leq \sigma < m$ by Proposition 2.2. Because $g(x)^{l-|\partial_j|}$ is a factor of $\partial_j(hg^l)$ and $\xi \in \mathrm{V}(g(x)), \partial_j[\xi](h(x)g(x)^l) = 0$. Since any $v \in D_\xi(\mathrm{I})$ is a linear combination of $\partial_j \in D_\xi(\mathrm{I}) = D_\xi^\sigma(\mathrm{I})$, we are done. $\qquad\square$

For a zero of a zero-dimensional triangular system, we have an intuitive definition of multiplicity as follows.

Definition 2.18. [Zhang *et al.* (2011)] For a zero-dimensional triangular system,

$$\begin{cases} f_1(x_1) = 0, \\ f_2(x_1, x_2) = 0, \\ \quad\vdots \\ f_n(x_1, \ldots, x_n) = 0, \end{cases}$$

and one of its zeros, $\xi = (\xi_1, \ldots, \xi_n)$, the *multiplicity* of ξ is defined to be $\prod_{i=1}^n m_i$, where m_i is the multiplicity of $x_i = \xi_i$ as a zero of the univariate polynomial $f_i(\xi_1, \ldots, \xi_{i-1}, x_i)$ for $i = 1, \ldots, n$.

Remark 2.7. It can be proved (see for example [Zhang *et al.* (2011)]) that Definition 2.18 is equivalent to Definition 2.16 for zero-dimensional triangular systems. So, the multiplicities of zeros of zero-dimensional triangular systems can be "read" out from the polynomials one by one. Take the following system (called DZ2 in [Dayton and Zeng (2005)]) for example:

$$\{x^4, x^2y + y^4, z + z^2 - 7x^3 - 8x^2\}.$$

The zeros $(0, 0, 0)$ and $(0, 0, -1)$ are both of multiplicity $4 \times 4 \times 1 = 16$.

2.6.2 *Zero Decomposition Keeping Multiplicity*

Lemma 2.8. *Assume* $I = \langle P \rangle$ *is a zero-dimensional ideal where* $P = \{f_1(\boldsymbol{x}), \ldots, f_s(\boldsymbol{x})\}$, *and* $C = [C_1(\boldsymbol{x}), \ldots, C_t(\boldsymbol{x})]$ *is a characteristic set of* P *such that (2.4) holds, i.e.*

$$V(P) = V(C \backslash J) \cup V(P, J),$$

where $J = \prod_{j=1}^{t} I_j$ *is the product of initials of all* C_i. *Then for each* $\xi \in V(C \backslash J)$, *the local multiplicity of* ξ *as a point in* $V(I)$ *and the local multiplicity of* ξ *as a point in* $V(\langle C \rangle)$ *are the same.*

Proof. Assume $\xi = (\xi_1, \ldots, \xi_n) \in V(C \backslash J)$, and let $M_\xi = \langle x_1 - \xi_1, \ldots, x_n - \xi_n \rangle$. According to Definition 2.16, the local multiplicity of ξ as a point in $V(I)$ is

$$\dim_{\mathbb{C}} \mathbb{C}[x_1, \ldots, x_n]_{M_\xi} / I\mathbb{C}[x_1, \ldots, x_n]_{M_\xi}$$

and the local multiplicity of ξ as a point in $V(\langle C \rangle)$ is

$$\dim_{\mathbb{C}} \mathbb{C}[x_1, \ldots, x_n]_{M_\xi} / \langle C \rangle \mathbb{C}[x_1, \ldots, x_n]_{M_\xi}.$$

It suffices to prove

$$I\mathbb{C}[x_1, \ldots, x_n]_{M_\xi} = \langle C \rangle \mathbb{C}[x_1, \ldots, x_n]_{M_\xi}.$$

On one hand, $\langle C \rangle \mathbb{C}[x_1, \ldots, x_n]_{M_\xi} \subseteq I\mathbb{C}[x_1, \ldots, x_n]_{M_\xi}$ since $\langle C \rangle \subseteq I$. On the other hand, $\mathrm{prem}(f_i(\boldsymbol{x}); C) = 0$ for each $f_i(\boldsymbol{x}) \in P$ since C is a characteristic set of P. Therefore, there exist $a_1, \ldots, a_t, q_1(\boldsymbol{x}), \ldots, q_t(\boldsymbol{x})$ such that

$$\prod_{j=1}^{t} I_j^{a_j} f_i = \sum_{j=1}^{t} q_j(\boldsymbol{x}) C_j(\boldsymbol{x}).$$

Because $\xi \in V(C \backslash J)$, $I_j(\xi) \neq 0$ for all $1 \leq j \leq t$. Thus, each $I_j(\boldsymbol{x})$ is invertible in $\mathbb{C}[x_1, \ldots, x_n]_{M_\xi}$. As a result, $f_j(\boldsymbol{x}) \in \langle C \rangle \mathbb{C}[x_1, \ldots, x_n]_{M_\xi}$. Thus,

$$I\mathbb{C}[x_1, \ldots, x_n]_{M_\xi} \subseteq \langle C \rangle \mathbb{C}[x_1, \ldots, x_n]_{M_\xi}.$$

That completes the proof. $\square$

For those points in $V(P, I_i)$, their local multiplicities as points of $V(P, I_i)$ may not be equal to their local multiplicities as points of $V(P)$. Intuitively, if $\xi \in V(P, I_i)$ and its local multiplicity as a point of $V(P)$ is m_1, its local multiplicity as a point of $V(P, I_i^{m_2})$ should also be m_1 provided that $m_2 \geq m_1$. And this is indeed true as implied by the following lemma.

Lemma 2.9. *Suppose* $\xi \in V(P, g(\boldsymbol{x}))$ *is an isolated zero of* $P = \{f_1(\boldsymbol{x}), \ldots, f_s(\boldsymbol{x})\}$ *and m is the local multiplicity of ξ as a point of* $V(P)$. *Then the local multiplicity of ξ as a point of* $V(P, g(\boldsymbol{x})^l)$ *is m if $l \geq m$.*

Proof. Let $I = \langle P \rangle$ and $I' = \langle P, g(\boldsymbol{x})^l \rangle$. According to Definition 2.17, it is sufficient to prove that

$$D_\xi(I) = D_\xi(I').$$

Obviously, $D_\xi(I) \supseteq D_\xi(I')$ according to the definition of dual space since $I \subseteq I'$.

On the other hand, for $f = \sum_{j=1}^{s} d_j(\boldsymbol{x}) f_j(\boldsymbol{x}) + h(\boldsymbol{x}) g(\boldsymbol{x})^l \in I'$ and any $v \in D_\xi(I)$,

$$v(f) = v(\sum_{j=1}^{s} d_j(\boldsymbol{x}) f_j(\boldsymbol{x})) + v(h(\boldsymbol{x}) g(\boldsymbol{x})^l) = v(h(\boldsymbol{x}) g(\boldsymbol{x})^l)$$

because $\sum_{j=1}^{s} d_j(\boldsymbol{x}) f_j(\boldsymbol{x}) \in I$. According to Lemma 2.7,

$$v(h(\boldsymbol{x}) g(\boldsymbol{x})^l) = 0$$

and thus $v \in D_\xi(I')$. That completes the proof. $\square$

By Lemma 2.8, Lemma 2.9 and Eq. (2.4), the following theorem is obvious.

Theorem 2.9 (Zero Decomposition with Multiplicity). *Suppose* P *is a finite non-empty set of non-zero polynomials in* $\mathbb{C}[\boldsymbol{x}]$ *and* $\langle P \rangle$ *is a zero-dimensional ideal. Assume that* $C = [C_1(\boldsymbol{x}), \ldots, C_n(\boldsymbol{x})]$ *is a characteristic set of* P, I_i *is the initial of* $C_i(\boldsymbol{x})$ *and* $J(i) = \prod_{k=1}^{i} I_k$ *for* $1 \leq i \leq n$. *There exists* $e_i \in \mathbb{N}$ *for* $2 \leq i \leq n$ *such that*

$$\mathrm{MZero}(P) = \mathrm{MZero}(C \backslash J(n)) \cup \bigcup_{i=2}^{n} \mathrm{MZero}(P_i \backslash J(i-1)), \qquad (2.7)$$

is a disjoint decomposition of $\mathrm{MZero}(P)$, *where* $P_i = P \cup \{I_i^{e_i}\}$.

Remark 2.8. The existence of such e_is is clear. For example, we can let e_i be the Bézout bound, or set $e_i \geq m$ where $m = \dim_{\mathbb{C}} \mathbb{C}[x_1, \ldots, x_n]/\langle P \rangle$ is the number (counted with multiplicity) of points in MZero(P), which can be obtained by computing the Gröbner basis of $\langle P \rangle$.

Remark 2.9. One may hope to apply `WuCharSet` to each P_i in Theorem 2.9 recursively to obtain a triangular decomposition of MZero(P). However, as pointed out in [Li *et al.* (2010)], only when some additional constraints are satisfied can the triangular decomposition be achieved.

Example 2.3. (Ojika2) [Ojika (1987); Dayton and Zeng (2005)] Consider a system of 3 polynomials

$$P = \{x^2 + y + z - 1, x + y^2 + z - 1, x + y + z^2 - 1\}.$$

We try to compute a zero decomposition with multiplicity of P. First, we use the Bézout bound on the number of zeros, $m = 8$, which is the product of all total degrees of polynomials in P. Next, we compute a characteristic set of P (to make the output more readable to the readers, we factor each polynomial):

$$C = [x^2(x^2 + 2x - 1)(x - 1)^2, x^2(x^2 + 2y - 1), x^2(x^2 - 1 + 2z)].$$

According to Theorem 2.9,

$$\text{MZero(P)} = \text{MZero}(C \backslash x^4) \cup \text{MZero}(P_1),$$

where $P_1 = P \cup C \cup \{x^{16}\}$. Because $\text{prem}(x^{16}; C) \neq 0$, we continue to compute a zero decomposition of MZero(P_1). The characteristic set of P_1 is:

$$C_1 = [x^2, y - x - y^2, 1 - z - y].$$

Therefore $\text{MZero}(P_1) = \text{MZero}(C_1)$. Finally, we get a zero decomposition with multiplicity of MZero(P) as follows:

$$\text{MZero(P)} = \text{MZero}(C \backslash x^4) \cup \text{MZero}(C_1).$$

Note that the decomposition is triangular.

To make the result more illustrative, we split the decomposition into 3 branches:

$$\begin{aligned}
T_1 &= [x^2 + 2x - 1, x^2 + 2y - 1, x^2 - 1 + 2z], \\
T_2 &= [(x - 1)^2, x^2 + 2y - 1, x^2 - 1 + 2z], \\
T_3 &= [x^2, y - x - y^2, 1 - z - y].
\end{aligned}$$

Obviously, the multiplicity of each zero in T_1 is 1, and the zeros in Zero(T_2) or Zero(T_3) all have multiplicity 2.

Chapter 3

Triangularization of Semi-Algebraic System

The following system
$$\begin{cases} p_1(\boldsymbol{u}, x_1, \ldots, x_n) = 0, \ldots, p_s(\boldsymbol{u}, x_1, \ldots, x_n) = 0, \\ g_1(\boldsymbol{u}, x_1, \ldots, x_n) \geq 0, \ldots, g_r(\boldsymbol{u}, x_1, \ldots, x_n) \geq 0, \\ g_{r+1}(\boldsymbol{u}, x_1, \ldots, x_n) > 0, \ldots, g_t(\boldsymbol{u}, x_1, \ldots, x_n) > 0, \\ h_1(\boldsymbol{u}, x_1, \ldots, x_n) \neq 0, \ldots, h_m(\boldsymbol{u}, x_1, \ldots, x_n) \neq 0, \end{cases}$$
is called a *Semi-Algebraic System*, or shortly SAS, and denoted as
$$[P, G_1, G_2, H], \tag{3.1}$$
where P, G_1, G_2 and H denote respectively
$$[p_1(\boldsymbol{u}, x_1, \ldots, x_n), \ldots, p_s(\boldsymbol{u}, x_1, \ldots, x_n)],$$
$$[g_1(\boldsymbol{u}, x_1, \ldots, x_n), \ldots, g_r(\boldsymbol{u}, x_1, \ldots, x_n)],$$
$$[g_{r+1}(\boldsymbol{u}, x_1, \ldots, x_n), \ldots, g_t(\boldsymbol{u}, x_1, \ldots, x_n)]$$
and
$$[h_1(\boldsymbol{u}, x_1, \ldots, x_n), \ldots, h_m(\boldsymbol{u}, x_1, \ldots, x_n)].$$
Herein, $n, s \geq 1$, $r, t, m \geq 0$, p_i, g_j, h_k are all polynomials in $\mathbb{Q}[\boldsymbol{u}, x_1, \ldots, x_n]$, and $\boldsymbol{u} = (u_1, \ldots, u_d)$ are real parameters.

For many cases considered in the rest of this book, the coefficient field $\mathbb{Q}$ in the above definition can be replaced by a real closed field or a computable ordered field. It will be stated clearly when necessary. Note that an SAS defined here contains at least one equation.

If an SAS contains no parameters, *i.e.* $d = 0$, it is called a *constant SAS*; otherwise, a *parametric SAS*. The *solution set* of an SAS S in the form of (3.1) is denoted as W(S), *i.e.*

$$\text{W(S)} = \{b \in \mathbb{R}^{d+n} \mid \text{all the equations and inequalities of S hold at } b\}.$$

3.1   Triangular SAS

If the equations of an SAS are in triangular form, it is called a *triangular SAS*, or shortly TSA. From Chapter 2, we know that regular chain is a special and important kind of triangular sets. In this section, we first define an analogue for triangular SAS.

Suppose $T : [F, G_1, G_2, H]$ is a TSA. For each $p_i \in F$ $(1 \le i \le s)$, set

$$\mathrm{BP}_{p_1} = \mathrm{res}(p_1, p_1', y_1);$$
$$\mathrm{BP}_{p_i} = \mathrm{res}(\mathrm{res}(p_i, p_i', y_i); p_{i-1}, \ldots, p_1), \quad 2 \le i \le s,$$

where $y_i = \mathrm{mvar}(p_i)$ and p_i' denotes the derivative of p_i with respect to y_i. Note that by Corollary 1.1,

$$\mathrm{res}(p_i, p_i', y_i) = (-1)^{\frac{a(a-1)}{2}} \mathrm{lc}(p_i)\mathrm{discrim}(p_i)$$

where $a = \deg(p_i)$, $\mathrm{lc}(p_i)$ is the leading coefficient and $\mathrm{discrim}(p_i)$ is the discriminant of p_i with respect to y_i. For each $q \in \{g_j \mid 1 \le j \le t\} \bigcup \{h_k \mid 1 \le k \le m\}$, set

$$\mathrm{BP}_q = \mathrm{res}(q; f_s, \ldots, f_1).$$

Definition 3.1. Let notations be as above. Define

$$\mathrm{BP(T)} = \prod_{1 \le i \le s} \mathrm{BP}_{p_i} \cdot \prod_{1 \le j \le t} \mathrm{BP}_{g_j} \cdot \prod_{1 \le k \le m} \mathrm{BP}_{h_k},$$

and call it the *border polynomial* of T.

Definition 3.2. If $\mathrm{BP(T)} \not\equiv 0$ for a TSA T, then T is said to be *regular*.

Definition 3.3. Given a TSA: $[F, G_1, G_2, H]$, if

$$F = [f_1(\boldsymbol{u}, x_1), f_2(\boldsymbol{u}, x_1, x_2), \ldots, f_n(\boldsymbol{u}, x_1, x_2, \ldots, x_n)],$$

i.e. the number of equations is the same as that of variables, it is said to be a *zero-dimensional TSA*.

Note that, for a regular zero-dimensional TSA, its BP is a non-zero polynomial in parameters. Then, the following proposition is almost obvious.

Proposition 3.1. *Suppose* $T : [F, G_1, G_2, H]$ *is a regular zero-dimensional TSA. Then the equation F is a squarefree regular chain and F is coprime with respect to any polynomial in* $G_1 \cup G_2 \cup H$ *(i.e.* g_j, h_k*) if specialized at any parameter value* $\boldsymbol{v}$ *such that* $\mathrm{BP}(\boldsymbol{v}) \ne 0$*. Furthermore,* T *is equivalent to* $[F, [\,], G_1 \cup G_2, [\,]]$ *if* $\mathrm{BP}(\boldsymbol{v}) \ne 0$*.*

Moreover, by Theorem 3.1, a regular zero-dimensional TSA has the same number of real solutions at any two parameter values in the same connected component of BP $\neq 0$ in $\mathbb{R}^d$. That is why BP is called border polynomial.

The following lemma is widely used, which can be seen as a corollary of some well-known results, for example, Proposition 3.11 in [Basu *et al.* (2003)].

Lemma 3.1. *Suppose*
$$f(a, x) = a_m x^m + a_{m-1} x^{m-1} + \cdots + a_0$$
is a univariate polynomial in x with real parametric coefficients, where a stands for $a_m, \ldots, a_0$. Let $R = \mathrm{res}(f, f', x)$. If c_1, c_2 are two points in the same connected component of $R \neq 0$ in parametric space $\mathbb{R}^{m+1}$, then $f(c_1, x)$ and $f(c_2, x)$ have the same number of real solutions.

Proof. By Corollary 1.1, $R = \mathrm{res}(f, f', x) = \pm \, \mathrm{lc}(f, x) \cdot \mathrm{discrim}(f, x)$. Since $R(c_1) \neq 0$, $f(c_1, x)$ is of degree m and has no multiple roots. Because roots depend continuously on coefficients (see for example Proposition 3.11 in [Basu *et al.* (2003)]), there exists a neighborhood of c_1, say $N(c_1; r_1)$ (the open ball centered at c_1 with radius r_1), contained in the connected component, such that $f(a, x)$ has the same number of real solutions at any points in $N(c_1; r_1)$.

Assume P is a path connecting c_1 and c_2 in the connected component. Define
$$C = \{c \in P \mid f(c, x) \text{ and } f(c_1, x) \text{ have different number of real solutions}\}.$$
Let d_c be the distance from $c \in C$ to c_1 along P. If C is not empty, d_c must have a greatest lower bound since there exists a lower bound r_1. Denote the greatest lower bound by d_0 and assume $d_{c_0} = d_0$ for a point $c_0 \in C$.

On the other hand, $R(c_0) \neq 0$ and thus $f(c_0, x)$ is of degree m and has no multiple roots. Similarly, there exists a neighborhood of c_0, say $N(c_0; r_0)$ (the open ball centered at c_0 with radius r_0), contained in the connected component, such that $f(a, x)$ has the same number of real solutions at any points in $N(c_0; r_0)$. However, by the property of c_0, we have $N(c_0; r_0) \cap C \neq \emptyset$ and $N(c_0; r_0) \cap (P \setminus C) \neq \emptyset$. This means there exist at least two points in $N(c_0; r_0)$ such that the numbers of real solutions of $f(a, x)$ at these two points are different. A contradiction. Therefore, C must be empty. That completes the proof. $\qquad\square$

Theorem 3.1. *Suppose* $\mathrm{T} : [F, G_1, G_2, H]$ *is a regular zero-dimensional TSA. Then the number of real solutions of* T *is invariant over each connected component of* $\mathrm{BP}(\mathrm{T}) \neq 0$ *in* $\mathbb{R}^d$.

Proof.   Assume C is a connected component of $\text{BP}(\text{T}) \neq 0$ in $\mathbb{R}^d$.

Obviously, $\text{lc}(f_1, x_1) \cdot \text{discrim}(f_1, x_1)$ is sign-invariant and non-zero in C. By Lemma 3.1, the number of real solutions of $f_1(\boldsymbol{u}, x_1)$ is invariant over C. We view $f_2(\boldsymbol{u}, x_1, x_2)$ as a polynomial in x_2. Since

$$f_1(\boldsymbol{u}, x_1) = 0 \quad \text{and} \quad \text{res}(\text{lc}(f_2, x_2)\text{discrim}(f_2, x_2), f_1, x_1) \neq 0$$

in C, $\text{lc}(f_2, x_2)\text{discrim}(f_2, x_2) \neq 0$ in $C \times \{x_1 \mid f_1(\boldsymbol{u}, x_1) = 0,\ \boldsymbol{u} \in C\}$. So, by Lemma 3.1 again, the number of real solutions of $f_2(\boldsymbol{u}, x_1, x_2)$ (with respect to x_2) is invariant in any component of $f_1(\boldsymbol{u}, x_1) = 0$ over C. Thus, the number of real solutions of $\{f_1 = 0, f_2 = 0\}$ is invariant over C. By similar induction, we know that the number of real solutions of $\{f_1 = 0, \ldots, f_s = 0\}$ is invariant over C.

For any $q \in G_1 \cup G_2 \cup H$, since $\text{res}(q; f_s, \ldots, f_1) \neq 0$ in C, q is sign-invariant and non-zero at any real solution component of $\{f_1 = 0, \ldots, f_s = 0\}$ over C. That completes the proof. $\square$

3.2   Triangular Decomposition of SASs

Similar to the case of polynomial system, transforming an SAS equivalently into some TSAs is called *triangular decomposition of SAS*.

Obviously, for an SAS S : $[P, G_1, G_2, H]$ in the form of (3.1), triangularizing the equations P will give a triangular decomposition of S. If S is a parametric SAS, one can view all the indeterminants ($\boldsymbol{x}$ and $\boldsymbol{u}$) as variables and get a zero decomposition for P over an extension field of $\mathbb{Q}$, say $\mathbb{C}$. That will lead to a triangular decomposition of S over the extension field of $\mathbb{Q}$. For more details of the method and recent advances in this direction, see [Chen *et al.* (2012b, 2013)].

Another way for triangular decomposition of SAS is the so-called "hierarchical strategy" introduced in [Xia (1998); Yang *et al.* (1999, 2001); Yang and Xia (2004, 2005)]. By this strategy, one first computes a generic regular decomposition for P over an extension field of $\mathbb{Q}(\boldsymbol{u})$, which leads to a generic regular decomposition for S. Some other decompositions with more properties such as "squarefree" or "disjoint" can be computed based on the generic regular decomposition. With these decompositions, one can then study the real solutions to S when the parameters are not on a variety V in the parametric space. So, after the first step of the hierarchical strategy, one gets hold of the real solutions to S when $V \neq 0$ in $\mathbb{R}^d$. For the case $V = 0$, one may add the equation to the system S and recursively call the above procedure to study the real solutions to S over a Zariski open set in $\mathbb{R}^{d-1}$. By repeating similar procedure, one can eventually get hold of all the real solutions to S. The details of the strategy will be elaborated in this section and Chapter 6.

The hierarchical strategy is very useful for computing real root classification of parametric SAS (see Chapter 6), which is a special kind of *quantifier elimination* (QE) problems. For some big SASs, the computation based on the strategy may stop at some stage due to huge consumption of space with outputs called *partial solution* to the problem, while other complete QE methods cannot get any outputs.

In this section, we introduce algorithms for computing regular, square-free or disjoint decomposition for SAS. The main content of the rest of this chapter is from [Tang *et al.* (2014); Chen *et al.* (2014, 2015)].

3.2.1 *Generic Zero-Dimensional Case*

For a parametric SAS S : $[P, G_1, G_2, H]$ in the form of (3.1), if the equations P is generic zero-dimensional (see Definition 2.10), S is said to be a *generic zero-dimensional SAS*. We discuss in this subsection the simple and basic case that S is generic zero-dimensional.

Firstly, we have an analogous of Definition 2.15 for SAS.

Definition 3.4. A pair $[\mathcal{T}, p]$ is called a *generic regular decomposition* of a given generic zero-dimensional SAS S : $[P, G_1, G_2, H]$ if

(1) $\mathcal{T} = \{T_i : [T_i, [\ ], G_{2i}, [\ ]] \mid 1 \leq i \leq l\}$ is a finite set of regular zero-dimensional TSAs and p is a polynomial in $\mathbb{Q}[\boldsymbol{u}]$ such that $[\{T_i \mid 1 \leq i \leq l\}, p]$ is a generic regular decomposition of P in $\mathcal{K}[\boldsymbol{u}, \boldsymbol{x}]$ where $\mathcal{K}$ is an extension field of $\mathbb{Q}$; and

(2) $\mathrm{BP}(T_i)$ divides p for all i and $\mathrm{W}(S(a)) = \cup_{i=1}^{l} \mathrm{W}(T_i(a))$ for any $a \in \mathbb{R}^d \backslash \mathrm{V}^{\boldsymbol{u}}(p)$.

To get a generic regular decomposition for an SAS, we may use Algorithm `GRDforZD`. However, Algorithm `GRDforZD` only outputs a generic regular decomposition for the equations of the SAS. The resulting regular chain is not necessarily square-free. So, we first need an algorithm for transforming regular chain to square-free regular chain.

Definition 3.5. A pair $[\mathcal{H}, p]$ is called a *generic square-free decomposition* of a given zero-dimensional regular chain T in $\mathcal{K}[\boldsymbol{u}, \boldsymbol{x}]$, if

(1) $\mathcal{H}$ is a finite set of zero-dimensional square-free regular chains in

$\mathcal{K}[\boldsymbol{u}, \boldsymbol{x}]$ such that $V_{\tilde{\mathcal{K}}}(\mathrm{T}) = \cup_{\mathrm{H} \in \mathcal{H}} V_{\tilde{\mathcal{K}}}(\mathrm{H})$;

(2) p is a polynomial in $\mathcal{K}[\boldsymbol{u}]$ such that for any $a \in \hat{\mathcal{K}}^d \backslash V^{\boldsymbol{u}}(p)$, $V(\mathrm{T}(a)) = \cup_{\mathrm{H} \in \mathcal{H}} V(\mathrm{H}(a))$ and $\mathrm{H}(a)$ is square-free and regular for any $\mathrm{H} \in \mathcal{H}$.

The polynomial p in Definition 3.5 is called the *square-free decomposition unstable polynomial* (SDU).

Algorithm 3.1 RCtoSqrfree

Input: A zero-dimensional regular chain $T = [f_1, \ldots, f_n]$, variables $\boldsymbol{x}$

Output: $[\mathcal{T}, p]$, a generic square-free decomposition of T

1:   $k \leftarrow$ the maximal integer such that $\mathrm{res}(\mathrm{discrim}(f_k); f_{k-1}, \ldots, f_1) = 0$;

2:   **if** k is undefined **then**

3:      **return** $[\{T\}, 1]$;

4: **end if**

5:   $\mathcal{T} \leftarrow \emptyset$; $p \leftarrow 1$; $T_{k-1} \leftarrow [f_1, \ldots, f_{k-1}]$;

6:   $[\mathcal{H}, \mathcal{G}, p'] \leftarrow \mathrm{WRSD}(T_{k-1}, \mathrm{discrim}(f_k), \boldsymbol{x})$;

7:   $\mathcal{T} \leftarrow \mathcal{T} \cup \bigcup_{G \in \mathcal{G}} \{\mathrm{RCtoSqrfree}([G, f_k, \ldots, f_n], \boldsymbol{x})_1\}$;

8:   $p \leftarrow p \cdot p' \cdot \prod_{G \in \mathcal{G}} \{\mathrm{RCtoSqrfree}([G, f_k, \ldots, f_n], \boldsymbol{x})_2$;

9:   $j \leftarrow$ the least nonnegative integer such that $\mathrm{prem}(R_j; T_{k-1}) \neq 0$, where $R_j = R_j(f_k, \frac{\partial}{\partial x_k} f_k)$ is the PSC corresponding to the jth subresultant $S_j(f_k, \frac{\partial}{\partial x_k} f_k)$ of f_k and $\frac{\partial}{\partial x_k} f_k$ with respect to x_k;

10: **if** $\mathrm{res}(R_j; T_{k-1}) \neq 0$ **then**

11:     compute, by Theorem 2.4, a common divisor with highest degree of f_k and $\frac{\partial}{\partial x_k} f_k$ in $\mathbb{Q}[x_1, \ldots, x_k]/\mathrm{sat}(T_{k-1})$ and denote it by f_{k1};

12:     $f_{k2} \leftarrow \mathrm{pquo}(f_k, f_{k1}, x_k)$;

13:     $[\mathcal{H}, q'] \leftarrow \mathrm{RCtoSqrfree}([f_1, \ldots, f_{k-1}, f_{k2}, f_{k+1}, \ldots, f_n], \boldsymbol{x})$;

14:     $\mathcal{T} \leftarrow \mathcal{T} \cup \mathcal{H}$; $p \leftarrow p \cdot \mathrm{res}(R_j; T_{k-1}) \cdot q'$;

15: **else**

16:     $[\mathcal{H}, \mathcal{G}, q'] \leftarrow \mathrm{WRSD}(T_{k-1}, R_j, \boldsymbol{x})$;

17:     $\mathcal{T} \leftarrow \mathcal{T} \cup \bigcup_{G \in \mathcal{H} \cup \mathcal{G}} \mathrm{RCtoSqrfree}([G, f_k, \ldots, f_n], \boldsymbol{x})_1$;

18:     $p \leftarrow p \cdot q' \cdot \prod_{G \in \mathcal{H} \cup \mathcal{G}} \mathrm{RCtoSqrfree}([G, f_k, \ldots, f_n], \boldsymbol{x})_2$;

19: **end if**

20: **return** $[\mathcal{T}, p]$

Theorem 3.2. *Algorithm 3.1 (*RCtoSqrfree*) terminates correctly.*

Proof.   By induction on the depth of recursion, we can easily obtain the termination of the algorithm. The correctness follows directly from the correctness of Algorithm WRSD and Theorem 2.4. $\square$

Algorithm 3.2 GRDforZDSAS

Input: A generic zero-dimensional SAS S : $[P, G_1, G_2, H]$ of the form (3.1), variables $\boldsymbol{x}$

Output: $[\mathcal{T}, p]$, a generic regular decomposition of S

1: $[O, p] \leftarrow$ GRDforZD$(P, \boldsymbol{x})$; $\mathcal{T} \leftarrow \emptyset$;
2: **for** T in O **do**
3:      T $\leftarrow \{[T, G_1, G_2, H]\}$;
4:      $[\text{T}, q] \leftarrow$ CoprimeH$(\text{T}, \boldsymbol{x})$; $p \leftarrow p \cdot q$;
5:      $[\text{T}, q] \leftarrow$ CoprimeG2$(\text{T}, \boldsymbol{x})$; $p \leftarrow p \cdot q$;
6:      $[\text{T}, q] \leftarrow$ CoprimeG1$(\text{T}, \boldsymbol{x})$; $p \leftarrow p \cdot q$;
7:      **for** F in T **do**
8:          $[\mathcal{H}, q] \leftarrow$ RCtoSqrfree$(F, \boldsymbol{x})$; $\mathcal{T} \leftarrow \mathcal{T} \bigcup \mathcal{H}$; $p \leftarrow p \cdot q$;
9:      **end for**
10: **end for**
11: **return** $[\mathcal{T}, p]$

Algorithm 3.3 CoprimeH

Input: T $= \{\text{T}_i : [F_i, G_1, G_2, H_i] \mid 1 \le i \le l\}$, a finite set of zero-dimensional TSAs where each F_i is a regular chain, and variables $\boldsymbol{x}$

Output: $[\text{T}', q]$ where $\text{T}' = \{\text{T}'_i : [F'_i, G_1, G_2, [\,]] \mid 1 \le i \le l'\}$, each F'_i is a regular chain and q is a polynomial in parameter such that $\text{W}(\text{T}) = \text{W}(\text{T}')$ over $q \ne 0$ in $\mathbb{R}^d$

1: $q \leftarrow 1$;
2: **for** T_i in T **do**
3:      **for** h in H_i **do**
4:          $H_i \leftarrow H_i \setminus \{h\}$; $\text{BP}_h \leftarrow \text{res}(h; F_i)$;
5:          **if** $\text{BP}_h \ne 0$ **then**
6:              $q \leftarrow q \cdot \text{BP}_h$;
7:          **else**
8:              $[\mathcal{H}, \mathcal{G}, p'] \leftarrow$ WRSD$(F_i, h, \boldsymbol{x})$;
9:              T $\leftarrow (\text{T} \setminus \{\text{T}_i\}) \cup \bigcup_{F'_i \in \mathcal{G}} \{[F'_i, G_1, G_2, H_i]\}$; $q \leftarrow q \cdot p'$; **break**;
10:          **end if**
11:      **end for**
12: **end for**
13: **return** $[\text{T}, q]$

Algorithm 3.4 `CoprimeG2`

Input: $T = \{T_i : [F_i, G_1, G_2, [\,]] \mid 1 \leq i \leq l\}$, a finite set of zero-dimensional TSAs where each F_i is a regular chain, and variables $\boldsymbol{x}$

Output: $[T', q]$ where $T' = \{T'_i : [F'_i, G_1, G_2, [\,]] \mid 1 \leq i \leq l'\}$, each F'_i is a regular chain and q is a polynomial in parameter such that $W(T) = W(T')$ over $q \neq 0$ in $\mathbb{R}^d$

1:   $q \leftarrow 1;\ G'_2 \leftarrow G_2;$
2: **for** T_i in T **do**
3:    **for** g in G'_2 **do**
4:      $G'_2 \leftarrow G'_2 \setminus \{g\};\ \mathrm{BP}_g \leftarrow \mathrm{res}(g; F_i);$
5:      **if** $\mathrm{BP}_g \neq 0$ **then**
6:        $q \leftarrow q \cdot \mathrm{BP}_g;$
7:      **else**
8:        $[\mathcal{H}, \mathcal{G}, p'] \leftarrow \mathrm{WRSD}(F_i, g, \boldsymbol{x});$
9:        $T \leftarrow (T \setminus \{T_i\}) \cup \bigcup_{F'_i \in \mathcal{G}}\{[F'_i, G_1, G_2, [\,]]\};\ q \leftarrow q \cdot p';$ **break;**
10:     **end if**
11:    **end for**
12: **end for**
13: **return** $[T, q]$

Theorem 3.3. *Algorithm 3.2 (`GRDforZDSAS`) terminates correctly.*

Proof.   Termination is obvious because those "for" loops are finite. Correctness follows directly from the correctness of Algorithm `WRSD`, Algorithm `GRDforZD` and Algorithm `RCtoSqrfree`. $\square$

Remark 3.1. Note that, if the input to Algorithm 3.2 is a constant zero-dimensional SAS, the output is indeed a set of constant zero-dimensional regular TSAs since the RDU p is a non-zero constant.

Let $\mathcal{T}$ is a generic regular decomposition of a generic zero-dimensional SAS S. Because any two TSAs in $\mathcal{T}$ may have common solutions, if we want to count the distinct real solutions of S by counting the real solutions of each TSA in $\mathcal{T}$, we have to make the decomposition *disjoint,* *i.e.* any two TSAs in $\mathcal{T}$ have no common solutions.

Definition 3.6. The *difference* of two triangular sets (regular chains) T_1 and T_2 is a set of triangular sets (regular chains) $\mathcal{T}$ such that $V(T_1) \setminus V(T_2) = \bigcup_{T \in \mathcal{T}} V(T)$. For two parametric systems T_1 and T_2, $[\mathcal{T}, p]$ is said to be a *generic difference* of T_1 and T_2 if $V(T_1(a)) \setminus V(T_2(a)) =$

Algorithm 3.5 `CoprimeG1`

Input: $\mathrm{T} = \{\mathrm{T}_i : [F_i, G_1, G_2, [\,]] \mid 1 \leq i \leq l\}$, a finite set of zero-dimensional TSAs where each F_i is a regular chain, and variables $\boldsymbol{x}$

Output: $[\mathrm{T}', q]$ where $\mathrm{T}' = \{\mathrm{T}_i' : [F_i', [\,], G_2', [\,]] \mid 1 \leq i \leq l'\}$, each F_i' is a regular chain and q is a polynomial in parameter such that $\mathrm{W}(\mathrm{T}) = \mathrm{W}(\mathrm{T}')$ over $q \neq 0$ in $\mathbb{R}^d$

1:   $q \leftarrow 1$; $G_1' \leftarrow G_1$;

2:   **for** T_i in T **do**

3:     **for** g in G_1' **do**

4:       $G_1' \leftarrow G_1' \setminus \{g\}$; $\mathrm{BP}_g \leftarrow \mathrm{res}(g; F_i)$;

5:       **if** $\mathrm{BP}_g \neq 0$ **then**

6:         $q \leftarrow q \cdot \mathrm{BP}_g$; $G_2 \leftarrow G_2 \cup \{g\}$;

7:       **else**

8:         $[\mathcal{H}, \mathcal{G}, p'] \leftarrow \mathrm{WRSD}(F_i, g, \boldsymbol{x})$;

9:         $q \leftarrow q \cdot p'$; $\mathrm{T} \leftarrow (\mathrm{T} \setminus \{\mathrm{T}_i\}) \cup \bigcup_{F_i' \in \mathcal{G}} \{[F_i', G_1', G_2 \cup \{g\}, [\,]]\} \cup \bigcup_{F_i' \in \mathcal{H}} \{[F_i', G_1', G_2, [\,]]\}$; **break**;

10:       **end if**

11:     **end for**

12:   **end for**

13:   **return** $[\mathrm{T}, q]$

$\bigcup_{\mathrm{T} \in \mathcal{T}} \mathrm{V}(\mathrm{T}(a))$ for all $a \in \mathbb{R}^d \setminus \mathrm{V}^{\boldsymbol{u}}(p)$. The polynomial p is called *difference decomposition unstable polynomial* (DDU).

For two zero-dimensional regular chains, it is not hard to give an algorithm for computing their difference based on Algorithm `WRSD`. Suppose $\mathrm{T}_1 = [T_{11}, \ldots, T_{1n}], \mathrm{T}_2 = [T_{21}, \ldots, T_{2n}]$. First, compute

$$[\mathcal{H}_1, \mathcal{G}_1, p_1] = \mathrm{WRSD}(T_{11}, T_{21}, x_1).$$

Obviously, $[G, T_{12}, \ldots, T_{1n}]$ belongs to the difference for any $G \in \mathcal{G}_1$. Then, for each $H \in \mathcal{H}_1$, compute

$$[\mathcal{H}_{2H}, \mathcal{G}_{2H}, p_{2H}] = \mathrm{WRSD}([H, T_{12}], T_{22}, [x_1, x_2]).$$

So, $[G, T_{13}, \ldots, T_{1n}]$ belongs to the difference for any $G \in \mathcal{G}_{2H}$. Continuing the similar process, we will finally get the difference and the DDU. For convenience, we denote the algorithm by $\mathrm{Difference}(\mathrm{T}_1, \mathrm{T}_2, \boldsymbol{x})$. We only give the specification of the algorithm here and omit the detail.

Based on Algorithm 3.6, we can easily obtain an algorithm for making a set of zero-dimensional regular chains (or zero-dimensional regular TSAs) disjoint. The specification of the algorithm is as follows.

Algorithm 3.6 `Difference`

Input: two zero-dimensional regular chains T_1, T_2, and variables $\boldsymbol{x}$
Output: $[\mathcal{T}, p]$, a generic difference of T_1 and T_2

Algorithm 3.7 `Disjoint`

Input: $\mathcal{T}$, a set of zero-dimensional regular chains (or zero-dimensional regular TSAs), variables $\boldsymbol{x}$
Output: $[\mathcal{T}', p]$, where $\mathcal{T}'$ is a set of zero-dimensional regular chains (or zero-dimensional regular TSAs) p is a polynomial in parameters such that
   (1) the regular chains (TSAs) in $\mathcal{T}'$ are pairwise disjoint;
   (2) $V(\mathcal{T}(a)) = V(\mathcal{T}'(a))$ (or $W(\mathcal{T}(a)) = W(\mathcal{T}'(a))$) for any $a \in \mathbb{R}^d \setminus V^{\boldsymbol{u}}(p)$;
   (3) $\mathcal{T}'(a)$ is still a set of zero-dimensional regular chains (or zero-dimensional regular TSAs) for any $a \in \mathbb{R}^d \setminus V^{\boldsymbol{u}}(p)$.

Remark 3.2. Combining Algorithm 3.7 and Algorithm 3.2, it is very easy to give an algorithm which, for an input generic zero-dimensional SAS S, computes a disjoint generic regular decomposition of S. We do not give the details of the algorithm here.

Especially, if the input is a constant zero-dimensional SAS, all the algorithms in this subsection (Algorithms 3.1-3.7) work and the output *decomposition-unstable* varieties are constants. So, a constant zero-dimensional SAS can always be decomposed into finitely many pairwise disjoint square-free zero-dimensional regular TSAs.

3.2.2 *Positive Dimensional Case*

Positive dimensional case is more complicated. We first need a generic regular decomposition algorithm for positive dimensional polynomial systems. Generic regular decompositions in positive dimensional case have to be expressed by *regular systems* instead of regular chains.

Definition 3.7. Let $T \subset \mathcal{K}[\boldsymbol{u}, \boldsymbol{x}]$ be a regular chain and $h \in \mathcal{K}[\boldsymbol{u}, \boldsymbol{x}]$. If $\operatorname{res}(h; T) \neq 0$, then $[T, h]$ is said to be a *regular system* in $\mathcal{K}[\boldsymbol{u}, \boldsymbol{x}]$.

The definition of regular system here is the same as that introduced in [Chen *et al.* (2007)] and different from that firstly proposed in [Wang

(2000)], see more details in [Chen *et al.* (2007)]. By Theorem 1.1, we have

Proposition 3.2. *If* $[\mathrm{T}, h]$ *is a regular system in* $\mathcal{K}[\boldsymbol{u}, \boldsymbol{x}]$ *and* $\tilde{\mathcal{K}}$ *is an algebraically closed extension field of* $\mathcal{K}(\boldsymbol{u})$, *then* $\mathrm{V}_{\tilde{\mathcal{K}}}(\mathrm{T}\backslash h) \neq \emptyset$.

For regular systems, we need some similar concepts which are defined for regular chains in Chapter 2.

Definition 3.8. Let P be a parametric system in $\mathcal{K}[\boldsymbol{u}, \boldsymbol{x}]$ and

$$\mathcal{TH} = \{[\mathrm{T}_1, h_1], \ldots, [\mathrm{T}_s, h_s]\}$$

be a set of regular systems in $\mathcal{K}[\boldsymbol{u}, \boldsymbol{x}]$. The set $\mathcal{TH}$ is said to be a *parametric regular system decomposition* of P in $\mathcal{K}[\boldsymbol{u}, \boldsymbol{x}]$, if

$$\mathrm{V}_{\tilde{\mathcal{K}}}(\mathrm{P}) = \cup_{i=1}^{s} \mathrm{V}_{\tilde{\mathcal{K}}}(\mathrm{T}_i \backslash h_i)$$

where $\tilde{\mathcal{K}}$ is an extension field of $\mathcal{K}(\boldsymbol{u})$.

Definition 3.9. Let $[\mathrm{T}, h]$ be a regular system in $\mathcal{K}[\boldsymbol{u}, \boldsymbol{x}]$ and $a \in \hat{\mathcal{K}}^d$ where $\hat{\mathcal{K}}$ is an extension field of $\mathcal{K}$. If
    (1) $\mathrm{T}(a)$ is a regular chain in $\hat{\mathcal{K}}[\boldsymbol{x}]$,
    (2) rank(T(a))= rank(T) and
    (3) res$(h(a); \mathrm{T}(a)) \neq 0$,
then we say that the regular system $[\mathrm{T}, h]$ *specializes well* at a.

Definition 3.10. Let P be a parametric system in $\mathcal{K}[\boldsymbol{u}, \boldsymbol{x}]$ and

$$\mathcal{TH} = \{[\mathrm{T}_1, h_1], \ldots, [\mathrm{T}_s, h_s]\}$$

be a parametric regular system decomposition of P in $\mathcal{K}[\boldsymbol{u}, \boldsymbol{x}]$. For any $a \in \hat{\mathcal{K}}^d$, if $[\mathrm{T}_i, h_i](1 \leq i \leq s)$ specializes well at a and

$$\mathrm{V}(\mathrm{P}(a)) = \cup_{i=1}^{s} \mathrm{V}(\mathrm{T}_i(a) \backslash h_i(a)),$$

$\mathcal{TH}$ is said to be *stable* at a.

Definition 3.11. Let $\mathcal{TH}$ be a parametric regular system decomposition of a given parametric system P in $\mathcal{K}[\boldsymbol{u}, \boldsymbol{x}]$. If there is an affine variety $\mathcal{V}$ in $\hat{\mathcal{K}}^d$ with dim$(\mathcal{V}) < d$ such that $\mathcal{TH}$ is stable at any $a \in \hat{\mathcal{K}}^d \backslash \mathcal{V}$, then $\mathcal{TH}$ is said to be a *generic regular system decomposition* of P and $\mathcal{V}$ is said to be a *regular-decomposition-unstable (RDU) variety* of P with respect to $\mathcal{TH}$.

3.2.2.1   *Wu's Decomposition Under Specification*

Definition 3.12. Let P_1 be a parametric system in $\mathcal{K}[u, x]$ and $\mathcal{S} = \{C_1, \ldots, C_m\}$ be a Wu's decomposition of P_1 in $\mathcal{K}[u, x]$. Suppose $\mathcal{L} = \{C_{l,1}, C_{l,2}, \ldots, C_{l,k}\}$ is a subset of $\mathcal{S}$ satisfying that

    (1) $C_{l,1}$ is a characteristic set of P_1;

    (2) If $k \geq 2$, $C_{l,i}$ $(2 \leq i \leq k)$ is a characteristic set of $P_i = P_{i-1} \cup C_{l,i-1} \cup \{\mathrm{lc}(h_{l,i-1})\}$ where $h_{l,i-1} \in C_{l,i-1}$;

    (3) If $k = 1$, $C_{l,1}$ is a contradictory ascending chain. Otherwise, $C_{l,i}$ $(1 \leq i \leq k-1)$ is a non-contradictory ascending chain and $C_{l,k}$ is a contradictory ascending chain.

    Then $\mathcal{L}$ is said to be a *line* of $\mathcal{S}$ and $P^{\mathcal{L}} = \{P_1, \ldots, P_k\}$ the corresponding systems.

Example 3.1. Consider the Wu's decomposition of
$$P = \{(x - 1)y^2 + x^2 - 1, x - 1\}.$$
Let $x \prec y$. The Wu's decomposition of P is $\mathcal{S} = \{C_1, C_2, C_3\}$, where
$$C_1 = \{(x - 1)y^2 + x^2 - 1, x - 1\}, C_2 = \{x - 1\}, C_3 = \{1\}.$$
Then $\{C_1, C_2, C_3\}$ is a line of $\mathcal{S}$. The corresponding systems is
$$\{P, P \cup C_1 \cup \{x - 1\}, P \cup C_1 \cup C_2 \cup \{1\}\}.$$
In this case, $k = 3$.

    $\{C_1, C_3\}$ is also a line of $\mathcal{S}$. The corresponding systems is $\{P, P \cup C_1 \cup \{1\}\}$. In this case, $k = 2$.

Example 3.2. Consider the Wu's decomposition of $P = \{x, x - 1\}$. The Wu's decomposition of P is $\mathcal{S} = \{C_1\}$, where $C_1 = \{1\}$. In this case, $k = 1$.

Lemma 3.2. *Let* $P_1 = \{f_1, ..., f_t\}$ *be a parametric system in* $\mathcal{K}[u, x]$ *and* $\mathcal{S}$ *be a Wu's decomposition of* P_1 *in* $\mathcal{K}[u, x]$. *Let* $\mathcal{L} = \{C_{l,1}, \ldots, C_{l,k}\}$ *be a line of* $\mathcal{S}$ *with corresponding systems* $\{P_1, \ldots, P_k\}$. *Then for any* $a \in \hat{\mathcal{K}}^d \backslash V^u(C_{l,k})$ *and any* $P_i (1 \leq i \leq k)$, *there exists a polynomial* $p_i \in P_i$ *such that* $p_i(a) \not\equiv 0$.

Proof.   We prove it by induction on the number k of elements of $\mathcal{L}$. If $k = 1$, it means $\mathcal{L}$ contains only one element. Since $C_{l,1}$ is a contradictory ascending chain, we can assume that $C_{l,1} = \{g_{l,1}\}$ where $g_{l,1} \in \mathcal{K}[u]$ and we know that $g_{l,1} \in \langle P_1 \rangle$. Then $g_{l,1}$ can be written as $g_{l,1} = \sum_{j=1}^{t} h_j f_j$ where $h_j \in \mathcal{K}[u, x]$ for any j $(1 \leq j \leq t)$. Since $a \in \hat{\mathcal{K}}^d \backslash V^u(g_{l,1})$, we have

$$g_{l,1}(a) = \sum_{j=1}^{t} h_j(a) f_j(a) \neq 0.$$

Thus there must exist some $f_{j_0} \in \mathrm{P}_1$ such that $f_{j_0}(a) \not\equiv 0$.

Now we assume that the conclusion holds when $k < N$ ($N > 1$). Suppose $k = N$. By Definition 3.12, it is evident that $\mathcal{L}_2 = \{\mathrm{C}_{l,2}, \ldots, \mathrm{C}_{l,k}\} \subsetneq \mathcal{L}$ is a line of Wu's decomposition of P_2. According to the induction hypothesis, for any $a \in \hat{\mathcal{K}}^d \backslash \mathrm{V}^{\boldsymbol{u}}(\mathrm{C}_{l,k})$ and any $\mathrm{P}_i (2 \leq i \leq k)$, there exists a polynomial $p_i \in \mathrm{P}_i$ such that $p_i(a) \not\equiv 0$.

Without loss of generality, suppose

$$\mathrm{P}_2 = \mathrm{P}_1 \cup \mathrm{C}_{l,1} \cup \{\mathrm{lc}(h_{l,1})\}$$

for some $h_{l,1} \in \mathrm{C}_{l,1}$.

If $p_2 \in \mathrm{P}_1$, the conclusion holds.

If $p_2 \in \mathrm{C}_{l,1} \subset \langle \mathrm{P}_1 \rangle$, then p_2 can be written as $p_2 = \sum_{j=1}^{t} h_j f_j$ where $h_j \in \mathcal{K}[\boldsymbol{u}, \boldsymbol{x}]$ for any j ($1 \leq j \leq t$). Thus $p_2(a) = \sum_{j=1}^{t} h_j(a) f_j(a)$ holds. Since $p_2(a) \not\equiv 0$, there must exist some $f_{j_1} \in \mathrm{P}_1$ such that $f_{j_1}(a) \not\equiv 0$.

If $p_2 \in \{\mathrm{lc}(h_{l,1})\}$, it implies $h_{l,1}(a) \not\equiv 0$. Since $h_{l,1}$ can be written as $h_{l,1} = \sum_{j=1}^{t} g_j f_j$ where $g_j \in \mathcal{K}[\boldsymbol{u}, \boldsymbol{x}]$ for any j ($1 \leq j \leq t$). Thus there must exist some $f_{j_2} \in \mathrm{P}_1$ such that $f_{j_2}(a) \not\equiv 0$. $\qquad\square$

Corollary 3.1. *Suppose* $\mathrm{P} = \{p\} \subset \mathcal{K}[\boldsymbol{u}, \boldsymbol{x}]$, $\mathcal{S}$ *is a Wu's decomposition of* P *in* $\mathcal{K}[\boldsymbol{u}, \boldsymbol{x}]$ *and* $\mathcal{L} = \{\mathrm{C}_{l,1}, \mathrm{C}_{l,2}, \ldots, \mathrm{C}_{l,k}\}$ *is a line of* $\mathcal{S}$. *Then* $p(a) \not\equiv 0$ *for any* $a \in \hat{\mathcal{K}}^d \backslash \mathrm{V}^{\boldsymbol{u}}(\mathrm{C}_{l,k})$.

Lemma 3.3. *Suppose* $\mathrm{C} = \{g_1, \ldots, g_t\}$ *is a non-contradictory characteristic set of parametric system* P *in* $\mathcal{K}[\boldsymbol{u}, \boldsymbol{x}]$. *For any* $a \in \hat{\mathcal{K}}^d$,

$$\mathrm{V}(\mathrm{P}(a)) = \mathrm{V}(\mathrm{C}(a) \backslash \mathrm{lc}(\mathrm{C})(a)) \cup \cup_{i=1}^{t} \mathrm{V}(\mathrm{P}(a) \cup \mathrm{C}(a) \cup \{\mathrm{lc}(g_i)(a)\}).$$

Proof. For any $a \in \hat{\mathcal{K}}^d$, if $\mathrm{lc}(\mathrm{C})(a) \equiv 0$, the conclusion obviously holds since $\mathrm{lc}(\mathrm{C})(a) = \prod_{i=1}^{t} \mathrm{lc}(g_i)(a)$. Now we prove the conclusion when $\mathrm{lc}(\mathrm{C})(a) \not\equiv 0$.

According to the definition of characteristic set, we know that $\mathrm{C} \subset \langle \mathrm{P} \rangle$ and for any $p \in \mathrm{P}$,

$$\mathrm{lc}(g_1)^{k_1} \cdots \mathrm{lc}(g_t)^{k_t} p = q_1 g_1 + \cdots + q_t g_t$$

where $q_i \in \mathcal{K}[\boldsymbol{u}, \boldsymbol{x}]$ for any i ($1 \leq i \leq t$). Then $\mathrm{C}(a) \subset \langle \mathrm{P}(a) \rangle$ and

$$\mathrm{lc}(g_1)(a)^{k_1} \cdots \mathrm{lc}(g_t)(a)^{k_t} p(a) = q_1(a) g_1(a) + \cdots + q_t(a) g_t(a).$$

Therefore

$$\mathrm{V}(\mathrm{C}(a) \backslash \mathrm{lc}(\mathrm{C})(a)) \subset \mathrm{V}(\mathrm{P}(a)) \subset \mathrm{V}(\mathrm{C}(a)) \text{ and}$$

$$\mathrm{V}(\mathrm{P}(a)) = \mathrm{V}(\mathrm{C}(a) \backslash \mathrm{lc}(\mathrm{C})(a)) \cup \mathrm{V}(\mathrm{P}(a) \cup \{\mathrm{lc}(\mathrm{C})(a)\}).$$

Since $C(a) \subset \langle P(a) \rangle$,

$$V(P(a) \cup \{lc(C)(a)\}) = V(P(a) \cup C(a) \cup lc(C)(a))$$
$$= \cup_{i=1}^{t} V(P(a) \cup C(a) \cup \{lc(g_i)(a)\}). \qquad \square$$

Corollary 3.2. *Let* P *be a parametric system in* $\mathcal{K}[\boldsymbol{u}, \boldsymbol{x}]$ *and* $\{C_1, \ldots, C_m\}$ *be a Wu's decomposition of* P *in* $\mathcal{K}[\boldsymbol{u}, \boldsymbol{x}]$*. Suppose*

$$\mathcal{S} = \{C_i | 1 \leq i \leq m \text{ and } C_i \text{ is a non-contradictory ascending chain}\}$$

and

$$\mathcal{CS} = \{C_i | 1 \leq i \leq m \text{ and } C_i \text{ is a contradictory ascending chain}\}.$$

Then for any $a \in \hat{\mathcal{K}}^d \backslash (\cup_{CS \in \mathcal{CS}} V^{\boldsymbol{u}}(CS))$,

$$V(P(a)) = \cup_{C \in \mathcal{S}} V(C(a) \backslash lc(C)(a)).$$

3.2.2.2   *Converting to Regular Systems*

According to Algorithm 2.3, the following proposition is clear.

Proposition 3.3. *Let* T *be a triangular set in* $\mathcal{K}[\boldsymbol{u}, \boldsymbol{x}]$ *and* $[\mathcal{G}, p] =$ $\mathtt{ZDtoRC}(T, \mathrm{mvar}(T))$*. If* $\mathcal{G} \neq \emptyset$*, then* $lc(T)(a) \neq 0$ *for* $a \notin V^{\boldsymbol{u}}(p)$*.*

Now suppose $T = \{T_1, \ldots, T_l\}$ is a triangular set in $\mathcal{K}[\boldsymbol{u}, \boldsymbol{x}]$ and $\mathrm{mvar}(T) \subsetneq \boldsymbol{x}$. Assume that $[\mathcal{G}, p] = \mathtt{ZDtoRC}(T, \mathrm{mvar}(T))$. It is interesting to show that we have two versions to interpret the relationship between T and $[\mathcal{G}, p]$. Let $\boldsymbol{u}' = \boldsymbol{u} \cup (\boldsymbol{x} \setminus \mathrm{mvar}(T))$ and recall that $\hat{\mathcal{K}}$ and $\widehat{\mathcal{K}(\boldsymbol{u}')}$ are extension fields of $\mathcal{K}$ and $\mathcal{K}(\boldsymbol{u}')$, respectively.

On one hand, T can be regarded as a triangular set in $\mathcal{K}[\boldsymbol{u}'][\mathrm{mvar}(T)]$. Then, according to Algorithm 2.3, we know that

(1) $V_{\widehat{\mathcal{K}(\boldsymbol{u}')}}(T \backslash lc(T)) = \cup_{G \in \mathcal{G}} V_{\widehat{\mathcal{K}(\boldsymbol{u}')}}(G)$,

(2) for any $a \in \hat{\mathcal{K}}^{d+n-l} \backslash V^{\boldsymbol{u}'}(p)$, $V(T(a) \backslash lc(T)(a)) = \cup_{G \in \mathcal{G}} V(G(a))$ and G specializes well at a for any $G \in \mathcal{G}$.

On the other hand, T can also be regarded as a triangular set in $\mathcal{K}[\boldsymbol{u}][\boldsymbol{x} \setminus \mathrm{mvar}(T)][\mathrm{mvar}(T)]$. Let $\mathcal{K}' = \mathcal{K}(\boldsymbol{u})$, $\mathcal{U} = \boldsymbol{x} \setminus \mathrm{mvar}(T)$ and $\boldsymbol{x}' = \mathrm{mvar}(T)$. Then according to Algorithm 2.3,

(3) $V_{\widetilde{\mathcal{K}'(\mathcal{U})}}(T \backslash lc(T)) = \cup_{G \in \mathcal{G}} V_{\widetilde{\mathcal{K}'(\mathcal{U})}}(G)$,

(4) for any $a \in \tilde{\mathcal{K}'}^{n-l} \backslash V^{\mathcal{U}}(p)$, $V_{\tilde{\mathcal{K}'}}(T(a) \backslash lc(T)(a)) = \cup_{G \in \mathcal{G}} V_{\tilde{\mathcal{K}'}}(G(a))$ and G specializes well at a for any $G \in \mathcal{G}$.

Remark that the above statements (1) and (3) are exactly the same since $\mathcal{K}(\boldsymbol{u}') = \mathcal{K}'(\mathcal{U})$. So, the following lemma is clear.

Lemma 3.4. *Suppose* $[\mathcal{G}, p] = \mathtt{ZDtoRC}(T, \mathrm{mvar}(T))$ *for a triangular set* T *in* $\mathcal{K}[\boldsymbol{u}, \boldsymbol{x}]$. *Then*

$$\mathrm{V}_{\widetilde{\mathcal{K}(\boldsymbol{u})}}(T\backslash\mathrm{lc}(T)\cdot p) = \cup_{G\in\mathcal{G}}\mathrm{V}_{\widetilde{\mathcal{K}(\boldsymbol{u})}}(G\backslash p)$$

and

$$\mathrm{V}(T(a)\backslash\mathrm{lc}(T)(a)\cdot p(a)) = \cup_{G\in\mathcal{G}}\mathrm{V}(G(a)\backslash p(a)), \ \forall a \in \hat{\mathcal{K}}^d\backslash\mathrm{V}^{\boldsymbol{u}}(p).$$

Algorithm 3.8 TSToRS

Input: A triangular set $T = \{T_1, \ldots, T_l\} \subset \mathcal{K}[\boldsymbol{u}, \boldsymbol{x}]$, variables $\boldsymbol{x} = \{x_1, \ldots, x_n\}$

Output: $[\mathcal{G}, p]$, where

(1) $\mathcal{G}$ is a finite set of regular systems in $\mathcal{K}[\boldsymbol{u}, \boldsymbol{x}]$ such that $\mathrm{V}_{\widetilde{\mathcal{K}(\boldsymbol{u})}}(T\backslash\mathrm{lc}(T)) = \cup_{[G,h]\in\mathcal{G}}\mathrm{V}_{\widetilde{\mathcal{K}(\boldsymbol{u})}}(G\backslash h)$;

(2) p is a polynomial in $\mathcal{K}[\boldsymbol{u}]$, for any $a \in \hat{\mathcal{K}}^d\backslash\mathrm{V}^{\boldsymbol{u}}(p)$, $\mathrm{V}(T(a)\backslash\mathrm{lc}(T)(a)) = \cup_{[G,h]\in\mathcal{G}}\mathrm{V}(G(a)\backslash h(a))$ and $[G, h]$ specializes well at a for any $[G, h] \in \mathcal{G}$.

1: $W \leftarrow \mathtt{ZDtoRC}(T, \mathrm{mvar}(T))$;

2: $\mathcal{G} \leftarrow \mathtt{map}(t \rightarrow [t, W_2], W_1)$;

3: **if** $W_2 \in \mathcal{K}[\boldsymbol{u}]$ **then return** $[\mathcal{G}, W_2]$; **end if**;

4: Compute a Wu's decomposition $\{C_1, C_2, \ldots, C_m\}$ of $\{W_2\}$ in $\mathcal{K}[\boldsymbol{u}, \boldsymbol{x}]$;

5: $p \leftarrow 1$;

6: **for** $i = 1$ to m **do**

7: **if** C_i is a contradictory ascending chain **then**

8:   $p \leftarrow p \cdot \mathtt{op}(C_i)$;

9: **else**

10:   $T_i \leftarrow T \cup C_i$; $RS \leftarrow \mathtt{TSToRS}(T_i, \boldsymbol{x})$;

11:   $\mathcal{G} \leftarrow \mathcal{G} \cup RS_1$; $p \leftarrow p \cdot RS_2$;

12: **end if**

13: **end for**

14: **return** $[\mathcal{G}, p]$

Theorem 3.4. *Algorithm 3.8 terminates correctly.*

Proof. For an input triangular set $T = \{T_1, \ldots, T_l\}$ in $\mathcal{K}[\boldsymbol{u}, \boldsymbol{x}]$, suppose

$$[\mathcal{G}_0, p_0] = \mathtt{ZDtoRC}(T, \mathrm{mvar}(T)).$$

Firstly, we prove Algorithm 3.8 terminates. If $p_0 \in \mathcal{K}[\boldsymbol{u}]$, the algorithm terminates at Line 3. Otherwise, $p_0 \in \mathcal{K}[\boldsymbol{u}'] \subset \mathcal{K}[\boldsymbol{u}, \boldsymbol{x}]$. Then we know that

$\operatorname{mvar}(T) \subsetneq \operatorname{mvar}(T_i)$ at Line 10. Therefore, the recursive call at Line 10 can occur only finitely many times since the algorithm will return at Line 3 if $\operatorname{mvar}(T_i) = \boldsymbol{x}$.

Now we prove the correctness by induction on the recursive depth t. If $t = 0$, $p_0 \in \mathcal{K}[\boldsymbol{u}]$ and the conclusion follows from Lemma 3.4. Assume that the conclusion holds for $t < N$ ($N > 0$).

When $t = N$, let
$$\mathcal{N} = \{i | 1 \le i \le m \text{ and } C_i \text{ is a non-contradictory ascending chain}\}$$
and
$$\mathcal{CN} = \{1, 2, \ldots, m\} \backslash \mathcal{N}.$$
Assume that for each $j \in \mathcal{CN}$, $C_j = \{C_j\}$ while for any $i \in \mathcal{N}$,
$$[\mathcal{G}_i, p_i] = \texttt{TSToRS}(T_i, \boldsymbol{x}).$$
Then according to Algorithm 3.8,
$$\mathcal{G} = \cup_{G \in \mathcal{G}_0} \{[G, p_0]\} \cup \cup_{i \in \mathcal{N}} \mathcal{G}_i$$
and
$$p = \prod_{j \in \mathcal{CN}} C_j \cdot \prod_{i \in \mathcal{N}} p_i.$$
By Lemma 3.4, Wu's method and the induction hypothesis, we get
$$\begin{aligned}
& V_{\widetilde{\mathcal{K}(\boldsymbol{u})}}(T \backslash \mathrm{lc}_T) \\
&= V_{\widetilde{\mathcal{K}(\boldsymbol{u})}}(T \backslash \mathrm{lc}_T \cdot p_0) \cup V_{\widetilde{\mathcal{K}(\boldsymbol{u})}}(T \cup \{p_0\} \backslash \mathrm{lc}_T) \\
&= \cup_{G \in \mathcal{G}_0} V_{\widetilde{\mathcal{K}(\boldsymbol{u})}}(G \backslash p_0) \cup (V_{\widetilde{\mathcal{K}(\boldsymbol{u})}}(T \backslash \mathrm{lc}_T) \cap V_{\widetilde{\mathcal{K}(\boldsymbol{u})}}(p_0)) \\
&= \cup_{G \in \mathcal{G}_0} V_{\widetilde{\mathcal{K}(\boldsymbol{u})}}(G \backslash p_0) \cup (V_{\widetilde{\mathcal{K}(\boldsymbol{u})}}(T \backslash \mathrm{lc}_T) \cap \cup_{i \in \mathcal{N}} V_{\widetilde{\mathcal{K}(\boldsymbol{u})}}(C_i \backslash \mathrm{lc}_{C_i})) \\
&= \cup_{G \in \mathcal{G}_0} V_{\widetilde{\mathcal{K}(\boldsymbol{u})}}(G \backslash p_0) \cup (\cup_{i \in \mathcal{N}} V_{\widetilde{\mathcal{K}(\boldsymbol{u})}}(T_i \backslash \mathrm{lc}_{T_i})) \\
&= \cup_{G \in \mathcal{G}_0} V_{\widetilde{\mathcal{K}(\boldsymbol{u})}}(G \backslash p_0) \cup (\cup_{i \in \mathcal{N}} \cup_{[G,h] \in \mathcal{G}_i} V_{\widetilde{\mathcal{K}(\boldsymbol{u})}}(G \backslash h))).
\end{aligned}$$
Therefore, the statement (1) in the specification of Algorithm 3.8 holds.

For any $a \in \hat{\mathcal{K}}^d \backslash V^{\boldsymbol{u}}(p)$, $C_j(a) \ne 0$ and $p_i(a) \ne 0$ for any $i \in \mathcal{N}$ and $j \in \mathcal{CN}$. Thus by Corollary 3.2, $V(p_0(a)) = \cup_{i \in \mathcal{N}} V(C_i(a) \backslash \mathrm{lc}_{C_i}(a))$. By Lemma 3.4 and the induction hypothesis, we get
$$\begin{aligned}
& V(T(a) \backslash \mathrm{lc}_T(a)) \\
&= V(T(a) \backslash \mathrm{lc}_T(a) \cdot p_0(a)) \cup V(T(a) \cup \{p_0(a)\} \backslash \mathrm{lc}_T(a)) \\
&= (\cup_{G \in \mathcal{G}_0} V(G(a) \backslash p_0(a))) \cup (V(T(a) \backslash \mathrm{lc}_T(a)) \cap V(p_0(a))) \\
&= (\cup_{G \in \mathcal{G}_0} V(G(a) \backslash p_0(a))) \cup (V(T(a) \backslash \mathrm{lc}_T(a)) \cap \cup_{i \in \mathcal{N}} V(C_i(a) \backslash \mathrm{lc}_{C_i}(a))) \\
&= (\cup_{G \in \mathcal{G}_0} V(G(a) \backslash p_0(a))) \cup (\cup_{i \in \mathcal{N}} V(T_i(a) \backslash \mathrm{lc}_{T_i}(a))) \\
&= (\cup_{G \in \mathcal{G}_0} V(G(a) \backslash p_0(a))) \cup (\cup_{i \in \mathcal{N}} \cup_{[G,h] \in \mathcal{G}_i} V(G(a) \backslash h(a))).
\end{aligned}$$

In addition, by Lemma 3.2 and Corollary 3.1, $p_0(a) \neq 0$. Thus $[G, p_0]$ specializes well at a for every $G \in \mathcal{G}_0$ according to Algorithm 2.3. By the induction hypothesis, we know that $[G, h]$ specializes well at a for every $[G, h] \in \mathcal{G}_i$ and any $i \in \mathcal{N}$. Therefore, the statement (2) in the specification of Algorithm 3.8 holds. $\square$

Remark 3.3. By Algorithm 3.8, for any regular system $[T_i, h_i]$ in the first output of Algorithm 3.8, $\mathrm{res}(\mathrm{lc}_{T_i}; T_i)$ is a factor of h_i. If $p \in \mathcal{K}[\boldsymbol{u}]$ is the second output of Algorithm 3.8, by Corollary 3.1, we know that

$$\mathbf{V}^{\boldsymbol{u}}(\mathrm{res}(\mathrm{lc}_{T_i}; T_i)) \subset \mathbf{V}^{\boldsymbol{u}}(h_i) \subset \mathbf{V}^{\boldsymbol{u}}(p).$$

3.2.2.3 *Generic Regular System Decomposition*

Algorithm 3.9 shows how to compute a generic regular system decomposition and the associated RDU of a given system simultaneously.

Algorithm 3.9 GRSD

Input: A parametric system P in $\mathcal{K}[\boldsymbol{u}, \boldsymbol{x}]$, variables $\boldsymbol{x}$

Output: $[\mathcal{TH}, p]$, where $\mathcal{TH}$ is a generic regular system decomposition for P and p is the corresponding RDU

  1: Compute a Wu's decomposition $\{C_1, \ldots, C_m\}$ of P in $\mathcal{K}[\boldsymbol{u}, \boldsymbol{x}]$;

  2: $p \leftarrow 1$; $\mathcal{TH} \leftarrow \emptyset$;

  3: **for** $i = 1$ to m **do**

  4:    **if** C_i is a contradictory ascending chain **then**

  5:        $p \leftarrow p \cdot \mathrm{op}(C_i)$;

  6:    **else**

  7:        $[W, q] \leftarrow \mathrm{TSToRS}(C_i, \boldsymbol{x})$;

  8:        $\mathcal{TH} \leftarrow \mathcal{TH} \cup W$; $p \leftarrow p \cdot q$;

  9:    **end if**

10: **end for**

11: **return** $[\mathcal{TH}, p]$;

Theorem 3.5. *Algorithm 3.9 terminates correctly.*

Proof.   The termination follows from the termination of Algorithm 3.8. We only need to show the correctness. In fact, $\mathcal{TH}$ being a parametric regular system decomposition for P follows from Wu's method and Algorithm 3.8. Moreover, by Corollary 3.2 and Algorithm 3.8, it is easy to see that

$p(\boldsymbol{u}) = 0$ is an affine variety (denoted by $\mathcal{V}$) in $\hat{\mathcal{K}}^d$ with $\dim(\mathcal{V}) < d$ and $\mathcal{TH}$ is stable at any $a \in \hat{\mathcal{K}}^d \backslash \mathcal{V}$. $\qquad \square$

Corollary 3.3. *Let* P *be a parametric polynomial system,* $\mathrm{GRSD}(\mathrm{P}, \boldsymbol{x}) = [\mathcal{TH}, p]$. *Then* $\mathcal{TH}$ *is stable at any* $a \in \hat{\mathcal{K}}^d \backslash \mathrm{V}^{\boldsymbol{u}}(p)$.

We use the following simple example to illustrate the main steps of Algorithm 3.9.

Example 3.3. Consider the system

$$\mathrm{P} = \begin{cases} (ux + 1)z^3 + (vy + 1)z^2 + wxz + 1 \\ ux + 1 \end{cases}$$

where x, y and z are variables ($x \prec y \prec z$) and u, v and w are parameters.

Step 1: According to the first step of Algorithm 3.9, we get Wu's decomposition $\mathcal{S} = \{\mathrm{C}_1, \mathrm{C}_2, \mathrm{C}_3\}$ of P in $\mathcal{K}[u, v, w][x, y, z]$ where

$$\mathrm{C}_1 = \{ux + 1, u + uvyz^2 + uz^2 - wz\},$$
$$\mathrm{C}_2 = \{ux + 1, vy + 1, u - wz\},$$
$$\mathrm{C}_3 = \{-v^2 u^5 w\}.$$

Step 2: Let $\mathcal{TH} = \emptyset$ and $p = 1$.

Step 3: Because C_1 and C_2 are both non-contradictory ascending chains and C_3 is a contradictory ascending chain, we need to execute $\mathtt{TSToRS}(\mathrm{C}_1, [x, y, z])$ and $\mathtt{TSToRS}(\mathrm{C}_2, [x, y, z])$.

Step 3.1: We execute $\mathtt{ZDtoRC}(\mathrm{C}_1, \mathrm{mvar}(\mathrm{C}_1))$ where $\mathrm{mvar}(\mathrm{C}_1) = \{x, z\}$. It returns

$$W = [\{\{ux + 1, u + uvyz^2 + uz^2u - wz\}\}, u(vy + 1)].$$

Since $W_2 = u(vy + 1) \notin \mathcal{K}[u, v, w]$, we compute Wu's zero decomposition for $u(vy + 1)$ and get $\mathrm{C}_{11} = \{vy + 1\}, \mathrm{C}_{12} = \{u\}$.

Step 3.2: Let $\mathrm{T}_1 = \mathrm{C}_1 \cup \{vy + 1\}$. Now we need to execute $\mathtt{TSToRS}(\mathrm{T}_1, [x, y, z])$. When we execute $\mathtt{ZDtoRC}(\mathrm{T}_1, \mathrm{mvar}(\mathrm{T}_1))$, it returns $[\emptyset, vu]$. Since $vu \in \mathcal{K}[u, v, w]$, the output of $\mathtt{TSToRS}(\mathrm{T}_1, [x, y, z])$ is

$$[\{[\{ux + 1, u + uvyz^2 + uz^2 - wz\}, u(vy + 1)]\}, uvw].$$

Step 3.3: Let

$$\mathcal{TH} \leftarrow \mathcal{TH} \cup \{[\{ux + 1, u + uvyz^2 + uz^2 - wz\}, u(vy + 1)]\}, p \leftarrow p \cdot uvw.$$

Step 3.4: We execute $\texttt{ZDtoRC}(C_2, \mathrm{mvar}(C_2))$ where $\mathrm{mvar}(C_2) = \{x, y, z\}$. It returns

$$W = [\{\{ux + 1, vy + 1, u - wz\}\}, -uvw].$$

Since $W_2 \in \mathcal{K}[u, v, w]$, the output of $\texttt{TSToRS}(C_2, [x, y, z])$ is

$$[\{[\{ux + 1, vy + 1, u - wz\}, -uvw]\}, uvw].$$

Step 3.5: Let $\mathcal{TH} \leftarrow \mathcal{TH} \cup \{[\{ux + 1, vy + 1, u - wz\}, -uvw]\}$ and $p \leftarrow p \cdot uvw$.

Step 3.6: Since C_3 is a contradictory ascending chain, we execute $p \leftarrow p \cdot (-v^2 u^5 w)$.

Step 4: Finally, we get $p = u^2 v^2 w^2$ and

$$\mathcal{TH} = \{[\{ux + 1, u + uvyz^2 + uz^2 - wz\}, u(vy + 1)],$$
$$[\{ux + 1, vy + 1, u - wz\}, -uvw]\}.$$

For the performance of Algorithm 3.9 on more examples, see [Chen *et al.* (2015)].

For an SAS S : $[P, G_1, G_2, H]$, computing a generic regular system decomposition of P by $\texttt{GRSD}(P, x)$ will lead to a triangular decomposition $\mathcal{TH}$ of S. Each TSA $T_i : [F_i, G_1, G_2, H]$ in $\mathcal{TH}$ is not necessarily regular because F_i may not be coprime with respect to the inequalities. It is not hard to modify Algorithms 3.3-3.5 to handle the positive dimensional case. The details are omitted. In Chapter 6, we will return to this topic from the viewpoint of real root classification.

Chapter 4

Real Root Counting

Automated inequality proving and discovering rely on the methods and tools of computational real algebra and computational real algebraic geometry. From this chapter on, we start to discuss the problems related to the real solutions of polynomials, polynomial systems and semi-algebraic systems. In this chapter, after a quick review of some classical results on real root counting for polynomials, we introduce our contribution along this direction.

4.1 Classical Results

When describing classical results, we follow the way and notations of [Gantmacher (1955)], starting from the Cauchy index.

Definition 4.1. For a real-valued rational function $R(x)$ and an interval (a, b), when x changes from a to b, the difference between the number of discontinuous points of $R(x)$ jumping from $-\infty$ to $+\infty$ and that of those jumping from $+\infty$ to $-\infty$ is called the *Cauchy index* and is denoted by $I_a^b R(x)$. Herein, a, b can be $-\infty$ and $+\infty$, respectively.

Suppose a real-valued rational function $R(x) = \sum_{i=1}^{k} \frac{c_i}{x - \alpha_i} + S(x)$, where $\alpha_i, c_i \in \mathbb{R}$, the field of real numbers, and $S(x)$ is a real-valued rational function such that any real point is not a *pole point* (the limit at this point is infinity) of $S(x)$. If the interval (a, b) contains exactly one α_i, then $I_a^b R(x) = \operatorname{sgn}(c_i)$. Herein, sgn is the sign function, which takes values $1, -1$ and 0, respectively, when c_i is positive, negative or zero. Therefore,

$$I_{-\infty}^{+\infty} R(x) = \sum_{i=1}^{k} \operatorname{sgn}(c_i).$$

In the following, as usual, f' denotes the derivative of a univariate function f.

Theorem 4.1.

(a) *The number of distinct real roots in (a, b) of a non-zero polynomial f equals $I_a^b \dfrac{f'(x)}{f(x)}$;*

(b) *For any two non-zero polynomials $f(x)$ and $g(x)$, $I_a^b \dfrac{f'(x)g(x)}{f(x)} = f_{g+} - f_{g-}$, where*

$$f_{g+} = \operatorname{card}(\{\alpha \in (a, b) | f(\alpha) = 0, g(\alpha) > 0\}),$$

$$f_{g-} = \operatorname{card}(\{\alpha \in (a, b) | f(\alpha) = 0, g(\alpha) < 0\}).$$

Proof.   (a) Write $f(x)$ as $f(x) = f_1(x) \prod_{i=1}^{m} (x - \alpha_i)^{j_i}$, where $\alpha_i \in \mathbb{R}$ and $f_1(x)$ has no real roots. Then

$$\frac{f'(x)}{f(x)} = \sum_{i=1}^{m} \frac{j_i}{x - \alpha_i} + \frac{f_1'}{f_1}.$$

Without loss of generality, suppose that the roots in (a, b) of f are $\alpha_1, \ldots, \alpha_k$, then

$$I_a^b \frac{f'(x)}{f(x)} = \sum_{i=1}^{k} \operatorname{sgn}(j_i) = k.$$

(b)

$$\frac{f'(x)g(x)}{f(x)} = \sum_{i=1}^{m} \frac{j_i g(x)}{x - \alpha_i} + h(x),$$

where $h(x) = f_1'g/f_1$ is a real-valued rational function without pole points. Without loss of generality, suppose the roots in (a, b) of $f(x)$ such that $g(\alpha_i) \neq 0$ are exactly the first n $(0 \leq n \leq m)$ α_is. It is easy to see that

$$I_a^b \frac{f'(x)g(x)}{f(x)} = \sum_{i=1}^{n} \operatorname{sgn}(g(\alpha_i)) = f_{g+} - f_{g-}.$$

$\square$

Assume $r(x)$ is the remainder of $f'(x)g(x)$ divided by $f(x)$, then obviously

$$I_a^b \frac{f'(x)g(x)}{f(x)} = I_a^b \frac{r(x)}{f(x)}.$$

In the following, a method for computing the Cauchy index is established so that, according to Theorem 4.1, a method for real root counting is also established. Let $a_1, \ldots, a_l$ be a sequence of non-zero real numbers. The *sign changes* of the sequence, denoted by $V(a_1, \ldots, a_l)$, is defined as the number of negative numbers in the set $\{a_i a_{i+1} \mid 1 \le i \le l - 1\}$, that is

$$\sum_{i=1}^{l-1} \frac{1 - \operatorname{sgn}(a_i a_{i+1})}{2}.$$

For a sequence of real numbers $b_1, \ldots, b_m$, deleting the zeros in the sequence gives a new sequence $b'_1, \ldots, b'_{m'}$. Then define $V(b_1, \ldots, b_m) = V(b'_1, \ldots, b'_{m'})$.

Definition 4.2. Suppose a sequence of non-zero polynomials

$$f_0(x), f_1(x), \ldots, f_s(x) \tag{4.1}$$

satisfy the following conditions in the interval (a, b):

(1) f_s does not have real roots in (a, b);
(2) if there exists $c \in (a, b)$ and some i $(0 < i < s)$ such that $f_i(c) = 0$, then $f_{i-1}(c) f_{i+1}(c) < 0$;

then the sequence (4.1) is called a *Sturm sequence* (starting from $f_0(x)$ and $f_1(x)$) in the interval (a, b). Without loss of generality, we can assume that $f_i(a) f_i(b) \neq 0$ $(0 \le i \le s)$ (otherwise, we can discuss equivalently in $(a + \epsilon, b - \epsilon)$ for some sufficiently small ϵ). Obviously, any two adjacent polynomials in the Sturm sequence have no common roots in (a, b).

If every polynomial in the Sturm sequence is multiplied by the same non-zero polynomial (assume neither a nor b is the root of this polynomial), the new sequence is called the *generalized Sturm sequence*.

For two polynomials $g_0(x), g_1(x) \in \mathbb{R}[x]$, we can construct a (generalized) Sturm sequence as follows: divide g_0 by g_1 and let $-g_2(x)$ be the remainder; generally, if $g_k(x)(\neq 0)$ and $g_{k-1}(x)$ are obtained, let $-g_{k+1}(x)$ be the remainder of $g_{k-1}(x)$ divided by $g_k(x)$; repeat this process until some polynomial is zero, *i.e.*

$$g_2(x) = -\mathrm{rem}(g_0(x), g_1(x)),$$

$$\cdots \cdots$$

$$g_{k+1}(x) = -\mathrm{rem}(g_{k-1}(x), g_k(x)),$$

$$\cdots \cdots$$

$$g_s(x) = -\mathrm{rem}(g_{s-2}(x), g_{s-1}(x)) \neq 0,$$

$$g_{s+1}(x) = -\mathrm{rem}(g_{s-1}(x), g_s(x)) = 0.$$

It is clear that $g_s(x) = \gcd(g_0(x), g_1(x))$. If $g_s(x)$ has no real roots in $[a, b]$, then $g_0, g_1, \ldots, g_s$ is a Sturm sequence; otherwise, $g_0, g_1, \ldots, g_s$ is a generalized Sturm sequence and dividing each g_i by g_s gives a Sturm sequence.

The *sign changes* at $x = r$ of a (generalized) Sturm sequence $f_0(x), f_1(x), \ldots, f_s(x)$, denoted by $V(f_0, f_1; r)$ or $V(r)$, is the sign changes of the real number sequence $f_0(r), f_1(r), \ldots, f_s(r)$.

Theorem 4.2. *Suppose* $f_0(x), f_1(x), \ldots, f_s(x)$ *is a Sturm sequence in* (a, b). *Then*

$$I_a^b \frac{f_1(x)}{f_0(x)} = V(a) - V(b).$$

Proof. Let $x_1, x_2 \in (a, b)$ and $x_1 < x_2$. If $[x_1, x_2]$ contains no roots of any f_i $(0 \leq i \leq s)$, it is clear that $V(x_1) - V(x_2) = 0$.

Suppose $a < c < b$ and $f_i(c) = 0$ $(1 \leq i < s)$. By the condition (2) of Sturm sequence, if $x \in (c - \epsilon, c + \epsilon)$, $f_{i-1}(x) f_{i+1}(x) < 0$. Thus $V(c - \epsilon) - V(c + \epsilon) = 0$.

Suppose $a < c < b$ and $f_0(c) = 0$. Then $f_1(c) \neq 0$. If $f_0(c - \epsilon) f_1(c) < 0$, $f_0(c + \epsilon) f_1(c) > 0$, then

$$I_{c-\epsilon}^{c+\epsilon} \frac{f_1(x)}{f_0(x)} = 1 = V(c - \epsilon) - V(c + \epsilon);$$

If $f_0(c - \epsilon) f_1(c) > 0$, $f_0(c + \epsilon) f_1(c) < 0$, then

$$I_{c-\epsilon}^{c+\epsilon} \frac{f_1(x)}{f_0(x)} = -1 = V(c - \epsilon) - V(c + \epsilon);$$

If $f_0(c - \epsilon) f_0(c + \epsilon) > 0$, then $I_{c-\epsilon}^{c+\epsilon} \dfrac{f_1(x)}{f_0(x)} = 0 = V(c - \epsilon) - V(c + \epsilon).$ $\square$

Multiply a Sturm sequence $f_0(x), f_1(x), \ldots, f_s(x)$ by $g(x)$ and let the new sequence be $g_0(x), g_1(x), \ldots, g_s(x)$. Obviously

$$I_a^b \frac{g_1(x)}{g_0(x)} = I_a^b \frac{f_1(x)}{f_0(x)}$$

$$= V(f_0, f_1; a) - V(f_0, f_1; b)$$

$$= V(g_0, g_1; a) - V(g_0, g_1; b).$$

The last equality is valid because we can assume that neither a nor b is a root of $g(x)$. Therefore, the above theorem is still valid for generalized Sturm sequence.

From Theorem 4.1 and Theorem 4.2, we have the following two important results.

Corollary 4.1 (Sturm's Theorem). *Suppose* $f(x) \in \mathbb{R}[x]$, $f(a)f(b) \neq 0$. *The number of distinct real roots in* (a,b) *of* $f(x)$ *is*

$$V(f, f'; a) - V(f, f'; b).$$

Especially, the number of distinct real roots of $f(x)$ *is* $V(f, f'; -\infty) - V(f, f'; +\infty)$.

Corollary 4.2 (Sturm-Tarski's Theorem). *Suppose* $f(x), g(x) \in \mathbb{R}[x]$, $f(a)f(b) \neq 0$ *and let* $r = \mathrm{rem}(f'g, f, x)$. *Then*

$$V(f, f'g; a) - V(f, f'g; b) = V(f, r; a) - V(f, r; b) = f_{g+} - f_{g-},$$

where

$$f_{g+} = \mathrm{card}(\{\alpha \in (a,b) | f(\alpha) = 0, g(\alpha) > 0\}),$$
$$f_{g-} = \mathrm{card}(\{\alpha \in (a,b) | f(\alpha) = 0, g(\alpha) < 0\}).$$

Obviously, the Sturm Theorem gives a way to compute the number of distinct real roots (in a given interval) of a univariate polynomial f: First, construct a (generalized) Sturm sequence (starting from f, f') by successive division as introduced above; Second, compute the difference of the sign changes of this sequence at $-\infty$ and $+\infty$ (or the endpoints of the given interval). This is a so-called "online" algorithm which is very efficient for polynomials with constant coefficients. It is not a so-called "explicit criterion" so that it is not very efficient for parametric polynomials in general.

In the rest of this section, we introduce two other classical results: Budan-Fourier's Theorem and Descartes' rule of signs. Although they do not give exact number of real roots in general, the theorems are concise and the algorithms based on them are very efficient.

Suppose $f(x) \in \mathbb{R}[x]$ is a polynomial of degree m. Consider its derivatives of different orders

$$f(x), f'(x), f''(x), \ldots, f^{(m-1)}(x), f^{(m)}(x). \tag{4.2}$$

The last polynomial of the sequence is obviously $m! \cdot \mathrm{lc}(f(x))$ (a constant in $\mathbb{R}$), and thus the sign of it is fixed. As usual, when discussing the number of real roots in $[a, b]$ of $f(x)$, we assume that $f(a)f(b) \neq 0$. Furthermore, we also assume that neither a nor b is a root of any polynomial in the sequence (4.2). Otherwise, if a (or b) is a root of some $f^{(i)}(x)$ $(1 \leq i < m)$, we can discuss equivalently in $[a + \epsilon, b - \epsilon]$ for a sufficiently small positive number ϵ.

Theorem 4.3 (Budan-Fourier's Theorem). *Suppose $f(x) \in \mathbb{R}[x]$, $a < b$ are two real numbers and $f(a)f(b) \neq 0$. The number (counted with multiplicity) of real roots in $[a, b]$ of $f(x)$ equals $V(a) - V(b) - 2k$ for some $k \in \mathbb{N}$. Herein $V(x)$ stands for the sign changes of sequence (4.2) at x.*

Proof.   See, for example, [Mignotte (1992)]. $\square$

Applying Budan-Fourier's Theorem to $(0, +\infty)$, we have

Theorem 4.4 (Descartes' rule of signs). *The number of positive roots (counted with multiplicity) of $f(x) = \sum_{i=0}^{m} a_i x^i \in \mathbb{R}[x]$ equals*

$$V(a_0, a_1, ..., a_m) - 2k$$

for some $k \in \mathbb{N}$.

Note that, in the Budan-Fourier Theorem and Descartes's rule of signs, only when the sign changes or the difference of sign changes is 1 or 0, is the answer determined. If we know in advance that all the roots of $f(x)$ are real, we have the following exact answer.

Theorem 4.5. *If all the roots of $f(x) \in \mathbb{R}[x]$ are real, the number of positive roots (counted with multiplicity) of $f(x) \in \mathbb{R}[x]$ equals the sign changes of the coefficients sequence of $f(x)$.*

By Theorem 4.5, it is easy to obtain the following proposition.

Proposition 4.1. *If all the roots of $g(x) = x^m + b_{m-1}x^{m-1} + \cdots + b_0 \in \mathbb{R}[x]$ are real, the necessary and sufficient condition for the roots of $g(x)$ to be all nonnegative is*

$$(-1)^{i+m} b_i \geq 0, \;\; i = 0, \ldots, m - 1.$$

Proof.   Necessity. It is easy to be deduced by the Viète theorem.

Sufficiency. If 0 is a root of g with multiplicity s (≥ 0) and the sign changes of the coefficients of g is t, by Theorem 4.5, the number of positive roots (counted with multiplicity) of $g(x)$ is exactly t. On the other hand, the sign changes of the coefficients of $g(-x)$ is 0. Thus, $n = t + s$. This means the roots of g are all nonnegative. $\square$

Similarly, we can also prove

Proposition 4.2. *If all the roots of* $g(x) = x^m + b_{m-1}x^{m-1} + \cdots + b_0 \in \mathbb{R}[x]$ *are real, the necessary and sufficient condition for the roots of* $g(x)$ *to be all positive is*

$$(-1)^{i+m}b_i > 0, \ i = 0, \ldots, m-1.$$

4.2 Discrimination Systems for Polynomials

It is well-known that polynomials of degrees no less than five cannot be solved by radicals in general. On the other hand, the sign of the discriminant of a polynomial of degree 2 determines the so-called *root classification* (*i.e.* the conditions in terms of polynomials in the coefficients on the numbers and multiplicities of real and complex roots). For a cubic polynomial, the signs of its discriminant and another polynomial in its coefficients can determine its root classification. A natural question is: Are there similar "explicit criteria" for polynomials of arbitrary degrees? In other words, can we have root classifications for polynomials of arbitrary degrees? The answer is positive. In this section, we introduce such a criterion presented in Yang *et al.* (1996a,b).

Given a univariate polynomial of real parametric coefficients

$$f(x) = a_0 x^m + a_1 x^{m-1} + \cdots + a_m \quad (a_0 \neq 0) \tag{4.3}$$

and another non-zero polynomial $g(x)$. Set

$$r(x) = \text{rem}(f'g, f) = b_1 x^{m-1} + \cdots + b_m. \tag{4.4}$$

The following $2m \times 2m$ matrix

$$\begin{pmatrix}
a_0 & a_1 & a_2 & \cdots & a_m & & & & \\
0 & b_1 & b_2 & \cdots & b_m & & & & \\
& a_0 & a_1 & \cdots & a_{m-1} & a_m & & & \\
& 0 & b_1 & \cdots & b_{m-1} & b_m & & & \\
& & & & \vdots & \vdots & & & \\
& & & a_0 & a_1 & a_2 & \cdots & a_m \\
& & & 0 & b_1 & b_2 & \cdots & b_m
\end{pmatrix}$$

is called the *discrimination matrix* of f with respect to g and denoted by Discr (f, g). If $g = 1$, we denote Discr $(f, 1)$ by Discr (f) for short and call it the *discrimination matrix* of f.

If $r(x)$ is viewed as a polynomial of degree m

$$r(x) = 0 \cdot x^m + b_1 x^{m-1} + \cdots + b_m,$$

the matrix above is the Sylvester matrix of $f(x)$ and $r(x)$ (up to some exchanges of rows. See Chapter 1).

Set $D_0 = 1$, and let

$$D_1(f,g), \, D_2(f,g), \ldots, D_m(f,g)$$

denote the even order principal minors of $\mathrm{Discr}\,(f,g)$, respectively. The following list

$$[D_0, D_1(f,g), \ldots, D_m(f,g)]$$

is called the *discriminant sequence* of f with respect to g, denoted by $\mathrm{GDL}(f,g)$. If $g = 1$,

$$[D_0, D_1(f,1), \ldots, D_m(f,1)]$$

denoted also by $\mathrm{DiscrList}\,(f)$, is called the *discriminant sequence* of f. That is $\mathrm{DiscrList}\,(f) = \mathrm{GDL}(f,1)$.

The following list

$$[\mathrm{sgn}(A_1), \ldots, \mathrm{sgn}(A_m)]$$

is called the *sign list* of a given list $[A_1, \ldots, A_m]$. For a sign list $[s_1, s_2, \ldots, s_m]$, its *revised sign list* $[t_1, t_2, \ldots, t_m]$ is constructed as follows:

- If $[s_i, s_{i+1}, \ldots, s_{i+j}]$ is a segment of the list and

$$s_i \neq 0, \, s_{i+1} = \cdots = s_{i+j-1} = 0, \, s_{i+j} \neq 0,$$

then replace the zeros

$$[s_{i+1}, \ldots, s_{i+j-1}]$$

with the first $j - 1$ elements of $[-s_i, -s_i, s_i, s_i, -s_i, -s_i, s_i, s_i, \ldots]$, *i.e.* set

$$t_{i+r} = (-1)^{\lfloor (r+1)/2 \rfloor} \cdot s_i, \quad r = 1, \ldots, j - 1,$$

where $\lfloor (r+1)/2 \rfloor$ is the maximal integer less than or equal to $(r+1)/2$.
- Otherwise, set $t_k = s_k$.

For example, the revised sign list of

$$[1, 1, -1, 0, 0, 0, 0, 0, 1, 0, 0, -1, 1, 0, 0]$$

is

$$[1, 1, -1, 1, 1, -1, -1, 1, 1, -1, -1, -1, 1, 0, 0].$$

Theorem 4.6 (Discrimination Theorem I). *For two polynomials $f = f(x)$ and $g = g(x)$ in $\mathbb{R}[x]$, if the sign changes of revised sign list of $\mathrm{GDL}(f,g) = [D_0, D_1(f,g), \ldots, D_m(f,g)]$ is ν, $D_\eta \neq 0$ and $D_t = 0$ $(t > \eta)$, then*

$$\eta - 2\nu = f_{g+} - f_{g-},$$

where

$$f_{g+} = \mathrm{card}(\{x \in \mathbb{R} \mid f(x) = 0,\, g(x) > 0\}),$$
$$f_{g-} = \mathrm{card}(\{x \in \mathbb{R} \mid f(x) = 0,\, g(x) < 0\}).$$

Theorem 4.7 (Discrimination Theorem II). *If the sign changes of the revised sign list of the discriminant sequence of $f(x)$ is ν, then $f(x)$ has ν pairs of distinct imaginary roots. Furthermore, if the number of non-zero elements in the revised sign list is $\eta + 1$, then $f(x)$ has $\eta - 2\nu$ distinct real roots.*

We will prove Theorem 4.6 in the next section and Theorem 4.7 is clearly a corollary of Theorem 4.6. By these two theorems, f_{x+}, the number of distinct positive roots of f, can be computed by solving the following simple linear equations (without loss of generality, assume $f(0) \neq 0$)

$$\begin{pmatrix} 1 & 1 \\ 1 & -1 \end{pmatrix} \begin{pmatrix} f_{x+} \\ f_{x-} \end{pmatrix} = \begin{pmatrix} k_1 \\ k_2 \end{pmatrix},$$

where k_1 and k_2 are given by Theorem 4.7 and Theorem 4.6, respectively.

Remark 4.1. Because $\mathrm{sgn}(D_1(F,1)) = 1$ if $a_0 \neq 0$, we usually call

$$[D_1, \ldots, D_m]$$

the discriminant sequence of f when its meaning is clear. Of course, in this situation, the number of non-zero elements in Discrimination Theorem II should be η.

Corollary 4.3. *[Weiss (1963)] If $f(x) \in \mathbb{R}[x]$ is a monic squarefree polynomial of degree d, the sign of its discriminant is $(-1)^{\frac{d-r}{2}}$, where r is the number of its real roots.*

Proof. Because $f(x)$ is monic, the sign list of discriminant sequence of $f(x)$ is

$$L = [\mathrm{sgn}(D_0), \mathrm{sgn}(D_1), \mathrm{sgn}(D_2), \ldots, \mathrm{sgn}(D_d)]$$
$$= [1, 1, \mathrm{sgn}(D_2), \ldots, \mathrm{sgn}(D_d)].$$

It is straightforward to verify that $\mathrm{sgn}(D_d) = \mathrm{sgn}(\mathrm{discrim}(f))$ since f is monic. Then f being squarefree implies $\mathrm{sgn}(D_d) \neq 0$ and thus the number of non-zero elements in the revised sign list of L is $d+1$. By Discrimination Theorem II (Theorem 4.7), the sign change v of the revised sign list is $\frac{d-r}{2}$. If v is odd, $\mathrm{sgn}(D_d)$ must be -1. If v is even, $\mathrm{sgn}(D_d)$ must be 1. That completes the proof. $\qquad\square$

It is clear that the conclusion of the above corollary still holds when $\mathrm{lc}(f, x)$ is positive.

The following Maple codes can compute discriminant sequences efficiently.

```
with(LinearAlgebra):
discrg := proc(poly1,poly2,var)
local f,g,tt,d,bz,i,ar,j,mm,dd;
      f := expand(poly1);
      g := expand(poly2*diff(f,var));
      d := degree(f,var);
      if d <= degree(g,var) then
         g := rem(g,f,x);
      fi;
      g := tt*var^d+g;
      bz := subs(tt=0,bezout(f,g,var));
      ar := [ ];
      for i to d do
          ar := [op(ar),row(bz,d+1-i)]
      od;
      mm := matrix(ar);
      dd := [1]
      for j to d do
          dd := [op(dd),det(submatrix(mm,1..j,1..j))]
      od;
end:
```

Remark 4.2. In the above program, we use Bezoutian matrix instead of Sylvester's matrix. Due to the relation between the two matrices (see for example Section 49 of [Yang *et al.* (1996b)]), the correctness of the program is guaranteed.

If we type in $\mathrm{discrg}(f, g, x)$ under Maple, the output of the function is the discriminant sequence of f with respect to g. Note that, if f is a parametric polynomial, we should assume the leading coefficient of f is not zero.

Example 4.1. Compute the root classification of

$$f = -2x^{16} + 4x^{15} + 2x^{14} - 4x^{13} - 2x^{12} + x^5 - 7x^4 + 9x^3 + 7x^2 - 9x - 5.$$

Call `discrg`$(f, 1, x)$ to compute the discriminant sequence of f, whose sign list is

$$[1, 1, 1, 1, 0, 0, 0, 0, 0, -1, 1, 1, -1, 1, 1, 0, 0],$$

and the revised sign list is

$$[1, 1, 1, 1, -1, -1, 1, 1, -1, -1, 1, 1, -1, 1, 1, 0, 0].$$

The sign changes is 6, and the number of non-zero elements is 15. By Discrimination Theorem II, f has 6 pairs of distinct imaginary roots and 2 distinct real roots. Furthermore, by computing the root classification of $\gcd(f, f') = x^2 - x - 1$, we know that each real root of f is of multiplicity 2.

For any $n_1 \times n_2$ $(n_1 \leq n_2)$ matrix M, denote by $M(k, l)$ the sub-matrix formed by the first k rows, the first $k - 1$ columns and the $(k + l)$th column of M. Let $f(x)$ be as in (4.3). Define

$$\Delta_k(f) = \Delta_k = \sum_{t=0}^{k} |\text{Discr}\,(f)(2\,(m - k), t)|\, x^{k-t}, \quad k = 0, 1, \ldots, m - 1,$$

and call $\Delta_0, \ldots, \Delta_{m-1}$ the *repeated factor sequence* of $f(x)$. Noticing that the slight difference between the discrimination matrix and the Sylvester matrix, we know that the difference between each Δ_i and the corresponding subresultant of $f(x)$ and $f'(x)$ is $\pm a_0$. Sometimes, this difference can be omitted and we may view the repeated factor sequence as the subresultant chain of $f(x)$ and $f'(x)$. The following lemma is a direct corollary of the theory of subresultants.

Lemma 4.1. *Suppose $f(x) \in \mathbb{R}[x]$ is a polynomial of degree d and its discriminant sequence is $[D_1, \ldots, D_d]$. If $D_k \neq 0$ and $D_j = 0$ for $k < j \leq d$, then*

$$\gcd(f(x), f'(x)) = \Delta_{d-k}.$$

This means $\gcd(f, f')$ is always contained in the repeated factor sequence of $f(x)$.

Definition 4.3. Let U denote the union of

$$\{f(x)\}, \ \{\Delta_k(f)\}, \ \{\Delta_j(\Delta_k(f))\}, \{\Delta_i(\Delta_j(\Delta_k(f)))\}, \ \ldots,$$

i.e. the union set of repeated factor sequences in different levels. Every polynomial in U has its own discriminant sequence and all these discriminant sequences form the so-called *complete discrimination system* of $f(x)$, denoted by $\mathrm{CDS}(f)$.

Definition 4.4. Let $\Delta(f) = \gcd(f(x), f'(x))$, and call it the *repeated part* of $f(x)$. Let

$$\Delta^0(f) = f, \quad \Delta^j(f) = \Delta(\Delta^{j-1}(f)), \quad j = 1, 2, \ldots$$

and call

$$\{\Delta^0(f), \quad \Delta^1(f), \quad \Delta^2(f), \quad \ldots\}$$

the Δ-*sequence* of $f(x)$.

In fact, to get the root classification of $f(x)$, not all the polynomials in $\mathrm{CDS}(f)$ are needed. The discriminant sequences of the polynomials in the Δ-sequence of f are enough.

Lemma 4.2. *If $\Delta^j(f)$ has k distinct real roots of multiplicities $n_1, n_2, \ldots, n_k$, respectively, and $\Delta^{j-1}(f)$ has m distinct real roots, then $\Delta^{j-1}(f)$ has k distinct real roots of multiplicities $n_1 + 1, n_2 + 1, \ldots, n_k + 1$, respectively, and $m - k$ distinct real roots of multiplicity 1. The case of imaginary roots is similar.*

We are now ready to give an algorithm for computing the numbers and multiplicities of the roots of a univariate polynomial $f(x) \in \mathbb{R}[x]$.

Step 1. Compute the discriminant sequence $f(x)$

$$[D_1(f), \ldots, D_n(f)]$$

and its revised sign list. Compute the sign changes of the list to determine the numbers of distinct real and complex roots by Discrimination Theorem II. If no zeros appear in the list, quit.

Step 2. If there are k zeros in the list, by Lemma 4.1, $\Delta(f) = \Delta_{n-k}(f)$, which can be constructed by its definition. Call this procedure recursively with input $\Delta(f)$, that is, go back to Step 1 with input $\Delta(f)$.

Step 3. Continue this recursively for $\Delta^2(f), \Delta^3(f), \ldots,$ until there is some j such that the revised sign list of $\Delta^j(f)$ does not contain 0.

Step 4. Compute the numbers of distinct real and complex roots of $\Delta^j(f)$ by Discrimination Theorem II. Then compute the numbers and multiplicities of real and complex roots of $\Delta^{j-1}(f)$ by Lemma 4.2. Continuing in this way, finally, we will get the numbers and multiplicities of distinct real and complex roots of $f(x)$.

Example 4.2. By Discrimination Theorem II, we can easily obtain a complete real root classification of a generic quintic equation

$$f = x^5 + px^3 + qx^2 + rx + s.$$

In the following table, the first column lists the condition the coefficients should satisfy, and the second column lists corresponding number of real roots under the condition, where, for example, $\{2, 2, 1\}$ means f has 3 distinct real roots of which two have multiplicity 2 and one has multiplicity 1.

$$(1)\ D_5 > 0 \wedge D_4 > 0 \wedge D_3 > 0 \wedge D_2 > 0, \qquad \{1, 1, 1, 1, 1\}$$

$$(2)\ D_5 > 0 \wedge (D_4 \le 0 \vee D_3 \le 0 \vee D_2 \le 0), \qquad \{1\}$$

$$(3)\ D_5 < 0, \qquad \{1, 1, 1\}$$

$$(4)\ D_5 = 0 \wedge D_4 > 0, \qquad \{2, 1, 1, 1\}$$

$$(5)\ D_5 = 0 \wedge D_4 < 0, \qquad \{2, 1\}$$

$$(6)\ D_5 = 0 \wedge D_4 = 0 \wedge D_3 > 0 \wedge E_2 \ne 0, \qquad \{2, 2, 1\}$$

$$(7)\ D_5 = 0 \wedge D_4 = 0 \wedge D_3 > 0 \wedge E_2 = 0, \qquad \{3, 1, 1\}$$

$$(8)\ D_5 = 0 \wedge D_4 = 0 \wedge D_3 < 0 \wedge E_2 \ne 0, \qquad \{1\}$$

$$(9)\ D_5 = 0 \wedge D_4 = 0 \wedge D_3 < 0 \wedge E_2 = 0, \qquad \{3\}$$

$$(10)\ D_5 = 0 \wedge D_4 = 0 \wedge D_3 = 0 \wedge D_2 \ne 0 \wedge F_2 \ne 0,\ \{3, 2\}$$

$$(11)\ D_5 = 0 \wedge D_4 = 0 \wedge D_3 = 0 \wedge D_2 \ne 0 \wedge F_2 = 0,\ \{4, 1\}$$

$$(12)\ D_5 = 0 \wedge D_4 = 0 \wedge D_3 = 0 \wedge D_2 = 0, \qquad \{5\}$$

where

$$D_2 = -p,$$
$$D_3 = 40rp - 12p^3 - 45q^2,$$
$$D_4 = -88r^2p^2 + 117prq^2 + 12p^4r - 4p^3q^2 - 40qp^2s + 125ps^2$$
$$\qquad - 27q^4 - 300qrs + 160r^3,$$

$$D_5 = 2000ps^2r^2 - 1600qsr^3 - 3750ps^3q + 560r^2p^2sq - 72p^4rsq$$
$$- 630prq^3s - 900rs^2p^3 - 4p^3q^2r^2 + 16p^3q^3s + 825q^2p^2s^2$$
$$+ 144pq^2r^3 + 2250q^2rs^2 + 256r^5 + 3125s^4 - 128r^4p^2$$
$$+ 16p^4r^3 + 108p^5s^2 - 27q^4r^2 + 108q^5s,$$
$$E_2 = 160r^2p^3 + 900q^2r^2 - 48ro^5 + 60q^2p^2r + 1500pqrs$$
$$+ 16q^2p^4 - 1100qp^3s + 625s^2p^2 - 3375q^3s,$$
$$F_2 = 3q^2 - 8rq.$$

In case (2) of the above table, f has a single real root and two distinct pairs of imaginary roots while in case (8), f has a single real root and one pair of imaginary roots of multiplicity 2. As for real root classification, these two cases have no difference.

The method given in this section for root classification of real parametric polynomials can be generalized to complex parametric polynomials. See, for example, [Liang and Zhang (1999)].

Example 4.3. What are the conditions on a, b, c such that
$$\forall x \, (x^6 + ax^2 + bx + c \geq 0)?$$
Let $f(x) = x^6 + ax^2 + bx + c$, then the problem is reduced to finding conditions on a, b, c such that $f(x)$ has no real roots or each real root is of even multiplicity. The discriminant sequence of $f(x)$ is
$$[1, 1, 0, 0, a^3, D_5, D_6],$$
where (up to a positive constant)
$$D_5 = 256\,a^5 + 1728\,c^2a^2 - 5400\,acb^2 + 1875\,b^4,$$
$$D_6 = -1024\,a^6c + 256\,a^5b^2 - 13824\,c^3a^3 + 43200\,c^2a^2b^2$$
$$- 22500\,b^4ca + 3125\,b^6 - 46656\,c^5.$$
By the method of constructing revised sign list and Theorem 4.7, the conditions on a, b, c are one of the following

    (1) $D_6 < 0 \wedge D_5 \geq 0$,

    (2) $D_6 < 0 \wedge a \geq 0$,

    (3) $D_6 = 0 \wedge D_5 > 0$,

    (4) $D_6 = 0 \wedge D_5 = 0 \wedge a > 0$,

    (5) $D_6 = 0 \wedge D_5 = 0 \wedge a < 0 \wedge E_2 > 0$,

    (6) $D_6 = 0 \wedge D_5 = 0 \wedge a = 0$,

where $E_2 = 25\,b^2 - 96\,ac$ is the discriminant of $\Delta_2(F) = 4\,ax^2 + 5\,bx + 6\,c$.

The following example originates from a model of chemical reaction, which has been studied extensively in the literature. See for example [Gatermann and Huber (2002)]. As for the application of CDS to other fields of technology, see, for example, [Wang and Hu (1999, 2000)].

Example 4.4. Given a system

$$\begin{cases} f_1 = k_{21}x_1 - k_{12}x_1^2 - k_{43}x_1x_2 + k_{34}x_3 = 0, \\ f_2 = -k_{43}x_1x_2 + (k_{34} + k_{54})x_3 - k_{45}x_2 = 0, \\ f_3 = x_2 + x_3 - c = 0, \end{cases} \tag{4.5}$$

where x_1, x_2, x_3 are variables and $c, k_{12}, k_{21}, k_{34}, k_{43}, k_{45}, k_{54}$ are parameters which are all positive in the real problem. We want to know the condition for the system to have 3 positive solutions.

We first compute Wu's zero decomposition of the system (4.5) in $\mathbb{Q}(c, k_{12}, k_{21}, k_{34}, k_{43}, k_{45}, k_{54})[x_1, x_2, x_3]$, and get

$$\begin{cases} g_1 = k_{12}k_{43}x_1^3 - u_2x_1^2 + u_3x_1 - ck_{34}k_{45} = 0, \\ g_2 = h_2x_2 - h_1 = 0, \\ g_3 = h_2x_3 - x_1h_3 = 0, \end{cases} \tag{4.6}$$

such that

$$V(\{f_1, f_2, f_3\}) = V(\{g_1, g_2, g_3\}) \setminus V(\{h_2\}),$$

where

$$\begin{aligned} u_2 &= k_{43}k_{21} - k_{12}k_{45} - k_{12}k_{34} - k_{12}k_{54}, \\ u_3 &= ck_{43}k_{54} - k_{21}k_{54} - k_{21}k_{45} - k_{21}k_{34}, \\ h_1 &= -k_{12}x_1^2 + k_{21}x_1 + ck_{34}, \\ h_2 &= k_{43}x_1 + k_{34}, \\ h_3 &= k_{12}x_1 + ck_{43} - k_{21}. \end{aligned}$$

Denote by D the last term in the discriminant sequence of g_1.

Proposition 4.3. *The system (4.5) has 3 positive solutions if and only if g_1 has 3 positive roots. And the latter is equivalent to*

$$u_2 > 0 \wedge u_3 > 0 \wedge D \geq 0. \tag{4.7}$$

Proof. Because h_2 must not be zero when $x_1 > 0$, the system (4.5) has 3 positive solutions if and only if system (4.6) has 3 positive solutions. Obviously, (4.6) has 3 positive solutions if and only if g_1 has 3 positive roots and $h_1 > 0 \wedge h_2 > 0 \wedge h_3 > 0$ or $h_1 < 0 \wedge h_2 < 0 \wedge h_3 < 0$. Because g_1 is of degree 3, by Discrimination Theorem, all the roots of g_1 are real if and only if $D \geq 0$. Then, by Descartes' rule of signs, g_1 has 3 positive roots if and only if $u_2 > 0 \wedge u_3 > 0 \wedge D \geq 0$, which is just formula (4.7).

Now we prove that (4.7) implies $h_1 > 0 \wedge h_2 > 0 \wedge h_3 > 0$.

First, since all parameters are positive, $h_2 = k_{43}x_1 + k_{34}$ must be positive when $x_1 > 0$. This means $h_1 < 0 \wedge h_2 < 0 \wedge h_3 < 0$ is impossible. Second, $u_3 = ck_{43}k_{54} - k_{21}k_{54} - k_{21}k_{45} - k_{21}k_{34} > 0$ implies $ck_{43}k_{54} - k_{21}k_{54} = (ck_{43} - k_{21})k_{54} > 0$, and the latter implies $ck_{43} - k_{21} > 0$. Thus if $x_1 > 0$, $h_3 = k_{12}x_1 + ck_{43} - k_{21} > 0$.

Third, we show that (4.7) implies $h_1 > 0$. Step 1, it is easy to verify that $h_1 = -k_{12}x_1^2 + k_{21}x_1 + ck_{34}$ has one negative root (denoted by α_1) and one positive root (denoted by α_2). Then $h_1(x_1) > 0$ for $x_1 \in (\alpha_1, \alpha_2)$. Step 2, let the three positive roots of g_1 be $\beta_1, \beta_2, \beta_3$, respectively. By the relations between roots and coefficients of polynomials, we have

$$\beta_1 + \beta_2 + \beta_3 = \frac{u_2}{k_{12}k_{43}}$$
$$= \frac{k_{21}}{k_{12}} - \frac{k_{45} + k_{34} + k_{54}}{k_{43}}$$
$$< \frac{k_{21}}{k_{12}}.$$

On the other hand,

$$\frac{k_{21}}{k_{12}} = \alpha_1 + \alpha_2 < \alpha_2.$$

That is to say, $\beta_1, \beta_2, \beta_3$ are all in the interval $(0, \alpha_2)$. That completes the proof. $\qquad\square$

Remark 4.3. In the above example, D is a polynomial with 81 terms, total degree 15, and the degrees with respect to $c, k_{12}, k_{21}, k_{34}, k_{43}, k_{45}, k_{54}$ are $3, 3, 4, 4, 4, 4, 4$, respectively. For more similar results, see [Gatermann and Xia (2003)].

4.3 Proof of Discrimination Theorem

Discrimination Theorem and its proof first appeared in [Yang *et al.* (1996a)]. In this section, we give a new proof based on the theory of subresultants. For all the concepts, notations and results related to subresultants, see Chapter 1 of this book. It should be pointed out that a result equivalent to Discrimination Theorem was proved in [Gonzalez *et al.* (1989)]. It was based on subresultant theory and described in a different way. See also Chapter 9 of [Basu *et al.* (2003)].

Let f, g, r be as in (4.3) and (4.4). We first establish the relation between the discriminant sequence of f with respect to g and the leading coefficients of the Sturm sequence starting from f and r.

According to the method of constructing Sturm sequence, we denote

$$
\begin{aligned}
T_0 &= f(x), \\
T_1 &= r(x), \\
T_2 &= -\mathrm{rem}(T_0, T_1), \\
&\;\;\vdots \\
T_{k+1} &= -\mathrm{rem}(T_{k-1}, T_k), \\
&\;\;\vdots
\end{aligned}
$$

Set

$$
s_{-1} = 0, \quad s_i = \deg(T_i, x) - \deg(T_{i+1}, x), \quad i = 0, 1, \ldots;
$$

$$
q_0 = 0, \quad q_j = \sum_{i=0}^{j-1} s_i, \quad j = 1, 2, \ldots;
$$

$$
\overline{T}_i = \mathrm{lc}(T_i, x), \quad i = 0, 1, \ldots.
$$

On the other hand, let

$$
[D_0 = 1, \; D_1, \ldots, D_m]
$$

be the discriminant sequence of f with respect to g,

$$
S_m = f, \; S_{m-1} = r, \; S_{m-2}, \ldots, S_0
$$

be the subresultant chain of f and r (where $\mu = m - 1$) and $R_m = 1$, $R_{m-1}, \ldots, R_0$ be the corresponding principal subresultant coefficients (psc). It is easy to obtain the following lemma by noticing the relation between the discrimination matrix and the Sylvester matrix of f and r.

Lemma 4.3.

$$
D_0 = 1 = R_m, \; D_i = (-1)^{\frac{(i-1)i}{2}} a_0 R_{m-i}, \quad (1 \le i \le m).
$$

Let $d_1, d_2, \ldots, d_v$ be the block indices of $S_m, \ldots, S_0$. It is clear that $d_i = m - q_{i-1}$ $(1 \le i \le v)$. By the definition of psc and the theorems of subresultant chain, we have

Lemma 4.4. *(a) If $0 < i \ne q_k$ $(k = 1, 2, \ldots)$, then $R_{m-i} = 0$ and thus $D_i = 0$. This means the elements between $D_{q_{i-1}}$ and D_{q_i} of the discriminant sequence of f with respect to g are all 0;*

(b) If there exists some k $(k > 0)$ such that $0 < i = q_k$, then

$$
D_i = (-1)^{\frac{(i-1)i}{2}} a_0 R_{d_{k+1}} = (-1)^{\frac{(i-1)i}{2}} a_0 \psi_{k+1}.
$$

The last equality is obtained directly from Theorem 1.7, where ψ_k is a quantity defined in Definition 1.6 (subresultant polynomial remainder sequence). We need to establish the relation between the leading coefficients of Sturm sequence $T_0, T_1, \ldots$ and discriminant sequence $[D_0, D_1, \ldots, D_m]$. The bridge is subresultant polynomial remainder sequence. So, for convenience, we copy Definition 1.6 here.

A sequence of non-zero polynomials $P_1, P_2, \ldots, P_v$ in $\mathcal{R}[x]$ is called the subresultant polynomial remainder sequence of P_1 and P_2 with respect to x where $\deg(P_1, x) \geq \deg(P_2, x)$, if

$$P_{i+2} = \mathrm{prem}(P_i, P_{i+1}, x)/\beta_{i+2}, \quad 1 \leq i \leq v - 2,$$
$$\mathrm{prem}(P_{v-1}, P_v, x) = 0,$$

where

$$\beta_3 = (-1)^{\delta_2}, \; \beta_{i+1} = (-1)^{\delta_i}\psi_{i-1}^{\delta_i - 1} I_{i-1}, \qquad i = 3, \ldots, v - 1,$$
$$I_1 = 1, \; I_i = \mathrm{lc}(P_i, x), \qquad\qquad\qquad i = 2, \ldots, v,$$
$$\delta_i = \deg(P_{i-1}) - \deg(P_i) + 1, \qquad\qquad i = 2, \ldots, v,$$
$$\psi_1 = 1, \; \psi_2 = I_2^{\delta_2 - 1}, \; \psi_i = \psi_{i-1}\left(\frac{I_i}{\psi_{i-1}}\right)^{\delta_i - 1}, \quad i = 3, \ldots, v.$$

Assume $T_0, T_1, \ldots, T_{v-1}$ is the Sturm sequence starting from f and r, and $P_1 = f, P_2 = r, \ldots, P_v$ is the subresultant polynomial remainder sequence of f and r. By Theorems of subresultant chain, each $\deg(P_i)$ corresponds to a block index d_i of $S_m, \ldots, S_0$, that is, $\deg(P_i) = d_i$ $(1 \leq i \leq v)$. Then, $\delta_i - 1 = s_{i-2}$.

Lemma 4.5. *Let notations be as above. For $0 \leq i < v$, denote*

$$\tau_i = (\delta_i - 1) + (\delta_{i-2} - 1) + \cdots + (\delta_{\lambda+1} - 1),$$
$$u_i = (-1)^{\tau_i} \cdot \frac{I_{i-1}I_{i-3}\cdots I_\lambda}{I_i I_{i-2}\cdots I_{\lambda+1}} \cdot \frac{\psi_{i-1}\psi_{i-3}\cdots\psi_\lambda}{\psi_i\psi_{i-2}\cdots\psi_{\lambda+1}}.$$

Herein, if i is even, $\lambda = 1$; Otherwise, $\lambda = 2$. And we stipulate that $\delta_i = 1$ for $i < 2$, $I_i = 1$ and $\psi_i = 1$ for $i < 1$. Then

$$T_i = u_i P_{i+1}. \tag{4.8}$$

Proof. We use induction. It is clear that (4.8) holds for $i = 0, 1$. Assume (4.8) holds for $k < i$. By the definition of subresultant polynomial remainder sequence, we have the following formula

$$I_i^{\delta_i} P_{i-1} = Q_{i+1}P_i + \beta_{i+1}P_{i+1},$$

where Q_{i+1} is the pseudo-quotient of P_{i-1} pseudo-divided by P_i. By the induction hypothesis, we have

$$I_i^{\delta_i} u_{i-2}^{-1} T_{i-2} = Q_{i+1} u_{i-1}^{-1} T_{i-1} + (-1)^{\delta_i} \psi_{i-1}^{\delta_i-1} I_{i-1} P_{i+1}.$$

Therefore

$$T_{i-2} = Q_{i+1} I_i^{-\delta_i} u_{i-2} u_{i-1}^{-1} T_{i-1} + I_i^{-\delta_i} u_{i-2} \cdot (-1)^{\delta_i} \psi_{i-1}^{\delta_i-1} I_{i-1} P_{i+1}.$$

By the method of constructing Sturm sequence, we need only to prove that

$$u_i = -I_i^{-\delta_i} u_{i-2} \cdot (-1)^{\delta_i} \psi_{i-1}^{\delta_i-1} I_{i-1}.$$

We compute as follows:

$$- I_i^{-\delta_i} u_{i-2} \cdot (-1)^{\delta_i} \psi_{i-1}^{\delta_i-1} I_{i-1}$$

$$= (-1)^{\delta_i-1} \cdot \frac{I_{i-1}}{I_i} \cdot \left(\frac{\psi_{i-1}}{I_i}\right)^{\delta_i-1} \cdot u_{i-2}$$

$$= (-1)^{\delta_i-1} \cdot \frac{I_{i-1}}{I_i} \cdot \frac{\psi_{i-1}}{\psi_i} \cdot u_{i-2}$$

$$= u_i$$

That completes the proof. $\qquad\qquad\square$

Lemma 4.6. *Let notation be as above. For $k \geq 1$,*

$$D_{q_{k+1}}/D_{q_k} = (-1)^{(s_k-1)s_k/2} (\overline{T}_k \overline{T}_{k+1})^{s_k}.$$

Proof. By Lemma 4.5,

$$\overline{T}_k \overline{T}_{k+1} = u_k I_{k+1} \cdot u_{k+1} I_{k+2}$$

$$= (-1)^{s_0+s_1+\cdots+s_{k-1}} \cdot \frac{I_{k+2}}{\psi_{k+1}}$$

$$= (-1)^{q_k} \cdot \frac{I_{k+2}}{\psi_{k+1}}.$$

On the other hand, by Lemma 4.4,

$$\frac{D_{q_{k+1}}}{D_{q_k}} = (-1)^{\frac{(q_k-1)q_k}{2} + \frac{(q_{k+1}-1)q_{k+1}}{2}} \frac{\psi_{k+2}}{\psi_{k+1}}$$

$$= (-1)^{\frac{(q_k-1)q_k}{2} + \frac{(q_{k+1}-1)q_{k+1}}{2}} \left(\frac{I_{k+2}}{\psi_{k+1}}\right)^{s_k}.$$

And

$$\frac{(q_k-1)q_k}{2} + \frac{(q_{k+1}-1)q_{k+1}}{2}$$

$$\equiv \frac{(q_k-1)q_k}{2} + \frac{(s_k+q_k-1)(s_k+q_k)}{2} \pmod 2$$

$$\equiv q_k s_k + \frac{(s_k-1)s_k}{2} \pmod 2.$$

That completes the proof. $\qquad\qquad\square$

Now, we are ready to complete the proof of Theorem 4.7. By Theorem 4.2 and Sturm-Tarski's Theorem, we only need to prove that

$$V(T_0, T_1; -\infty) - V(T_0, T_1; +\infty) = \eta - 2\nu.$$

The signs of $T_0, T_1, \ldots$ at $-\infty$ and $+\infty$ are:

$$-\infty: \ (-1)^{m-q_i} \operatorname{sgn}(\overline{T}_i); \quad +\infty: \ \operatorname{sgn}(\overline{T}_i) \ (i = 0, 1, \ldots).$$

Then,

$$V(T_0, T_1; -\infty) - V(T_0, T_1; +\infty)$$

$$= \sum_{i=0}^{k-1} \frac{1}{2} \left[1 - \operatorname{sgn}\left((-1)^{2m - q_i - q_{i+1}} \overline{T}_i \overline{T}_{i+1}\right)\right]$$

$$- \sum_{i=0}^{k-1} \frac{1}{2} \left[1 - \operatorname{sgn}\left(\overline{T}_i \overline{T}_{i+1}\right)\right]$$

$$= \sum_{i=0}^{k-1} \frac{1}{2}[1 - (-1)^{s_i}] \operatorname{sgn}(\overline{T}_i \overline{T}_{i+1})$$

$$= \sum_{i=0, \, 2 | s_i + 1}^{k-1} \operatorname{sgn}(\overline{T}_i \overline{T}_{i+1}).$$

Denote D_{q_i} by σ_i and let the revised sign list of the discriminant sequence of f with respect to g be $[\epsilon_0, \ldots, \epsilon_m]$, where

$$\epsilon_j = 0, \quad \text{if } j > \eta = q_k;$$

$$\epsilon_{q_k} = \operatorname{sgn}(\sigma_k) = \operatorname{sgn}(D_\eta);$$

$$\epsilon_{q_i + p_i} = (-1)^{p_i(p_i+1)/2} \operatorname{sgn}(\sigma_i),$$

$$p_i = 0, \ldots, s_i - 1, \ i = 0, \ldots, k - 1.$$

Therefore

$$\eta - 2v = \eta - 2 \sum_{i=0}^{\eta-1} \frac{1}{2}[1 - \operatorname{sgn}(\epsilon_i \epsilon_{i+1})] = \sum_{i=0}^{\eta-1} \operatorname{sgn}(\epsilon_i \epsilon_{i+1})$$

$$= \sum_{i=0}^{k-1} \left[\sum_{j=0, \, s_i > 1}^{s_i - 2} \operatorname{sgn}(\epsilon_{q_i+j} \epsilon_{q_i+j+1}) + \operatorname{sgn}(\epsilon_{q_i+s_i-1} \epsilon_{q_{i+1}}) \right]$$

$$= \sum_{i=0}^{k-1} \Bigg[\sum_{j=0, s_i>1}^{s_i-2} (-1)^{\frac{j(j+1)}{2} + \frac{(j+1)(j+2)}{2}} \operatorname{sgn}(\sigma_i^2)$$

$$+ (-1)^{\frac{(s_i-1)s_i}{2}} \operatorname{sgn}(\sigma_i \sigma_{i+1}) \Bigg]$$

$$= \sum_{i=0}^{k-1} \Bigg[\sum_{j=0,\, s_i>1}^{s_i-2} (-1)^{j+1} + (-1)^{(s_i-1)s_i} \operatorname{sgn}((\overline{T}_i \overline{T}_{i+1})^{s_i} \sigma_i^2) \Bigg]$$

$$= \sum_{i=0}^{k-1} \left\{ \frac{1}{2}[(-1)^{s_i-1} - 1] + \operatorname{sgn}((\overline{T}_i \overline{T}_{i+1})^{s_i}) \right\}$$

$$= \sum_{i=0,\, 2|s_i+1}^{k-1} \operatorname{sgn}(\overline{T}_i \overline{T}_{i+1}).$$

That completes the proof of Discrimination Theorem I.

4.4 Properties of Discrimination Matrix

Discrimination Theorem II indicates that the even order principal minors of the discrimination matrix of a polynomial $f(x)$ (*i.e.* discriminant sequence) determine the numbers of distinct real and complex roots of $f(x)$. Then, what is the property of the odd order principal minors of the discrimination matrix? We will prove in this section that the odd order principal minors, together with the even order principal minors, can determine the number of negative (positive) roots of $f(x)$.

Proposition 4.4. *Let f, g, r be as in (4.3) and (4.4) and*

$$[H_0, H_1, \ldots, H_m]$$

the discriminant sequence of f with respect to g. The other notations are as in Lemma 4.4.

(a) If $H_i = 0$ and $H_{i-1} \cdot H_{i+1} \neq 0$ for some $i = 1, \ldots, m-1$, then $H_{i-1} \cdot H_{i+1} < 0$; If for some $i = 2, \ldots, m-2$,

$$H_{i-1} = H_i = H_{i+1} = 0, \quad H_{i-2} \cdot H_{i+2} \neq 0,$$

then $H_{i-2} \cdot H_{i+2} > 0$.
(b) Denote by

$$h_1, h_2, \ldots, h_{2m-1}, h_{2m}$$

the sequence of principal minors of the discrimination matrix of f and g. Of course we have $H_i = h_{2i}$ $(i = 1, \ldots, m)$. If $h_{2n} = h_{2n+2} = 0$ for some n $(1 \le n \le m - 1)$, then $h_{2n+1} = 0$.

Proof. (a). Let H_{i-1} be the jth non-zero element of the sequence $[H_1, \ldots, H_m]$, then $i - 1 = q_j$, $i + 1 = q_{j+1}$. So, $s_j = q_{j+1} - q_j = 2$. By Lemma 4.4, we have

$$\frac{H_{i+1}}{H_{i-1}} = (-1)^{(s_j-1)s_j/2}(\overline{T_j T_{j+1}})^{s_j} = -(\overline{T_j T_{j+1}})^2 < 0.$$

Similarly, we know that, if

$$H_{i-1} = H_i = H_{i+1} = 0, \ H_{i-2} \cdot H_{i+2} \ne 0,$$

then $H_{i-2} \cdot H_{i+2} > 0$.

(b). A direct corollary of subresultant chain theorem (Theorems 1.5 and 1.6), or see [Xia and Yang (2003)]. $\qquad\square$

For a given real parametric polynomial $f(x) = a_0 x^m + a_1 x^{m-1} + \ldots + a_m$ $(a_0 \ne 0)$, construct a new matrix by adding one row and one column to Discr (f) as follows.

$$\begin{bmatrix}
a_0 & a_1 & a_2 & \cdots & a_m & & & & \\
0 & ma_0 & (m-1)a_1 & \cdots & a_{m-1} & & & & \\
& a_0 & a_1 & \cdots & a_{m-1} & a_m & & & \\
& 0 & ma_0 & \cdots & 2a_{m-2} & a_{m-1} & & & \\
& & \cdots & \cdots & & & & & \\
& & \cdots & \cdots & & & & & \\
& & & a_0 & a_1 & \cdots & \cdots & a_m & \\
& & & 0 & ma_0 & \cdots & \cdots & a_{m-1} & \\
& & & & a_0 & a_1 & \cdots & \cdots & a_m
\end{bmatrix}$$

For convenience, we call the above matrix the *extended discrimination matrix* of $f(x)$, denoted by EDiscr (f). In this section, the sequence of principal minors of EDiscr(f) is denoted by $\{d_1, d_2, \ldots, d_{2m+1}\}$. We have an analogue of Proposition 4.4 as follows.

Proposition 4.5.

(a) *If $d_{2i+1} = 0$ and $d_{2i-1} \cdot d_{2i+3} \ne 0$ for some $i\,(1 \le i \le m - 1)$, then $d_{2i-1} \cdot d_{2i+3} < 0$;*

(b) *If $d_{2n-1} = d_{2n+1} = 0$ for some $n\,(1 \le n \le m)$, then $d_{2n} = 0$.*

Definition 4.5. The sequence $[d_1 d_2, d_2 d_3, \ldots, d_{2m} d_{2m+1}]$ is called the *negative root discriminant sequence* of $f(x)$, denoted by nrd(f).

Denote by $f_{(a,b)}$ the number of distinct real roots in (a,b) of $f(x)$. Let

$$\tilde{h}(x) = f(x^2), \quad h(x) = f(-x^2)$$

and assume $f(0) \neq 0$, then

$$f_{(0,\infty)} = \frac{1}{2}\tilde{h}_{(-\infty,\infty)}, \quad f_{(-\infty,0)} = \frac{1}{2}h_{(-\infty,\infty)}.$$

Theorem 4.8. *Let notations be as above. The discriminant sequence of* $h(x)$, $[D_1(h),\ldots,D_{2m}(h)]$, *equals* $\mathrm{nrd}(f)\,[d_1 d_2,\ldots,d_{2m}d_{2m+1}]$, *that is,*

$$D_k(h) = d_k d_{k+1}, \quad k = 1, 2, \ldots, 2m,$$

up to a factor with the same sign of a_0.

Proof. (1) Assume k is even. Let $k = 2j$ $(1 \leq j \leq m)$ and $t_j = m - j$, then

$$D_k(h) =$$

$$\begin{vmatrix} (-1)^m a_0 & 0 & (-1)^{t_1} a_1 & 0 & \cdots & 0 \\ 0 & (-1)^m 2m a_0 & 0 & (-1)^{t_1} 2t_1 a_1 & \cdots & (-1)^{t_{2j-1}} 2t_{2j-1} a_{2j-1} \\ & (-1)^m a_0 & 0 & (-1)^{t_1} a_1 & \cdots & (-1)^{t_{2j-1}} a_{2j-1} \\ & & (-1)^m 2m a_0 & 0 & \cdots & 0 \\ \vdots & \vdots & \vdots & \vdots & \ddots & 0 \\ 0 & \cdots & (-1)^m a_0 & \cdots & \cdots & (-1)^{t_j} a_j \\ 0 & \cdots & 0 & (-1)^m 2m a_0 & \cdots & 0 \end{vmatrix}_{4j \times 4j}$$

$$= (-1)^m 2^k a_0 \times$$

$$\begin{vmatrix} (-1)^m m a_0 & 0 & (-1)^{t_1} t_1 a_1 & 0 & \cdots & (-1)^{t_{2j-1}} t_{2j-1} a_{2j-1} \\ (-1)^m a_0 & 0 & (-1)^{t_1} a_1 & 0 & \cdots & (-1)^{t_{2j-1}} a_{2j-1} \\ 0 & (-1)^m m a_0 & 0 & (-1)^{t_1} t_1 a_1 & \cdots & 0 \\ 0 & (-1)^m a_0 & 0 & (-1)^{t_1} a_1 & \cdots & 0 \\ \vdots & \vdots & \vdots & \vdots & \ddots & \vdots \\ \cdots & \cdots & (-1)^m a_0 & \cdots & \cdots & (-1)^{t_j} a_j \\ \cdots & \cdots & 0 & (-1)^m m a_0 & \cdots & 0 \end{vmatrix}$$

In the above determinant, move one by one the 2nd, 4th, 6th, $\cdots$, and the $(4j-2)$th columns to the first $(2j-1)$ columns, and then, move two by two the 3rd and 4th rows, the 7th and 8th rows, $\cdots$, and the $(4j-5)$th and $(4j-4)$th rows to the first $(2j-1)$ rows. Then we get that

$$D_k(h) = (-1)^\delta \cdot (-1)^m \cdot 2^k \cdot a_0 \cdot \begin{vmatrix} A & 0 \\ 0 & B \end{vmatrix}$$

where

$$\begin{aligned} \delta &= (2-1) + (4-2) + (6-3) + \cdots + (4j-2-2j+1) + \\ & \quad (3-1) + (4-2) + (7-3) + (8-4) + \cdots + (4j-1-2j+1) \\ &\equiv 1 + 2 + 3 + \cdots + (2j-1) \mod 2 \\ &\equiv j \mod 2, \end{aligned}$$

$$A = \begin{vmatrix} (-1)^m ma_0 & (-1)^{t_1} t_1 a_1 & \cdots & & \cdots & (-1)^{t_{2j-2}} t_{2j-2} a_{2j-2} \\ (-1)^m a_0 & (-1)^{t_1} a_1 & \cdots & & \cdots & (-1)^{t_{2j-2}} a_{2j-2} \\ & & (-1)^m ma_0 & \cdots & & \cdots \vdots \\ & & (-1)^m a_0 & \cdots & & \cdots \vdots \\ \vdots & \vdots & & \ddots & & \ddots \vdots \\ & & & & (-1)^m ma_0 & \cdots \; (-1)^{t_{j-1}} t_{j-1} a_{j-1} \end{vmatrix}_{(2j-1)\times(2j-1)},$$

and

$$B = \begin{vmatrix} (-1)^m ma_0 & \cdots & & \cdots & (-1)^{t_{2j-1}} t_{2j-1} a_{2j-1} \\ (-1)^m a_0 & \cdots & & \cdots & (-1)^{t_{2j-1}} a_{2j-1} \\ \vdots & & \vdots & \ddots & \vdots \\ \cdots & & (-1)^m a_0 & \cdots & (-1)^{t_j} a_j \end{vmatrix}_{2j\times 2j}.$$

If m is even, multiply the odd columns of A and B by -1; Otherwise, if m is odd, multiply the even columns of A and B by -1. After that, multiply the 1st and 2nd rows, the 5th and 6th rows, the 9th and 10th rows and etc. of A and B by -1. We have

$$A = (-1)^{2j} A^* = A^*, \quad B = (-1)^j B^*, \text{ if } m \equiv 0 \mod 2,$$

$$A = (-1)^{2j-1} A^* = (-1) A^*, \; B = (-1)^j B^*, \text{ if } m \equiv 1 \mod 2,$$

where

$$A^* = \begin{vmatrix} ma_0 & t_1 a_1 & \cdots & & \cdots & t_{2j-2} a_{2j-2} \\ a_0 & a_1 & \cdots & & \cdots & a_{2j-2} \\ & & ma_0 & \cdots & & \cdots \vdots \\ & & a_0 & \cdots & & \cdots \vdots \\ \vdots & \vdots & & \ddots & & \ddots \vdots \\ & & & & ma_0 & \cdots \; t_{j-1} a_{j-1} \end{vmatrix}_{(2j-1)\times(2j-1)},$$

$$B^* = \begin{vmatrix} ma_0 & \cdots & & \cdots & t_{2j-1} a_{2j-1} \\ a_0 & \cdots & & \cdots & a_{2j-1} \\ \vdots & & \vdots & \ddots & \vdots \\ \cdots & & a_0 & \cdots & a_j \end{vmatrix}_{2j\times 2j}.$$

So, no matter what the parity of m is, we have

$$\begin{aligned} D_k(h) &= (-1)^\delta \cdot (-1)^m \cdot 2^k \cdot a_0 \cdot A \cdot B \\ &= (-1)^j \cdot (-1)^j \cdot 2^k \cdot a_0 \cdot A^* \cdot B^* \\ &= 2^k \cdot a_0 \cdot A^* \cdot B^*. \end{aligned}$$

Noticing that

$$A^* = \frac{1}{a_0} \begin{vmatrix} a_0 & 0 \\ 0 & A^* \end{vmatrix} = \frac{1}{a_0} d_{2j}, \quad B^* = \frac{1}{a_0} \begin{vmatrix} a_0 & 0 \\ 0 & B^* \end{vmatrix} = \frac{1}{a_0} d_{2j+1},$$

we obtain that

$$D_k(h) = \frac{2^k}{a_0} \cdot d_{2j} \cdot d_{2j+1}.$$

Because $k = 2j$,

$$D_k(h) = d_k \cdot d_{k+1},$$

up to a factor with the same sign of a_0.

(2) The proof in the case that k is odd is similar. □

Theorem 4.9. *Let* $[D_1(\tilde{h}), \ldots, D_{2m}(\tilde{h})]$ *be the discriminant sequence of* $\tilde{h}(x)$*, then*

$$D_k(\tilde{h}) = (-1)^{\lfloor \frac{k}{2} \rfloor} d_k d_{k+1}, \quad k = 1, \ldots, 2m,$$

up to a factor with the same sign of a_0*.*

Proof. Similar to the proof of Theorem 4.8. □

Theorem 4.10. *Let notations be as above. Assume* $a_0 \neq 0, a_m \neq 0$*, and the revised sign list of* $\mathrm{nrd}(f)$ *has* μ *sign changes and* $2l$ *non-zero elements, then the number of distinct negative roots of* $f(x)$ *is* $l - \mu$*.*

Proof. A direct corollary of Theorem 4.7 and Theorem 4.8. □

Theorem 4.11. *Let notations be as above. If the revised sign list of* $[d_1, d_3, \ldots, d_{2m+1}]$ *has* v *sign changes and* $l + 1$ *non-zero elements, that is,* $d_{2l+1} \neq 0, d_{2t+1} = 0 \ (t > l)$*, then*

$$l - 2v = f_{(-\infty,0)} - f_{(0,\infty)}.$$

Proof. First, if $t_0, t_1, \ldots, t_m$ is a sequence of non-zero real numbers, the sign changes of the sequence is

$$\sum_{i=0}^{m-1} \frac{1}{2}(1 - \mathrm{sgn}(t_i t_{i+1})).$$

Let $[H_0, H_1, \ldots, H_m]$ be the discriminant sequence of f with respect to $g(x) = x$. Assume $[\epsilon_0, \epsilon_1, \ldots, \epsilon_m]$ is the revised sign list of $[H_0, H_1, \ldots, H_m]$, which has v_1 sign changes and $\epsilon_{l_1} \neq 0, \ \epsilon_t = 0 \ (t > l_1)$. By Discrimination Theorem I,

$$l_1 - 2v_1 = f_{(0,\infty)} - f_{(-\infty,0)}.$$

Assume $[\epsilon'_0, \epsilon'_1, \ldots, \epsilon'_m]$ is the revised sign list of $[d_1, d_3, \ldots, d_{2m+1}]$. We need to show that

$$l - 2v = -(l_1 - 2v_1).$$

Suppose $l_1 = q_k$. Then

$$l_1 - 2v_1 = l_1 - 2 \sum_{i=0}^{l_1-1} \tfrac{1}{2}(1 - \mathrm{sgn}(\epsilon_i \epsilon_{i+1})) = \sum_{i=0}^{l_1-1} \mathrm{sgn}(\epsilon_i \epsilon_{i+1})$$

$$= \sum_{i=0}^{k-1} \left[\sum_{j=0, s_i > 1}^{s_i-2} \mathrm{sgn}(\epsilon_{q_i+j} \epsilon_{q_i+j+1}) + \mathrm{sgn}(\epsilon_{q_i+s_i-1} \epsilon_{q_{i+1}}) \right]$$

$$= \sum_{i=0}^{k-1} \left[\sum_{j=0, s_i > 1}^{s_i-2} (-1)^{\frac{j(j+1)}{2} + \frac{(j+1)(j+2)}{2}} \cdot \mathrm{sgn}(\sigma_i^2) + (-1)^{(s_i-1)s_i/2} \cdot \mathrm{sgn}(\sigma_i \sigma_{i+1}) \right]$$

$$= \sum_{i=0}^{k-1} \left[\sum_{j=0, s_i > 1}^{s_i-2} (-1)^{j+1} + (-1)^{(s_i-1)s_i} \cdot \mathrm{sgn}((\overline{T_i T_{i+1}})^{s_i} \cdot \sigma_i^2) \right]$$

$$= \sum_{i=0}^{k-1} \left[\tfrac{1}{2}((-1)^{s_i-1} - 1) + \mathrm{sgn}((\overline{T_i T_{i+1}})^{s_i}) \right]$$

$$= \sum_{i=0, s_i \text{ odd}}^{k-1} \mathrm{sgn}(\overline{T_i T_{i+1}}).$$

Herein, $\overline{T_i}$ stands for the leading coefficient of T_i. By the relations

$$d_{2i+1} = (-1)^i \cdot a_0 \cdot H_i, \ (0 \le i \le m)$$

we know that $l = q_k$ and

$$\epsilon'_{q_i} = (-1)^{q_i} \epsilon_{q_i}, \ 0 \le i \le k.$$

Therefore, by similar way, we know

$$l - 2v = \sum_{i=0}^{k-1} \left[\sum_{j=0, s_i > 1}^{s_i-2} \mathrm{sgn}(\epsilon'_{q_i+j} \epsilon'_{q_i+j+1}) + \mathrm{sgn}(\epsilon'_{q_i+s_i-1} \epsilon'_{q_{i+1}}) \right] \cdot$$

For every i $(0 \le i \le k-1)$, if
(i) q_i is odd and s_i is odd, then $q_{i+1} = q_i + s_i$ is even, and thus

$$\sum_{j=0, s_i > 1}^{s_i-2} \mathrm{sgn}(\epsilon'_{q_i+j} \epsilon'_{q_i+j+1}) + \mathrm{sgn}(\epsilon'_{q_i+s_i-1} \epsilon'_{q_{i+1}})$$

$$= \tfrac{1}{2}((-1)^{s_i-1} - 1) - \mathrm{sgn}((\overline{T_i T_{i+1}})^{s_i})$$

$$= -\mathrm{sgn}(\overline{T_i T_{i+1}}).$$

(ii) q_i is odd and s_i is even, then $q_{i+1} = q_i + s_i$ is odd, and thus

$$\sum_{j=0,s_i>1}^{s_i-2} \mathrm{sgn}(\epsilon'_{q_i+j}\epsilon'_{q_i+j+1}) + \mathrm{sgn}(\epsilon'_{q_i+s_i-1}\epsilon'_{q_{i+1}})$$

$$= \tfrac{1}{2}((-1)^{s_i-1} - 1) + \mathrm{sgn}((\overline{T_iT_{i+1}})^{s_i})$$

$$= 0.$$

The other two cases (q_i even s_i odd; q_i even s_i even) can be handled similarly. Finally, we have

$$l - 2v = -\sum_{i=0,s_i \text{ odd}}^{k-1} \mathrm{sgn}(\overline{T_iT_{i+1}})$$

$$= -(l_1 - 2v_1).$$

That completes the proof. $\qquad\square$

Theorem 4.12. *Let $[d_1, d_2, \ldots, d_{2m}, d_{2m+1}]$ be the sequence of principal minors of* $\mathrm{EDiscr}\,(f)$ *where $f(x) = a_0 x^m + a_1 x^{m-1} + \ldots + a_m$ ($a_0 \neq 0$, $a_m \neq 0$). If l_1, v_1; l_2, v_2; l, v are the sign changes and the number of non-zero elements of*

$$[d_2, d_4, \ldots, d_{2m}],$$

$$[d_1, d_3, \ldots, d_{2m+1}],$$

and

$$[d_1 d_2, d_2 d_3, \ldots, d_{2m} d_{2m+1}],$$

respectively, then $l = l_1 + l_2 - 1$, $v = v_1 + v_2$.

Proof. By Theorem 4.10, the number of distinct negative roots of $f(x)$ equals $l/2 - v$. On the other hand, by Discrimination Theorem II and Theorem 4.11, we know that the number also equals $(l_1 + l_2 - 1)/2 - (v_1 + v_2)$. Therefore, $v = v_1 + v_2$ if $l = l_1 + l_2 - 1$. By Proposition 4.4(b) and Proposition 4.5(b), we have $|2l_1 - (2l_2 - 1)| = 1$. Obviously, l must be even and thus $l = 2l_1$ and $2l_1 < (2l_2 - 1)$. So

$$2l_2 - 1 - 2l_1 = 1, \quad l_2 = l_1 + 1.$$

Finally, we get that $l = 2l_1 = l_1 + l_2 - 1$. $\qquad\square$

Remark 4.4. The nrd can be used to give an explicit criterion on the number of real roots in an interval of a polynomial. Suppose $f(x)$ is a polynomial of degree m and $f(a)f(b) \neq 0$. Then the numbers of real roots of $f(x)$ in (a, b), $(-\infty, a)$ and $(b, +\infty)$ equal respectively to a half of the numbers of non-zero roots of the following polynomials

$$(x^2 + 1)^m f\left(\frac{ax^2 + b}{x^2 + 1}\right), \quad f(a - x^2), \quad f(x^2 + b).$$

So we can apply Discrimination Theorem II to these polynomials and get corresponding explicit criteria. However, computing the discriminant sequences of these composite polynomials can be very time-consuming since the degrees are doubled.

We show by an example an application of the property of discrimination matrices.

Proposition 4.6. *[Yang and Xia (2006)] Suppose*

$$Q(\lambda) = \lambda^4 + p\,\lambda^3 + q\,\lambda^2 + r\lambda + s$$

where $s \neq 0$, is a quartic polynomial with real coefficients. Then

$$\forall \lambda > 0 \ (Q(\lambda) > 0)$$

if and only if

$$
\begin{aligned}
s > 0 \wedge (&(p \geq 0 \wedge q \geq 0 \wedge r \geq 0) \vee \\
&(d_8 > 0 \wedge (d_6 \leq 0 \vee d_4 \leq 0)) \vee \\
&(d_8 < 0 \wedge d_7 \geq 0 \wedge (p \geq 0 \vee d_5 < 0)) \vee \\
&(d_8 < 0 \wedge d_7 < 0 \wedge p > 0 \wedge d_5 > 0) \vee \\
&(d_8 = 0 \wedge d_6 < 0 \wedge d_7 > 0 \wedge (p \geq 0 \vee d_5 < 0)) \vee \\
&(d_8 = 0 \wedge d_6 = 0 \wedge d_4 < 0))
\end{aligned}
\tag{4.9}
$$

where

$$
\begin{aligned}
d_4 =\ & -8\,q + 3\,p^2, \\
d_5 =\ & 3\,r\,p + q\,p^2 - 4\,q^2, \\
d_6 =\ & 14\,q\,r\,p - 4\,q^3 + 16\,s\,q - 3\,p^3\,r + p^2\,q^2 - 6\,p^2\,s - 18\,r^2, \\
d_7 =\ & 7\,r\,p^2\,s - 18\,q\,p\,r^2 - 3\,q\,p^3\,s - q^2\,p^2\,r + 16\,s^2\,p + 4\,r^2\,p^3 + 12\,q^2\,p\,s \\
& + 4\,r\,q^3 - 48\,r\,s\,q + 27\,r^3, \\
d_8 =\ & p^2\,q^2\,r^2 + 144\,q\,s\,r^2 - 192\,r\,s^2\,p + 144\,q\,s^2\,p^2 - 4\,p^2\,q^3\,s + 18\,q\,r^3\,p \\
& - 6\,p^2\,s\,r^2 - 80\,r\,p\,s\,q^2 + 18\,p^3\,r\,s\,q - 4\,q^3\,r^2 + 16\,q^4\,s - 128\,s^2\,q^2 \\
& - 4\,p^3\,r^3 - 27\,p^4\,s^2 - 27\,r^4 + 256\,s^3.
\end{aligned}
$$

Proof. We need to find the necessary and sufficient condition such that $Q(\lambda)$ does not have positive roots. First of all, by Descartes' rule of signs we have the following results:

(1) $s > 0$ must hold. Otherwise, the sequence $[1, p, q, r, s]$ will have an odd number of sign changes which implies $Q(\lambda)$ has at least one positive root.

(2) If the roots of $Q(\lambda)$ are all real, $Q(\lambda)$ does not have positive roots if and only if $s > 0$ and p, q, r are all non-negative.

Therefore, in the following we always assume $s > 0$ and do not consider the case when $Q(\lambda)$ has four real roots (counted with multiplicity).

Let $P(\lambda) = Q(-\lambda)$, then we discuss the condition such that $P(\lambda)$ does not have negative roots. We compute the principal minors d_i $(1 \leq i \leq 9)$ of $\mathrm{Discr}(P)$ and consider the following two lists:

$$L_1 = [1, d_4, d_6, d_8] \quad \text{and} \quad L_2 = [1, d_3, d_5, d_7, d_9]$$

where $d_3 = -p$, $d_9 = sd_8$ and d_i $(4 \leq i \leq 8)$ are showed above in the statement of this proposition. In the following, we denote the numbers of non-zero elements and sign changes of the revised sign list of L_i by l_i and v_i $(i = 1, 2)$, respectively.

Case **I**. $d_8 > 0$.

In this case, by Theorem 4.7 $P(\lambda)$ has either four imaginary roots or four real roots. $P(\lambda)$ has four imaginary roots if and only if $d_6 \leq 0 \vee d_4 \leq 0$ by Theorem 4.7. As stated above, we need not consider the case when $P(\lambda)$ has four real roots. Thus,

$$d_8 > 0 \ \wedge \ (d_6 \leq 0 \ \vee \ d_4 \leq 0)$$

must be satisfied under Case I.

Case **II**. $d_8 < 0$.

In this case, L_1 becomes $[1, d_4, d_6, -1]$ with $l_1 = 4, v_1 = 1$ which implies by Theorem 4.7 that $P(\lambda)$ has two imaginary roots and two distinct real roots.

If $d_7 > 0$, L_2 becomes $[1, -p, d_5, 1, -1]$. By Theorem 4.12, v_2 should be 3 which is equivalent to $p \geq 0 \ \vee \ d_5 \leq 0$.

If $d_7 = 0$, L_2 becomes $[1, -p, d_5, 0, -1]$. By Theorem 4.12, v_2 should be 3 which is equivalent to $p \geq 0 \ \vee \ d_5 < 0$.

To combine the above two conditions, we perform pseudo-division of d_7 and d_5 with respect to r and obtain that

$$27p^3 d_7 = F d_5 + 12G^2 \tag{4.10}$$

where F, G are polynomials in p, q, r, s. It is easy to see that p should be non-negative if $d_7 > 0$ and $d_5 = 0$. Thus, we may combine the above two sub-cases into

$$d_7 \geq 0 \ \wedge \ (p \geq 0 \ \vee \ d_5 < 0).$$

If $d_7 < 0$, L_2 becomes $[1, -p, d_5, -1, -1]$. By (4.10) we know that $p = 0 \ \wedge \ d_5 > 0$ and $p > 0 \ \wedge \ d_5 = 0$ are both impossible. Thus, v_2 is 3 if and only if $p > 0 \ \wedge \ d_5 > 0$.

In Case II, we conclude that

$$d_8 < 0 \wedge \ [(d_7 \geq 0 \wedge (p \geq 0 \vee d_5 < 0)) \vee (d_7 < 0 \wedge p > 0 \wedge d_5 > 0)]$$

must be satisfied.

Case **III**. $d_8 = 0$.

If $d_6 > 0$, $P(\lambda)$ has four real roots (counted with multiplicity) and this is a case we have already discussed.

If $d_6 < 0$, then $l_1 = 3$ and $v_1 = 1$. We need to find the condition for $l_2/2 = v_2$ by Theorem 4.12. Obviously, l_2 must be an even integer. We consider the sign of d_7. First, $d_7 < 0$ implies $l_2/2 = 2$ and v_2 is an odd integer and thus $l_2/2 = v_2$ cannot be satisfied. Second, if $d_7 = 0$, by Theorem 4.12 $d_5 \neq 0$ since $d_6 < 0$. That means l_2 is odd which is impossible. Finally, if $d_7 > 0$, v_2 must be 2 and this is satisfied by $p \geq 0 \ \vee \ d_5 < 0$.

If $d_6 = 0$, L_1 becomes $[1, d_4, 0, 0]$. And $d_4 \geq 0$ implies $P(\lambda)$ has four real roots (counted with multiplicity) and this is the case we have already discussed. If $d_4 < 0$, $P(\lambda)$ has four imaginary roots (counted with multiplicity) and thus no negative roots.

In Case III, we conclude that

$$d_8 = 0 \ \wedge \ [(d_6 < 0 \wedge d_7 > 0 \wedge (p \geq 0 \vee d_5 < 0)) \ \vee \ (d_6 = 0 \wedge d_4 < 0)]$$

must be satisfied. That completes the proof. $\qquad\qquad\qquad\Box$

By similar discussion, we have

Proposition 4.7. *[Yang and Xia (2006)] Given a quartic polynomial of real coefficients,*

$$Q(\lambda) = \lambda^4 + p\,\lambda^3 + q\,\lambda^2 + r\lambda + s,$$

with $s \neq 0$, then

$$\forall \lambda \geq 0 \, (Q(\lambda) \geq 0)$$

is equivalent to

$$s > 0 \wedge ((p \geq 0 \wedge q \geq 0 \wedge r \geq 0) \vee$$
$$(d_8 > 0 \wedge (d_6 \leq 0 \vee d_4 \leq 0)) \vee$$
$$(d_8 < 0 \wedge d_7 \geq 0 \wedge (p \geq 0 \vee d_5 < 0)) \vee \qquad (4.11)$$
$$(d_8 < 0 \wedge d_7 < 0 \wedge p > 0 \wedge d_5 > 0) \vee$$
$$(d_8 = 0 \wedge d_6 < 0) \vee$$
$$(d_8 = 0 \wedge d_6 > 0 \wedge d_7 > 0 \wedge (p \geq 0 \vee d_5 < 0)) \vee$$
$$(d_8 = 0 \wedge d_6 = 0 \wedge (d_4 \leq 0 \vee E_1 = 0)))$$

where d_i $(4 \leq i \leq 8)$ are defined as in Proposition 4.6 and

$$E_1 = 8\,r - 4\,p\,q + p^3.$$

The above two propositions provide an easy-to-use tool for solving related problems. We take for example an application of Proposition 4.6 in program termination verification.

In the field of program verification, termination analysis is an important topic [Cousot (2001)]. It is well known that program termination is generally undecidable. However, for special kinds of programs in practice, it is expected that their termination can be proved and some computable explicit conditions for the termination of some special classes of programs can be established so that, for a concrete program in a special class, one can use directly the conditions to verify its termination.

Linear programs are a class of programs which have been studied extensively in the literature [Besson *et al.* (1999); Cousot and Halbwachs (1978); Halbwachs *et al.* (1997)]. A great deal of *reactive systems* can be described by linear programs exactly or approximately [Henzinger and Ho (1995)]. Unfortunately, the termination of linear programs is generally undecidable [Tiwari (2004)]. On the other hand, Tiwari (2004) proved that the termination of a class of linear programs of the following form is decidable.

$$\mathbf{P_1} : \ \mathbf{while} \ \ Bx > b \ \{x := Ax + c\},$$

where x is a vector of N program variables, b and c are vectors of real numbers, A and B are $N \times N$ and $M \times N$ real matrices, respectively, $Bx > b$ stands for the conjunction of M linear inequalities and $x := Ax + c$ means linear assignments to the variables.

Theorem 4.13. *[Tiwari (2004)] The termination of linear program $\mathbf{P_1}$ is decidable.*

If b and c are both zero vector, the linear program is called the *homogeneous version* of $\mathbf{P}_1$, denoted by

$$\mathbf{P}_2: \quad \textbf{while } (Bx > 0) \{x := Ax\}.$$

Theorem 4.14. *[Tiwari (2004)] If the program $\mathbf{P}_2$ is nonterminating, then there is a real eigenvector v of A, corresponding to a positive eigenvalue, such that $Bv \geq 0$.*

Definition 4.6. If matrix A has no positive eigenvalues, then the assignment $x := Ax$ in $\mathbf{P}_2$ is said to be a *terminating assignment*.

Obviously, if the assignment $x := Ax$ in $\mathbf{P}_2$ is a terminating assignment, then, for any matrix B, $\mathbf{P}_2$ is terminating. So, as a corollary of Proposition 4.6, we have the following theorem.

Theorem 4.15. *[Yang et al. (2005)] Suppose $A = (a_{ij})$ is a 4×4 matrix. $x := Ax$ is a terminating assignment if and only if the condition (4.9) holds, where*

$$p = -a_{11} - a_{22} - a_{33} - a_{44},$$

$$\begin{aligned}
q = {} & a_{33}a_{44} + a_{11}a_{22} - a_{41}a_{14} - a_{31}a_{13} - a_{32}a_{23} - a_{34}a_{43} + a_{22}a_{44} \\
& + a_{22}a_{33} - a_{21}a_{12} - a_{42}a_{24} + a_{11}a_{44} + a_{11}a_{33},
\end{aligned}$$

$$\begin{aligned}
r = {} & -a_{32}a_{24}a_{43} + a_{11}a_{34}a_{43} - a_{11}a_{33}a_{44} - a_{21}a_{42}a_{14} + a_{11}a_{32}a_{23} \\
& + a_{21}a_{12}a_{33} + a_{42}a_{24}a_{33} + a_{11}a_{42}a_{24} - a_{31}a_{12}a_{23} + a_{22}a_{34}a_{43} \\
& - a_{11}a_{22}a_{33} + a_{31}a_{13}a_{44} - a_{11}a_{22}a_{44} - a_{42}a_{23}a_{34} - a_{22}a_{33}a_{44} \\
& - a_{41}a_{12}a_{24} + a_{32}a_{23}a_{44} - a_{41}a_{13}a_{34} + a_{41}a_{14}a_{33} + a_{21}a_{12}a_{44} \\
& + a_{41}a_{22}a_{14} - a_{31}a_{14}a_{43} + a_{31}a_{22}a_{13} - a_{21}a_{32}a_{13},
\end{aligned}$$

$$\begin{aligned}
s = {} & -a_{11}a_{22}a_{34}a_{43} - a_{21}a_{32}a_{14}a_{43} - a_{21}a_{42}a_{13}a_{34} + a_{11}a_{32}a_{24}a_{43} \\
& + a_{21}a_{42}a_{14}a_{33} + a_{41}a_{12}a_{24}a_{33} + a_{31}a_{12}a_{23}a_{44} - a_{31}a_{12}a_{24}a_{43} \\
& + a_{11}a_{22}a_{33}a_{44} - a_{21}a_{12}a_{33}a_{44} + a_{21}a_{12}a_{34}a_{43} - a_{31}a_{22}a_{13}a_{44} \\
& - a_{41}a_{12}a_{23}a_{34} + a_{31}a_{22}a_{14}a_{43} - a_{31}a_{42}a_{14}a_{23} - a_{11}a_{32}a_{23}a_{44} \\
& + a_{41}a_{22}a_{13}a_{34} + a_{11}a_{42}a_{23}a_{34} - a_{11}a_{42}a_{24}a_{33} + a_{41}a_{32}a_{14}a_{23} \\
& + a_{21}a_{32}a_{13}a_{44} - a_{41}a_{22}a_{14}a_{33} - a_{41}a_{32}a_{13}a_{24} + a_{31}a_{42}a_{13}a_{24}.
\end{aligned}$$

Proof.   Notice that the characteristic polynomial of A is $\lambda^4 + p\lambda^3 + q\lambda^2 + r\lambda + s$, where p, q, r, s are defined as above. By the definition of terminating assignment and Proposition 4.6, the result is clear. $\square$

Chapter 5

Real Root Isolation

A univariate polynomial has only finitely many real roots. If a constant polynomial system or semi-algebraic system is given in this chapter, it is assumed to have only finitely many real solutions, too. Suppose a univariate polynomial (polynomial system / semi-algebraic system) has k distinct real solutions, roughly speaking, *real root isolation* means computing k disjoint regions to isolate all the distinct real solutions, with only one solution in each region.

Real root isolation for univariate polynomials (polynomial systems / semi-algebraic systems) with integer coefficients plays a significant role in many algorithms concerning computational real algebra and real algebraic geometry. In this chapter, we first introduce well-known algorithms for isolating real roots of polynomials. Then we present our contribution to real root isolation for constant semi-algebraic systems (SAS) based on triangularization of SAS (see Chapter 3) and interval arithmetic. Finally, we introduce our work on real root isolation for polynomial systems based on a combination of homotopy continuation method and interval arithmetic.

5.1 Real Root Isolation for Polynomials

Suppose $f(x) \in \mathbb{Z}[x]$ has k distinct real roots $\alpha_1, ..., \alpha_k$. Isolating the real roots of $f(x)$ is to compute k pairwise disjoint intervals $[a_1, b_1], ..., [a_k, b_k]$ with rational endpoints such that $\alpha_i \in [a_i, b_i]$ for $i = 1, ..., k$.

Real root isolation (RRI) for polynomials is a fundamental operation in computational real algebraic geometry. The research on algorithms for RRI has been a focus topic in the field of symbolic computation for many years. There are many well-known algorithms and tools for computing RRI for polynomials. In many computer algebra systems (CAS), one can find implementations of algorithms for real root isolation based on different principles. The `realroot` function in Maple and the `RealRootIntervals` function in Mathematica are such examples.

For the purpose of this book, it is sufficient to admit that there are efficient algorithms for computing RRI. So, we only introduce the basic idea and framework of the well-known algorithm for RRI based on Descartes' rule of signs in this section. For details and recent advances, see for example [Collins and Akritas (1976); Collins and Loos (1982); Collins and Johnson (1989); Akritas *et al.* (1994); Johnson and Krandick (1997); Rouillier and Zimmermann (2004); Akritas and Strzeboński (2005); Eigenwillig *et al.* (2006); Sharma (2007); Mehlhorn and Sagraloff (2009); Emiris *et al.* (2010a); Garcia and Galligo (2012); Sagraloff (2012); Sharma and Yap (2012); Sharma and Batra (2015)]. RRI for other kinds of functions can be found in [Achatz *et al.* (2008); Strzeboński (2008); Xu *et al.* (2015)].

A natural idea for real root isolation is bisection. Suppose B is a root bound of $f(x)$ and we have an effective rule $\mathcal{M}$ to determine the number of roots of $f(x)$ in an interval, then we may bisect $(-B, B)$ (or $(0, B)$) repeatedly and apply $\mathcal{M}$ each time to rule out intervals not containing roots. Sturm's theorem, Budan-Fourier's theorem and Descartes' rule of signs are instances of such rules.

Algorithm 5.1 RRI–Sturm

Input: A squarefree polynomial $f(x) \in \mathbb{Z}[x]$
Output: Real root isolation of $f(x)$

 1: $B \leftarrow$ a root bound of $f(x)$;
 2: Compute the Sturm sequence of $f(x)$ and denote it by Θ;
 3: $L \leftarrow \emptyset$; $W \leftarrow \{(-B, B)\}$;
 4: **for** $(a, b) \in W$ **do**
 5: $v \leftarrow V(\Theta; a) - V(\Theta; b)$;
 6: **if** $v = 1$ **then** $L \leftarrow L \cup \{(a, b)\}$; **end if**
 7: **if** $v \geq 2$ **then**
 8:   $W \leftarrow W \cup \{(a, \frac{a+b}{2}), (\frac{a+b}{2}, b)\}$;
 9:   **if** $f(\frac{a+b}{2}) = 0$ **then**
 10:    $L \leftarrow L \cup \{[\frac{a+b}{2}, \frac{a+b}{2}]\}$; $f \leftarrow \frac{f}{x-(a+b)/2}$; Re-compute Θ;
 11:   **end if**
 12: **end if**
 13: **end for**
 14: **return** L

Algorithm 5.1 based on Sturm's theorem is very simple but may illustrate the idea of bisection well. The correctness and termination of Algorithm 5.1 are clearly by Sturm's Theorem.

Algorithm 5.2 [Collins and Loos (1982)] based on Descartes' rule of signs is also a bisection method but is tricky and more efficient. For its correctness and termination, see for example [Collins and Loos (1982)].

Algorithm 5.2 RRI-Descartes

Input: A squarefree polynomial $f(x) \in \mathbb{Z}[x]$
Output: Real root isolation of positive roots of $f(x)$
  1: $B \leftarrow$ a root bound of $f(x)$;
  2: $g \leftarrow f(Bx)$;
  3: Compute $\mathtt{subRRI}(g)$ and assume the output is $\{(a_1, b_1), \ldots, (a_k, b_k)\}$;
  4: **return** $\{(Ba_1, Bb_1), \ldots, (Ba_k, Bb_k)\}$

Algorithm 5.3 subRRI

Input: A squarefree polynomial $g(x) \in \mathbb{Z}[x]$ whose positive roots are all in $(0, 1)$
Output: Real root isolation of positive roots of $g(x)$
  1: $m \leftarrow \deg(g)$; $L \leftarrow \emptyset$;
  2: $g^* \leftarrow (x + 1)^m g(1/(x + 1))$;
  3: Let v be the sign changes of the coefficients of g^*;
  4: **if** $v = 0$ **then return** L **end if**;
  5: **if** $v = 1$ **then** $L \leftarrow L \cup \{(0, 1)\}$; **return** L **end if**;
  6: **if** $g(1/2) = 0$ **then**
  7:    $L \leftarrow L \cup \{[\frac{1}{2}, \frac{1}{2}]\}$; $g \leftarrow \frac{g}{x - 1/2}$;
  8: **end if**
  9: $g_1 \leftarrow 2^m g(x/2)$;
 10: Compute $\mathtt{subRRI}(g_1)$ and assume the output is $\{(a_1, b_1), \ldots, (a_i, b_i)\}$;
 11: $L \leftarrow L \cup \{(\frac{a_1}{2}, \frac{b_1}{2}), \ldots, (\frac{a_i}{2}, \frac{b_i}{2})\}$;
 12: $g_2 \leftarrow 2^m g((x + 1)/2)$;
 13: Compute $\mathtt{subRRI}(g_2)$ and assume the output is $\{(c_1, d_1), \ldots, (c_j, d_j)\}$;
 14: $L \leftarrow L \cup \{(\frac{c_1+1}{2}, \frac{d_1+1}{2}), \ldots, (\frac{c_j+1}{2}, \frac{d_j+1}{2})\}$;
 15: **return** L

Remark 5.1. It is not difficult to know that, based on real root isolation algorithms, we may design an algorithm which, for two input polynomials

$p(x), q(x) \in \mathbb{Z}[x]$ without common roots, computes a real root isolation, say $[a_1, b_1], ..., [a_k, b_k]$, of $p(x)$ such that for any $i (1 \leq i \leq k)$ and any $\alpha \in [a_i, b_i]$, $q(\alpha) \neq 0$. We denote the algorithm by `RRI-TwoPoly`.

Remark 5.2. There are some algorithms for real root isolation for *polynomial equations* using different theories and techniques with focuses on complexity and/or solving benchmarks, see for example, [Xia and Yang (2002); Lu *et al.* (2004); Xia and Zhang (2006); Cheng *et al.* (2007); Boulier *et al.* (2009); Cheng *et al.* (2009); Emiris *et al.* (2010b); Mantzaflaris *et al.* (2011); Zhang *et al.* (2011); Hauenstein and Sottile (2012); Strzeboński and Tsigaridas (2012); Cheng *et al.* (2012)].

Real root isolation algorithms are usually symbolic but not numeric methods. An advantage of symbolic method is that exact results can be obtained. However, there are also some disadvantages. The major disadvantage is that symbolic methods are in general time-consuming and can hardly handle big problems. For example, some real root isolation algorithms need to triangularize the system first, which is very time-consuming when the scale of the system is big. While some methods that do not use triangularization have to give a huge initial interval to include all the real roots [Zhang (2004); Zhang *et al.* (2005)], which is extremely inefficient.

One interesting idea is to design a real root isolation algorithm for polynomial equations by combining numerical and symbolic computation. See for example [Shen (2012); Shen *et al.* (2014)].

We do not introduce such algorithms in this book. Instead, we present a real root isolation algorithm for constant SASs [Xia and Zhang (2006)] in the next section.

5.2　Real Root Isolation for Constant Semi-Algebraic Systems

When talking about real root isolation for a constant SAS, we will assume, of course, that the SAS is zero-dimensional, *i.e.* it has only a finite number of complex solutions.

Suppose S is a constant zero-dimensional SAS. By Algorithm 3.2, S can be decomposed equivalently to a set of constant zero-dimensional regular TSAs. Furthermore, those TSAs can be made pairwise disjoint by Algorithm 3.7. So, isolating the real solutions to S is essentially isolating the real solutions to a constant zero-dimensional regular TSA.

A key step for isolating the real solutions to a constant zero-dimensional regular TSA is to isolate the real solutions to a zero-dimensional regular chain. Assume the regular chain is $f_1(x_1), f_2(x_1, x_2), ..., f_s(x_1, ..., x_s)$. A natural idea is: isolate the real roots of f_1 first; then substitute the real roots of f_1 one by one for the x_1 in f_2 and isolate the real roots of resulting polynomials with respect to x_2; and so on. This way, one has to deal with polynomials with algebraic coefficients directly. Our idea is to bound f_2 from above and below by interval arithmetic so that we only deal with polynomials with integer (rational) coefficients.

5.2.1 *Interval Arithmetic*

Interval arithmetic is an important method in the field of numerical computation. For convenience, we introduce here some basic concepts and results of interval arithmetic which may be useful in subsequent subsections. For more details on interval arithmetic and interval algorithms for solving equations, see for example [Alefeld and Herzberger (1983); Moore *et al.* (2009)].

A subset of $\mathbb{R}$ of the form

$$X = [x_1, x_2] = \{x \mid x_1 \le x \le x_2\}, x_1, x_2 \in \mathbb{R}$$

is called an *interval*. The set of all intervals is denoted by $I(\mathbb{R})$. If $x_1 = x_2$, X is called a *point interval*. A subset of $\mathbb{R}$ of the form

$$X = [-\infty, a] = \{x \mid x \le a\}, a \in \mathbb{R}$$

or

$$X = [b, +\infty] = \{x \mid b \le x\}, b \in \mathbb{R}$$

is called a *semi-infinity interval*. The set of all semi-infinity intervals is denoted by $SI(\mathbb{R})$. Note that $I(\mathbb{R})$ and $SI(\mathbb{R})$ are disjoint sets.

Definition 5.1. For $X = [a, b] \in I(\mathbb{R})$, the *width*, the *radius*, the *midpoint* and the *sign* of X are defined, respectively, as $\mathrm{wid}(X) = b - a$, $\mathrm{rad}(X) = \frac{1}{2}\mathrm{wid}(X), \mathrm{mid}(X) = \dfrac{a + b}{2}$ and

$$\mathrm{sign}(X) = \begin{cases} -1, & b < 0; \\ 0, & a \le 0 \le b; \\ 1, & a > 0. \end{cases}$$

Obviously we have $X = [\mathrm{mid}(X) - \mathrm{rad}(X), \mathrm{mid}(X) + \mathrm{rad}(X)]$. An interval is usually expressed by its midpoint and radius. For example, if $m = \mathrm{mid}(X)$, $r = \mathrm{rad}(X)$, then we can write the above formula as $X = \mathrm{midrad}(m, r)$.

A vector is called an *interval vector* if all its components are intervals. *Interval matrix* can be similarly defined. For interval vectors and interval matrices, the concepts such as midpoint, width, radius, etc. and the arithmetic operations are defined in components.

Definition 5.2. For $X, Y \in I(\mathbb{R}) \cup SI(\mathbb{R})$ and $\diamond \in \{+, -, \cdot\}$, define $X \diamond Y = \{x \diamond y \mid x \in X, y \in Y\}$. For $X = [a, b] \in I(\mathbb{R})$, if $\text{sign}(X) \neq 0$, define

$$X^{-1} = 1/X = [1/b, 1/a];$$

if $\text{sign}(X) = 0$ and $\text{wid}(X) \neq 0$, define

$$X^{-1} = 1/X = \begin{cases} [-\infty, 1/a], & b = 0; \\ [1/b, +\infty], & a = 0; \\ [-\infty, 1/a] \cup [1/b, +\infty], & a < 0 < b; \end{cases} \tag{5.1}$$

if $X = [0, 0]$, X^{-1} is undefined. And then Y/X is defined to be $Y \cdot X^{-1}$. Note that

$$Y/X = Y \cdot [-\infty, 1/a] \cup Y \cdot [1/b, +\infty] \quad \text{if} \quad a < 0 < b.$$

For $a \in \mathbb{R}$, $X \in I(\mathbb{R})$ *and* $\diamond \in \{+, -, \cdot, /\}$, define $a \diamond X = [a, a] \diamond X$ and $X \diamond a = X \diamond [a, a]$.

Definition 5.3. Let f be an arithmetic expression of a polynomial in $\mathbb{R}[x_1, ..., x_n]$. We replace all operands of f as intervals and replace all operations of f as interval operations and denote the result by F. Then, the function $F : I(\mathbb{R})^n \to I(\mathbb{R})$ is called an *interval expansion*.

Let F be an interval expansion in $D \in I(\mathbb{R})^n$. If for all $X, Y \subseteq D$, $X \subseteq Y$ implies $F(X) \subseteq F(Y)$, we call F a *monotonic interval expansion*.

Theorem 5.1. *An interval expansion of any polynomial in $\mathbb{R}[x_1, ..., x_n]$ is a monotonic interval expansion.*

5.2.2   *Algorithm*

Obviously, a constant zero-dimensional regular TSA can be viewed as a system in the following form

$$\{f_1 = 0, ..., f_s = 0, g_1 > 0, ..., g_t > 0\} \tag{5.2}$$

where s is the number of variables.

In the following, we only discuss how to isolate the real solutions to a regular TSA in the form of (5.2).

Let a regular TSA T in the form of (5.2) be given. There exist some efficient methods to isolate the real roots of a univariate polynomial (see the first part of this chapter). To isolate the real solutions of T, as stated above, we first isolate the real roots of the first equation of the system and substitute each resulting interval for the variable in the rest of the equations and then repeat the above computation. Of course, we have to deal with polynomials with "interval coefficients".

Definition 5.4. Let a polynomial $q \in \mathbb{Z}[x_1, ..., x_{i+1}]$ be represented as

$$q = q_l(x_1, ..., x_i)x_{i+1}^l + \cdots + q_1(x_1, ..., x_i)x_{i+1} + q_0(x_1, ..., x_i),$$

where $q_l(x_1, ..., x_i) \not\equiv 0$. For any $X = ([a_1, b_1], ..., [a_i, b_i]) \in I(\mathbb{R})^i$, let Q_j ($0 \leq j \leq l$) be an interval expansion of q_j in X and

$$\begin{aligned} Q &= Q_l([a_1, b_1], ..., [a_i, b_i])x_{i+1}^l + ... + Q_0([a_1, b_1], ..., [a_i, b_i]) \\ &= [c_l, d_l]x_{i+1}^l + ... + [c_0, d_0]. \end{aligned}$$

We call

$$ {}_-q = c_l x_{i+1}^l + ... + c_0 \quad \text{and} \quad {}^+q = d_l x_{i+1}^l + ... + d_0 \tag{5.3}$$

the *lower bound polynomial* and *upper bound polynomial* of q in X, respectively.

Let $[q(x)]^{(n)}$ ($n \in \mathbb{N}$) denote the n order derivative of $q(x)$ with respect to x and $[q(x)]^{(0)} = q(x)$.

Proposition 5.1. *Suppose* $X = ([a_1, b_1], ..., [a_i, b_i])$ *is an isolating cube of some zero,* $\boldsymbol{x}_i^*$, *of* $\{f_1 = 0, ..., f_i = 0\}$ *in the system* T *and* $_-f_{i+1}$ *and* $^+f_{i+1}$ *are the lower bound and upper bound polynomials of* f_{i+1} *in* X, *respectively, then for all* $n \in \mathbb{N}$ *and all* $x_{i+1} \in (0, +\infty)$

$$[_-f_{i+1}]^{(n)} \leq [f_{i+1}(\boldsymbol{x}_i^*, x_{i+1})]^{(n)} \leq [\, ^+f_{i+1}]^{(n)}. \tag{5.4}$$

Proof. Suppose $_-f_{i+1} = c_l x_{i+1}^l + ... + c_0$, $^+f_{i+1} = d_l x_{i+1}^l + ... + d_0$, and $f_{i+1}(\boldsymbol{x}_i^*, x_{i+1}) = e_l x_{i+1}^l + ... + e_0$. From Definition 5.4, it is easy to see that $c_j \leq e_j \leq d_j$ for $0 \leq j \leq l$. Therefore, the relations (5.4) hold for all $x_{i+1} \in (0, +\infty)$. $\qquad\square$

In fact, the relations (5.4) hold not only for $\boldsymbol{x}_i^*$ but also for any $\boldsymbol{x}_i \in X$.

Now, suppose $\boldsymbol{x}_i^*$ is a real solution of $\{f_1 = 0, ..., f_i = 0\}$ and X is an isolating cube such that $\boldsymbol{x}_i^* \in X$. Let $_-f_{i+1}(x_{i+1})$ and $^+f_{i+1}(x_{i+1})$ be the lower bound and upper bound polynomials of f_{i+1} in X, respectively. By using

Proposition 5.1, we want to obtain the isolating intervals of $f_{i+1}(\boldsymbol{x}_i^*, x_{i+1})$ by isolating the real roots of $_-f_{i+1}$ and $^+f_{i+1}$.

From Proposition 5.1 (by letting $n = 0$), we have

$$-f_{i+1} \leq f_{i+1}(\boldsymbol{x}_i^*, x_{i+1}) \leq {}^+f_{i+1}$$

for $x_{i+1} > 0$. So, we first shift the real roots of $f_{i+1}(\boldsymbol{x}_i^*, x_{i+1})$ to the positive real roots of $\overline{f_{i+1}}(\boldsymbol{x}_i^*, x_{i+1}) = f_{i+1}(\boldsymbol{x}_i^*, x_{i+1} - B)$, where B satisfies that any real root of $f_{i+1}(\boldsymbol{x}_i^*, x_{i+1})$ is greater than B.

To determine the value of B, we let

$$\widetilde{f_{i+1}}(x_1, ..., x_i, x_{i+1}) = f_{i+1}(x_1, ..., x_i, -x_{i+1}).$$

Then the negative roots of $f_{i+1}(\boldsymbol{x}_i^*, x_{i+1})$ correspond to the positive roots of $\widetilde{f_{i+1}}(\boldsymbol{x}_i^*, x_{i+1})$ and thus the positive-root-bound of $\widetilde{f_{i+1}}(\boldsymbol{x}_i^*, x_{i+1})$ is the negative-root-bound of $f_{i+1}(\boldsymbol{x}_i^*, x_{i+1})$. We shrink X repeatedly until $\mathrm{lc}(_-f_{i+1}) \cdot \mathrm{lc}(^+f_{i+1}) > 0$ (this inequality must hold at a certain step because the TSA T being regular implies $\mathrm{lc}(f_{i+1}, x_{i+1})(\boldsymbol{x}_i^*) \neq 0$) which guarantees that the greatest real root of $f_{i+1}(\boldsymbol{x}_i^*)$ is smaller than that of $_-f_{i+1}(\boldsymbol{x}_i^*)$ or $^+f_{i+1}(\boldsymbol{x}_i^*)$. Then, let $_-\widetilde{f_{i+1}}$ and $^+\widetilde{f_{i+1}}$ be the lower bound and upper bound polynomials of $\widetilde{f_{i+1}}$ in X, respectively and $B > 0$ the maximum of the root-bounds of $_-\widetilde{f_{i+1}}$ and $^+\widetilde{f_{i+1}}$. We define

$$\overline{f_{i+1}}(x_1, ..., x_i, x_{i+1}) = f_{i+1}(x_1, ..., x_i, x_{i+1} - B).$$

Obviously, all the real roots of $f_{i+1}(\boldsymbol{x}_i^*, x_{i+1})$ are shifted to the real roots of $\overline{f_{i+1}}(\boldsymbol{x}_i^*, x_{i+1})$, which are all in $(0, +\infty)$. Therefore, without loss of generality, we only consider the positive roots of $f_{i+1}(\boldsymbol{x}_i^*, x_{i+1})$ in our algorithm.

Suppose

$$S_j = [[\alpha_1^{(j)}, \beta_1^{(j)}], ..., [\alpha_{m_j}^{(j)}, \beta_{m_j}^{(j)}]], \ j = 1, 2,$$

and S_1 and S_2 isolate all positive roots of $_-f_{i+1}(x_{i+1})$ and $^+f_{i+1}(x_{i+1})$, respectively. Because T is regular, $f_{i+1}(\boldsymbol{x}_i^*, x_{i+1}) = 0$ has no repeated roots. So, by Proposition 5.1, if X is small enough, $m_1 = m_2$ and we can define that

$$S = [[\alpha_1, \beta_1], ..., [\alpha_{m_1}, \beta_{m_1}]], \tag{5.5}$$

where for $1 \leq k \leq m_1$,

$$\alpha_k = \min(\alpha_k^{(1)}, \alpha_k^{(2)}), \ \beta_k = \max(\beta_k^{(1)}, \beta_k^{(2)}). \tag{5.6}$$

If the diameters of X, S_1 and S_2 are all small enough, any two adjacent intervals of S do not intersect. Furthermore, $_-f_{i+1}(x_{i+1})$, $^+f_{i+1}(x_{i+1})$ and

f_{i+1} are all monotonic in each $[\alpha_k, \beta_k]$ of S. In this case, we write $S = S_1 \triangle S_2$ and it is easy to prove that S isolates all positive roots of f_{i+1}.

Remark 5.3. Whenever we use the notion $S = S_1 \triangle S_2$ (or $S \leftarrow S_1 \triangle S_2$), we mean that, given S_1 and S_2, S is defined by (5.5) and (5.6) and the following three conditions are satisfied:

(1) $m_1 = m_2$;
(2) $f_{i+1}(x_{i+1})$ is monotonic when x_{i+1} is in each (α_k, β_k);
(3) any two adjacent intervals of S do not intersect.

To make the three conditions hold, we may have to shrink X repeatedly (finitely many times, of course).

Now, we can describe our algorithm as Algorithm 5.4. The termination

---

Algorithm 5.4 RRI-TSA

Input: A regular TSA T in the form of (5.2)
Output: A list of isolating cubes of the positive real solutions to T
 1: $L_1 \leftarrow$ the isolating intervals of the positive roots of f_1; $L_2 \leftarrow \emptyset$;
 2: **if** $L_1 = \emptyset$ **then return** $\emptyset$ **end if**;
 3: **for** $i = 1$ **to** $s - 1$ **do**
 4:     **for** X $= ([a_1, b_1], ..., [a_i, b_i]) \in L_1$ **do**
 5:         Compute $_-f_{i+1}$ and $^+f_{i+1}$ in X;
 6:         $S_1 \leftarrow$ the isolating intervals of the positive roots of $_-f_{i+1}$;
 7:         $S_2 \leftarrow$ the isolating intervals of the positive roots of $^+f_{i+1}$;
 8:         $S \leftarrow S_1 \triangle S_2$;
 9:         $L_2 \leftarrow L_2 \cup \{([a_1, b_1], ..., [a_i, b_i], [c, d]) | [c, d] \in S\}$;
10:     **end for**
11:     **if** $L_2 = \emptyset$ **then**
12:         **return** $\emptyset$;
13:     **else**
14:         $L_1 \leftarrow L_2$; $L_2 \leftarrow \emptyset$;
15:     **end if**
16: **end for**
17: For each X $\in L_1$, compute $G_j(X)$ $(1 \le j \le t)$, where G_j is an interval expansion of g_j. If $\text{sign}(G_{j_0}(X)) < 0$ for some j_0 $(1 \le j_0 \le t)$, delete X from L_1; If $\text{sign}(G_{j_1}(X)) = 0$ for some j_1 $(1 \le j_1 \le t)$, shrink X repeatedly until either $\text{sign}(G_{j_1}(X)) < 0$ or $\text{sign}(G_{j_1}(X)) > 0$.
18: **return** L_1

---

and correctness of the algorithm are guaranteed by the above discussion.

In Line 8 ($S \leftarrow S_1 \triangle S_2$) and Line 17, we may have to shrink X repeatedly. The following sub-algorithm is for this end.

Algorithm 5.5 SHR

Input: A cube $X = ([a_1, b_1], ..., [a_i, b_i])$ and the regular TSA T

Output: A cube $X' \subset X$ such that $x_i^* \in X'$, where $x_i^* = (x_1^*, ..., x_i^*)$ is the only solution of $\{f_1 = 0, ..., f_i = 0\}$ in X

1: By the intermediate value theorem, we obtain an interval $[a_1', b_1'] \subset [a_1, b_1]$ such that $x_1^* \in [a_1', b_1']$ and $\mathrm{wid}([a_1', b_1']) \leq \frac{1}{2}\mathrm{wid}([a_1, b_1])$.

2: $X' \leftarrow ([a_1', b_1'])$;

3: **for** $j = 1$ **to** $i - 1$ **do**

4:    Compute $_-f_{j+1}$ and $^+f_{j+1}$ with respect to X';

5:    By the intermediate value theorem, compute an interval $[a, b] \subset [a_{j+1}, b_{j+1}]$ such that $[a, b]$ contains the root of $_-f_{j+1}$ in $[a_{j+1}, b_{j+1}]$ and $\mathrm{wid}([a, b]) = \frac{1}{8}\mathrm{wid}([a_{j+1}, b_{j+1}])$;

6:    Similarly, compute an interval $[c, d] \subset [a_{j+1}, b_{j+1}]$ such that $[c, d]$ contains the root of $^+f_{j+1}$ in $[a_{j+1}, b_{j+1}]$ and $\mathrm{wid}([c, d]) = \frac{1}{8}\mathrm{wid}([a_{j+1}, b_{j+1}])$;

7:    $a_{j+1}' \leftarrow \min(a, c)$, $b_{j+1}' \leftarrow \max(b, d)$; $X' \leftarrow (X', [a_{j+1}', b_{j+1}'])$;

8: **end for**

9: **return** X'

Let us prove the correctness of the algorithm **SHR**. Only the loop of **SHR** needs some further description. Let us denote $_-f_{j+1}$ and $^+f_{j+1}$ with respect to X' by $_-f_{j+1}(X')$ and $^+f_{j+1}(X')$, respectively. By Definition 5.4 and Theorem 5.1, the following relations hold for all $x_{j+1} \in (0, +\infty)$,

$$_-f_{j+1}(X) \leq_- f_{j+1}(X') \leq f_{j+1}(x_1^*, ..., x_j^*, x_{j+1}) \leq {}^+f_{j+1}(X') \leq {}^+f_{j+1}(X),$$

$$_-f_{j+1}'(X) \leq_- f_{j+1}'(X') \leq f_{j+1}'(x_1^*, ..., x_j^*, x_{j+1}) \leq {}^+f_{j+1}'(X') \leq {}^+f_{j+1}'(X).$$

On the other hand, from Algorithm **RRI–TSA**, $_-f_{j+1}(X)$ is monotonic on $[a_{j+1}, b_{j+1}]$ and has only one root in it. So does $^+f_{j+1}(X)$. Then, by the above relations, $_-f_{j+1}(X')$ and $^+f_{j+1}(X')$ are both monotonic on $[a_{j+1}, b_{j+1}]$ and each of them has only one root in the interval. The correctness of Algorithm **SHR** is thus proved.

Remark 5.4. In the loop of Algorithm **SHR**, we use an empirical factor $1/8$. Theoretically speaking, the factor can be any rational number between zero and one.

5.2.3 *Examples*

Algorithm RRI-TSA has been implemented as a function in our program DISCOVERER, which has been integrated into the Maple package RegularChains since Maple 13. The calling sequence of the function in Maple for an SAS in the form of (3.1) is:

$$\text{RealRootIsolate}(P,\ G_1,\ G_2,\ H,\ R,\ method = \ 'Discoverer');$$

where R is the ordered variables. For more details of RealRootIsolate, see the Maple help page.

Example 5.1. [Xia and Yang (2002)]

$$\begin{cases} p_1 = x^2 + y^2 - xy - 1 = 0, \\ p_2 = y^2 + z^2 - yz - a^2 = 0, \\ p_3 = z^2 + x^2 - zx - b^2 = 0, \\ p_4 = a^2 - 1 + b - b^2 = 0, \\ p_5 = 3b^6 + 56b^4 - 122b^3 + 56b^2 + 3 = 0, \\ x > 0, y > 0, z > 0, a - 1 \geq 0, b - a \geq 0, a + 1 - b > 0, \end{cases}$$

which is a special case of the "P3P" problem in computer vision.

To isolate the real roots of the above SAS by Maple, we may execute the following:

```
> with(RegularChains):
> with(SemiAlgebraicSetTools):
> p1 := x^2+y^2-x*y-1:
> p2 := y^2+z^2-y*z-a^2:
> p3 := z^2+x^2-z*x-b^2:
> p4 := a^2-1+b-b^2:
> p5 := 3*b^6+56*b^4-122*b^3+56*b^2+3:
> R := PolynomialRing([b,a,x,y,z]):
> P := [p1,p2,p3,p4,p5]:
> G1 := [a-1,b-a]: G2 := [x,y,z,a+1-b]: H := [ ]:
> RealRootIsolate(P, G1, G2, H, R, method='Discoverer');
```

The program outputs an isolating box:

$$b = [1, \tfrac{11}{8}], a = [\tfrac{9}{8}, \tfrac{5}{4}], x = [\tfrac{9}{8}, \tfrac{5}{4}], y = [\tfrac{20379}{32768}, \tfrac{326397}{524288}],$$
$$z = [\tfrac{2373099311743 0801}{18014398509481984}, \tfrac{47461986234861603}{36028797018963968}].$$

The running time on a laptop (2.26Ghz CPU, 2G memory, windows 7, Maple 17) is 1.37 seconds.

By the function, we computed all the six examples in [Xia and Yang (2002)]. For readers' convenience, we list the six systems in the following but refer the reader to [Xia and Yang (2002)] for detail.

Example 5.2. (Chemical Reaction)

$$\begin{cases} h_1 = 2 - 7x_1 + x_1^2 x_2 - \dfrac{1}{2}(x_3 - x_1) = 0, \\[2mm] h_2 = 6x_1 - x_1^2 x_2 - 5(x_4 - x_2) = 0, \\[2mm] h_3 = 2 - 7x_3 + x_3^2 x_4 - \dfrac{1}{2}(x_1 - x_3) = 0, \\[2mm] h_4 = 6x_3 - x_3^2 x_4 + 1 + \dfrac{1}{2}(x_2 - x_4) = 0. \end{cases}$$

4 real solutions, 0.11 seconds.

Example 5.3. (Neural Network)

$$\begin{cases} f_1 = 1 - cx - xy^2 - xz^2 = 0, \\ f_2 = 1 - cy - yx^2 - yz^2 = 0, \\ f_3 = 1 - cz - zx^2 - zy^2 = 0, \\ f_4 = 8c^6 + 378c^3 - 27 = 0, \\ c > 0, 1 - c > 0. \end{cases}$$

4 real solutions, 1.045 seconds.

Example 5.4. (Cyclic 5)

$$\begin{cases} p_1 = a + b + c + d + e = 0, \\ p_2 = ab + bc + cd + de + ea = 0, \\ p_3 = abc + bcd + cde + dea + eab = 0, \\ p_4 = abcd + bcde + cdea + deab + eabc = 0, \\ p_5 = abcde - 1 = 0. \end{cases}$$

10 real solutions, 1.17 seconds.

Example 5.5.

$$\begin{cases} p_1 = 2x_1(2 - x_1 - y_1) + x_2 - x_1 = 0, \\ p_2 = 2x_2(2 - x_2 - y_2) + x_1 - x_2 = 0, \\ p_3 = 2y_1(5 - x_1 - 2y_1) + y_2 - y_1 = 0, \\ p_4 = y_2(3 - 2x_2 - 4y_2) + y_1 - y_2 = 0, \\ x_1 \geq 0, x_2 \geq 0, y_1 \geq 0, y_2 \geq 0. \end{cases}$$

4 real solutions, 0.171 seconds.

Example 5.6. (solving geometric constraints)

$$\begin{cases} f_1 = 1/100 - 4s(s-1)(s-b)(s-c) = 0, \\ f_2 = 1/5 - bc = 0, \\ f_3 = 2s - 1 - b - c = 0, \\ b > 0, c > 0, b + c - 1 > 0, 1 + c - b > 0, 1 + b - c > 0. \end{cases}$$

4 real solutions, 0.063 seconds.

5.3 Real Root Counting for Constant Semi-Algebraic Systems

As in the last section, we assume any given SAS has only a finitely many real solutions. For a given SAS S, if we use `RealRootIsolate`, we may obtain the number of distinct real solutions to S. In this section, we introduce a different algorithm [Xia and Hou (2002)] to determine the number of distinct real solutions to a given constant SAS, which is based on real root isolation for polynomials and triangularization of SAS.

By the method of Chapter 3 (see Remark 3.2), any given constant zero-dimensional SAS can be triangularized into a finite set of pairwise disjoint square-free zero-dimensional regular TSAs. So, by counting distinct real solutions to every TSA in the set, we can get the number of distinct real solutions to the original SAS. Therefore we only discuss how to count the real solutions to a given zero-dimensional regular constant TSA in this section.

Suppose $T : [F, [\ ], [G], [\ ]]$ is a zero-dimensional regular constant TSA, where

$$F = \{f_1(x_1), f_2(x_1, x_2), ..., f_s(x_1, ..., x_s)\}, G = \{g_1(x_1, ..., x_s), ..., g_t(x_1, ..., x_s)\}.$$

Define

$$\begin{aligned} Bf_2 &= \operatorname{res}(f_2, f_2', x_2), \\ Bf_i &= \operatorname{res}(\operatorname{res}(f_i, f_i', x_i); f_{i-1}, f_{i-2}, \ldots, f_2), \text{ for } i > 2, \\ Bg_j &= \operatorname{res}(g_j; f_s, f_{s-1}, \ldots, f_2), \text{ for } 1 \le j \le t, \end{aligned}$$

and

$$Bfg = \prod_{2 \le i \le s} Bf_i \cdot \prod_{1 \le j \le t} Bg_j.$$

Clearly, Bfg is a polynomial in x_1 and

$$\operatorname{res}(Bfg(x_1), f_1(x_1), x_1) \ne 0$$

because T is regular.

Definition 5.5. Let notations be as above. Suppose all the distinct real roots of $f_1(x_1)$ are $\alpha_1, \ldots, \alpha_k$. If a list of real numbers, $[r_1, \ldots, r_k]$, satisfies that for any $i(1 \leq i \leq k)$,

1) if $s > 1$, the system

$$\begin{cases} f_2(\alpha_i, x_2) = 0, \\ \cdots, \\ f_s(\alpha_i, x_2, ..., x_s) = 0, \\ g_1(\alpha_i, x_2, ..., x_s) > 0, ..., g_t(\alpha_i, x_2, ..., x_s) > 0, \end{cases}$$

and the system

$$\begin{cases} f_2(r_i, x_2) = 0, \\ \cdots, \\ f_s(r_i, x_2, ..., x_s) = 0, \\ g_1(r_i, x_2, ..., x_s) > 0, ..., g_t(r_i, x_2, ..., x_s) > 0, \end{cases}$$

have the same number of distinct real solutions and,

2) if $s = 1$, $\mathrm{sgn}(g_j(\alpha_i)) = \mathrm{sgn}(g_j(r_i))$ for any $g_j(1 \leq j \leq t)$,

then $[r_1, \ldots, r_k]$ is called a list of *near roots* of $f_1(x_1)$ with respect to T. If $f_1(x_1)$ has no real roots, the list of *near roots* is defined to be the empty list $[\,]$.

It is easy to prove

Theorem 5.2. *Algorithm 5.6 is correct.*

Proof. Note that every output number r_i corresponds to a root α_i of $p(x)$ such that they are both in one of those $k + 1$ intervals. Then, it is obvious that $\mathrm{sgn}(q(r_i)) = \mathrm{sgn}(q(\alpha_i))$. $\qquad\qquad\square$

Theorem 5.3. *Let notations be as above. The output of*

$$\mathtt{NearRoots}(f_1, BP(x_1))$$

is a list of near roots of $f_1(x_1)$ with respect to system T.

Proof. If $s = 1$, the conclusion is obvious. So, suppose $s > 1$. Because system T is regular, f_1 has no common roots with $Bfg(x_1)$. So, by $\mathtt{NearRoots}(f_1, Bfg(x_1))$, we get a list of rational numbers $[r_1, \ldots, r_k]$. We prove that this is indeed a list of near roots of $f_1(x_1)$ with respect to system T.

Algorithm 5.6 NearRoots

Input: two polynomials $p(x), q(x) \in \mathbb{Z}[x]$ which have no common roots
Output: a list of near roots of $p(x)$ w.r.t. system $\{p(x) = 0, q(x) > 0\}$

1: If $p(x)$ has no real roots, return an empty list $[\,]$. If $q(x)$ has no real roots and $p(x)$ has u distinct real roots, return a list $[0, \ldots, 0]$ in which the number of 0 is u.

2: By RRI-TwoPoly(q, p) (see Remark 5.1), compute a real root isolation, say $[a_1, b_1], \ldots, [a_k, b_k]$, of $q(x)$ such that for any $i(1 \leq i \leq k)$ and any $\alpha \in [a_i, b_i]$, $p(\alpha) \neq 0$.

3: Now we've got a sequence of intervals

$$(-\infty, a_1), \ldots, (b_i, a_{i+1}), \ldots, (b_k, \infty).$$

For every interval in this sequence, determine the number of distinct real roots in it of $p(x)$. If in $(-\infty, a_1)$, $p(x)$ has $u_0(> 0)$ distinct roots, choose u_0 rational numbers from $(-\infty, a_1)$, say $a_1 - 1$ (u_0 times). If in (b_i, a_{i+1}), $p(x)$ has $u_i(> 0)$ distinct roots, choose u_i rational numbers $(b_i + a_{i+1})/2$. If in (b_k, ∞), $p(x)$ has $u_k(> 0)$ distinct roots, choose u_k rational numbers $b_k + 1$. Finally, sort and return all the numbers we have chosen.

For any $i_0(1 \leq i_0 \leq k)$, we know from Step 3 in Algorithm NearRoots that α_{i_0} and r_{i_0} both lie between some two consecutive roots (say, β_{i_0} and β_{i_0+1}) of $Bfg(x_1) = \prod_{2 \leq i \leq s} Bf_i \cdot \prod_{1 \leq j \leq t} Bg_j$. Clearly, the sign of each Bf_i and Bg_j is invariant on the interval $(\beta_{i_0}, \beta_{i_0+1})$.

First of all, $Bf_2 = \mathrm{res}(f_2, f_2', x_2) \neq 0$ on the interval $(\beta_{i_0}, \beta_{i_0+1})$ implies the number of distinct real roots of f_2 is invariant on $(\beta_{i_0}, \beta_{i_0+1})$. Furthermore, $Bf_3 \neq 0$ on $(\beta_{i_0}, \beta_{i_0+1})$ implies, if $f_2 = 0$, $\mathrm{res}(f_3, f_3', x_3) \neq 0$ on $(\beta_{i_0}, \beta_{i_0+1})$, which means the number of distinct real solutions of equations $\{f_2 = 0, f_3 = 0\}$ is invariant on $(\beta_{i_0}, \beta_{i_0+1})$. Continuing similar discussions, we get that the number of distinct real solutions of equations $\{f_2 = 0, \ldots, f_s = 0\}$ is invariant on $(\beta_{i_0}, \beta_{i_0+1})$. Secondly, $Bg_j \neq 0$ on $(\beta_{i_0}, \beta_{i_0+1})$ implies, if $\{f_2 = 0, \ldots, f_s = 0\}$, $g_j \neq 0$ on $(\beta_{i_0}, \beta_{i_0+1})$, which means the number of distinct real solutions of system T without f_1 is invariant on $(\beta_{i_0}, \beta_{i_0+1})$. That completes the proof. $\qquad\square$

Theorem 5.4. *Algorithm 5.7 is correct.*

Proof. If equations $\{f_1 = 0, \ldots, f_s = 0\}$ has no real solutions, pts_s is obviously an empty list $[\,]$. If equations $\{f_1 = 0, \ldots, f_s = 0\}$ have k

---

Algorithm 5.7 `RealRootCount`

Input:　A zero-dimensional regular constant TSA T as above

Output:　the number of distinct real solutions of system T

1:　$i \leftarrow 1;\ rs_1 \leftarrow \text{NearRoots}(f_1(x_1), Bfg(x_1));$

2:　**if** $rs_1 = [\,]$ **then**

3:　　$pts_1 \leftarrow [\,];$

4:　**else**

5:　　Assume $rs_1 = [r_{11}, \ldots, r_{k_1 1}]$. Let $pts_1 \leftarrow [[r_{11}], \ldots, [r_{k_1 1}]];$

6:　**end if**

7:　**for** i from 1 to $s - 1$ **do**

8:　　**if** $pts_i = [\,]$ **then break; end if**

9:　　Assume $pts_i = [[r_{11}, \ldots, r_{1i}], \ldots, [r_{k_i 1}, \ldots, r_{k_i i}]].$

10:　　**for** j from 1 to k_i **do**

11:　　　Substitute $x_1 = r_{j1}, \ldots, x_i = r_{ji}$ into system T in which $f_1, \ldots, f_i$ have been deleted. For this new system, by **NearRoots** we get a list of near roots of $f_{i+1}(x_{i+1})$, say rs^j_{i+1}. If $rs^j_{i+1} = [\,]$, let $pts^j_{i+1} \leftarrow [\,]$. Otherwise, if $rs^j_{i+1} = [r_1, \ldots, r_{l_j}]$, let

$$pts^j_{i+1} \leftarrow [[r_{j1}, \ldots, r_{ji}, r_1], \ldots, [r_{j1}, \ldots, r_{ji}, r_{l_j}]];$$

12:　　**end for**

13:　　Let $pts_{i+1} \leftarrow \bigcup_{1 \leq j \leq k_i} pts^j_{i+1};$

14:　**end for**

15:　**if** $pts_s = [\,]$ **then return** 0; **end if**

16:　Assume

$$pts_s = [v_1, \ldots, v_k] = [[r_{11}, \ldots, r_{1s}], \ldots, [r_{k1}, \ldots, r_{ks}]].$$

If all $g_j > 0$ hold at u members of pts_s, return u.

---

distinct real solutions, $[w_1, \ldots, w_k]$, then by Theorem 5.3, pts_s must be

$$[v_1, \ldots, v_k] = [[r_{11}, \ldots, r_{1s}], \ldots, [r_{k1}, \ldots, r_{ks}]],$$

in which

$$\forall i(1 \leq i \leq k)\ \forall j(1 \leq j \leq t)\ \text{sgn}(g_j(w_i)) = \text{sgn}(g_j(v_i)).$$

That ends the proof.　　　　　　　　　　　　　　　　　　　　　　$\square$

Based on Algorithm 5.7, we develop a function in our tool DISCOVERER [Xia (2000, 2007)] (see Section 6.4) using Maple which may count the real solutions to any zero-dimensional constant SAS. The tool DISCOVERER has been integrated into Maple package `RegularChains` since Maple 13. The function in `RegularChains` is `RealRootCounting`.

Example 5.7.

$$\begin{cases} p_1(x,y,z) = x^2 + y^2 - xy - 1 = 0, \\ p_2(x,y,z) = y^2 + z^2 - yz - 4 = 0, \\ p_3(x,y,z) = 100z^2 + 100x^2 - 100zx - 441 = 0, \\ x > 0, y > 0, z > 0. \end{cases}$$

To count the real solutions to the above system by Maple, we may execute the following:

```
> with(RegularChains):
> with(SemiAlgebraicSetTools):
> p1 := x^2+y^2-x*y-1:
> p2 := y^2+z^2-y*z-4:
> p3 := 100*z^2+100*x^2-100*z*x-441:
> R := PolynomialRing([x,y,z]):
> P := [p1,p2,p3]:
> G1 := [ ]: G2 := [x,y,z]: H := [ ]:
> RealRootCounting(P, G1, G2, H, R);
      The answer is 1.
```

5.4 Termination of Linear Programs

(Tiwari, 2004) proved that the termination of a class of linear programs is decidable. The decision procedure proposed therein depends on the computation of *Jordan forms*. Thus, one may draw a wrong conclusion from this procedure, if they simply apply floating-point computation to compute Jordan forms. (Xia *et al.*, 2011) proposed a symbolic implementation of the decision procedure based on real root counting and isolation for SAS. In this section, we first use an example to explain the problem caused by floating-point computation and then present one special result in [Xia *et al.* (2011)].

(Tiwari, 2004) proved that the termination of the following loops on the reals is decidable.

$$P_1: \qquad \textbf{while} \quad (Bx > b) \quad \{x \leftarrow Ax + c\},$$

where A is an $n \times n$ matrix, B is an $m \times n$ matrix, and x, b and c are vectors. $Bx > b$ is a conjunction of strict linear inequalities which is the loop condition, while $x \leftarrow Ax + c$ is interpreted as updating the values of x by $Ax + c$ simultaneously and not in any sequential order. We say P_1 *terminates* if it terminates on all initial values.

The termination problem of P_1 is reduced to that of the following homogeneous loop in [Tiwari (2004)]

$$\mathsf{P}_2 : \qquad \textbf{while} \quad (Bx > 0) \quad \{\, x \leftarrow Ax \,\}.$$

A key step of the decision procedure in [Tiwari (2004)] is to compute the *Jordan form* of the matrix A so that one can have a diagonal description of A^n. In [Tiwari (2004)] it was proved that if the Jordan form of A is $A^* = Q^{-1}AQ$ and set $B^* = BQ$, then P_2 terminates if and only if

$$\mathsf{P}_2^* : \qquad \textbf{while } (B^*x > 0) \{\, x \leftarrow A^*x \,\}$$

terminates. This idea is natural, but, if we use floating-point computation routines to calculate the Jordan form in a conventional way, the errors of floating-point computation may lead to a wrong conclusion. To see this, let us consider the following example.

Example 5.8. Let

$$A = \begin{bmatrix} 2 & -3 \\ -1 & 2 \end{bmatrix}, \qquad B = \begin{bmatrix} 1 & b \\ -1 & b \end{bmatrix},$$

where

$$b = -\frac{1127637245}{651041667} = -\sqrt{3} + \epsilon \approx -1.732050807$$

with $\epsilon = \sqrt{3} - \dfrac{1127637245}{651041667} > 0$. Determine whether

$$\mathsf{Q}_1 : \qquad \textbf{while } (Bx > 0) \{\, x \leftarrow Ax \,\}$$

is terminating.

According to the conventional method, in order to compute the Jordan form of A we have to calculate the eigenvalues of A by using floating-point computation, say, through the package **LinearAlgebra** in Maple. The approximate eigenvalues of A are 3.732050808 and 0.267949192 (both take 10 decimal digits of precision). Hence, the Jordan form of A is

$$A^* = Q^{-1}AQ = \begin{bmatrix} 3.732050808 & 0 \\ 0 & 0.267949192 \end{bmatrix}$$

where

$$Q = \begin{bmatrix} 0.5 & 0.5 \\ -0.2886751347 & 0.2886751347 \end{bmatrix}.$$

Use the same package of Maple to calculate

$$B^* = BQ = \begin{bmatrix} 1.0 & 0.0 \\ 0.0 & -1.0 \end{bmatrix}.$$

Then, the loop Q_1 is terminating if and only if the following loop Q_2 terminates,

$$\mathsf{Q}_2: \qquad \textbf{while } (B^*\boldsymbol{x} > 0) \ \{\boldsymbol{x} \leftarrow A^*\boldsymbol{x}\}.$$

Obviously, $(A^*)^n = \begin{bmatrix} 3.732050808^n & 0 \\ 0 & 0.267949192^n \end{bmatrix}$. If we let $\boldsymbol{x} = [1, -1]^{\mathrm{T}}$, after n times of iteration,

$$(A^*)^n\boldsymbol{x} = [3.732050808^n, -0.267949192^n],$$

where $\boldsymbol{v}^{\mathrm{T}}$ stands for the transpose of the vector $\boldsymbol{v}$. And the loop condition is

$$B^*(A^*)^n\boldsymbol{x} = [3.732050808^n, 0.267949192^n] > [0, 0]$$

which is always true for all n. Therefore, Q_1 is not terminating.

However, this conclusion is not correct. Let us see how the floating-point computation leads us to the wrong result. The Jordan form of A is indeed (by symbolic computation)

$$J = P^{-1}AP = \begin{bmatrix} 2 + \sqrt{3} & 0 \\ 0 & 2 - \sqrt{3} \end{bmatrix},$$

where

$$P = \begin{bmatrix} \dfrac{1}{2} & \dfrac{1}{2} \\ -\dfrac{1}{6}\sqrt{3} & \dfrac{1}{6}\sqrt{3} \end{bmatrix}$$

and, in order to obtain B^*, we should compute BP instead of BQ symbolically as

$$BP = \begin{bmatrix} \dfrac{1}{2} - \dfrac{b}{6}\sqrt{3} & \dfrac{1}{2} + \dfrac{b}{6}\sqrt{3} \\ -\dfrac{1}{2} - \dfrac{b}{6}\sqrt{3} & -\dfrac{1}{2} + \dfrac{b}{6}\sqrt{3} \end{bmatrix} = \begin{bmatrix} 1 - \dfrac{\epsilon}{6}\sqrt{3} & \dfrac{\epsilon}{6}\sqrt{3} \\ -\dfrac{\epsilon}{6}\sqrt{3} & -1 + \dfrac{\epsilon}{6}\sqrt{3} \end{bmatrix} = \begin{bmatrix} m_{11} & m_{12} \\ m_{21} & m_{22} \end{bmatrix}.$$

Therefore $m_{12} > 0$, $m_{21} < 0$. However, when we use floating-point computation, these two entries (m_{12} and m_{21}) are both 0 (in Maple with Digits $= 10$). That is why we obtain wrong result by floating-point computation.

One may guess that if we evaluate BP rather than BQ through floating-point computation routines, we may obtain a more precise approximation.

Unfortunately, it is not true. In fact using floating-point computation to compute BP, we will still get some strange results. For example, computing BP by Maple with Digits $= 10$ outputs the following matrix

$$\begin{bmatrix} 1.0 & -1.0 \cdot 10^{-10} \\ 1.0 \cdot 10^{-10} & -1.0 \end{bmatrix}.$$

It is totally wrong, because, comparing with the signs of m_{12} and m_{21} in the above, m_{12} is negative and m_{21} positive.

To handle the above problem, (Xia *et al.*, 2011) developed a symbolic decision procedure for the termination of linear programs P_1. The general framework of the procedure is quite similar to that of [Tiwari (2004)], but we re-implement the two key steps, *i.e.* computing Jordan normal form of A and generating linear constraints. In Tiwari's decision procedure, the two steps are implemented numerically while the new procedure is based on symbolic computation and thus can avoid errors caused by floating-point computation. We do not elaborate the details of the procedure here but introduce a concise result of [Xia *et al.* (2011)] on the case when the characteristic polynomial of A is irreducible, which uses real root counting (isolation) as a tool for termination analysis of linear programs defined above.

Theorem 5.5. *[Tiwari (2004)] If the program P_2 is nonterminating, then there must be a real eigenvector v of A corresponding to a positive eigenvalue such that $Bv \geq 0$.*

A direct corollary of Theorem 5.5 is

Corollary 5.1. *Assume that for every real eigenvector v of A corresponding to a positive eigenvalue, every element of Bv is not zero. Then, program P_2 is nonterminating if and only if there is a real eigenvector v of A corresponding to a positive eigenvalue such that $Bv > 0$.*

Theorem 5.6. *Suppose A and B are both matrices with rational entries and the characteristic polynomial $D(\lambda)$ of A is irreducible in $\mathbb{Q}[\lambda]$. The program P_2 is nonterminating if and only if there is a real eigenvector v of A corresponding to a positive eigenvalue such that $Bv > 0$.*

Proof. The sufficiency is obvious, so we only prove the necessity. The irreducibility of the characteristic polynomial $D(\lambda)$ of A implies that the eigenvalues of A are pairwise distinct. Otherwise, set

$$D_1 = \gcd(D(\lambda), \frac{\mathrm{d}}{\mathrm{d}\lambda}D(\lambda)), \ D_2 = \frac{D(\lambda)}{\gcd(D(\lambda), \frac{\mathrm{d}}{\mathrm{d}\lambda}D(\lambda))},$$

then $D(\lambda) = D_1 D_2$ is reducible over $\mathbb{Q}$.

Suppose $\lambda_1, \ldots, \lambda_n$ are the eigenvalues of A. Set

$$A(\lambda) = A - \lambda I = \begin{bmatrix} a_{11} - \lambda & a_{12} & \cdots & a_{1n} \\ a_{21} & a_{22} - \lambda & \cdots & a_{2n} \\ \vdots & \vdots & \cdots & \vdots \\ a_{n1} & a_{n2} & \cdots & a_{nn} - \lambda \end{bmatrix},$$

and denote the (i, j) algebraic complement minor of $A(\lambda)$ by $A_{ij}(\lambda)$ for $i = 1, \ldots, n$ and $j = 1, \ldots, n$.

Obviously, $A_{11}(\lambda)$ is a nonzero polynomial in λ of degree $n - 1$ because its leading monomial is $(-\lambda)^{n-1}$. Then for any eigenvalue λ_β $(\beta = 1, \ldots, n)$, $A_{11}(\lambda_\beta) \neq 0$ since λ_β is a root of $D(\lambda)$ which is an irreducible polynomial of degree n. For each $\beta = 1, \ldots, n$, set

$$\boldsymbol{v}_\beta = [A_{11}(\lambda_\beta), A_{12}(\lambda_\beta), \ldots, A_{1n}(\lambda_\beta)]^{\mathrm{T}}.$$

It is clear that each $\boldsymbol{v}_\beta$ is nonzero and $A\boldsymbol{v}_\beta - \lambda_\beta \boldsymbol{v}_\beta = A(\lambda_\beta)\boldsymbol{v}_\beta = \boldsymbol{0}$. Thus $\boldsymbol{v}_\beta$ is exactly the unique eigenvector (up to a scale multiplier) of A, related to λ_β.

If P_2 is nonterminating, by Theorem 5.5, there exists a real eigenvector $\boldsymbol{v}_\alpha$ of A corresponding to a positive eigenvalue such that $B\boldsymbol{v}_\alpha \geq 0$. That is to say, if we denote the kth row of B by $\boldsymbol{b}_k = (b_{k1}, \ldots, b_{kn})$ and set $u_k = \boldsymbol{b}_k \cdot \boldsymbol{v}_\alpha$, then $u_k \geq 0$ $(k = 1, \ldots, m)$. We need only to show that each u_k is nonzero. Set

$$\boldsymbol{v}(\lambda) = [A_{11}(\lambda), A_{12}(\lambda), \cdots, A_{1n}(\lambda)]^{\mathrm{T}},$$

and $u_k(\lambda) = B_k \cdot \boldsymbol{v}(\lambda)$, $(k = 1, \ldots, m)$. Then $u_k = u_k(\lambda_\alpha)$.

Note that, unless $u_k(\lambda)$ is a zero polynomial, $u_k(\lambda)$ does not equal to zero at λ_α because it is a polynomial in λ of degree at most $n - 1$ but λ_α is a root of an irreducible polynomial of degree n. We continue to show that $u_k(\lambda)$ is not a zero polynomial.

If some $u_k(\lambda)$ is a zero polynomial, then for all $\beta = 1, \ldots, n$, $u_k(\lambda_\beta) = 0$, *i.e.* $\boldsymbol{b}_k \cdot \boldsymbol{v}_1 = 0$, $\boldsymbol{b}_k \cdot \boldsymbol{v}_2 = 0, \ldots, \boldsymbol{b}_k \cdot \boldsymbol{v}_n = 0$. Thus $\boldsymbol{v}_1, \ldots, \boldsymbol{v}_n$ are linear dependent. However, these eigenvectors must be linear independent because their eigenvalues are pairwise distinct. This is a contradiction. $\square$

Corollary 5.2. *If A and B are both matrices on a field of numbers (e.g., the second extension of the field of rational numbers) and the characteristic polynomial of A is irreducible on this field, then the program P_2 is nonterminating if and only if there exists a real eigenvector $\boldsymbol{v}$ of A, corresponding to a positive eigenvalue λ, such that $B\boldsymbol{v} > 0$. In other words, if and only if*

$$\exists \lambda \exists \boldsymbol{v}(\lambda \in \mathbb{R}^+ \wedge \boldsymbol{v} \in \mathbb{R}^n \wedge A\boldsymbol{v} = \lambda \boldsymbol{v} \wedge B\boldsymbol{v} > 0). \tag{5.7}$$

Based on Theorem 5.6, we design the following Algorithm `TermIrr` to determine the termination of P_2 with the assumption that the characteristic polynomial of A is irreducible.

Algorithm 5.8 `TermIrr`

Input: an $n \times n$ matrix A and an $m \times n$ matrix B of linear program P_2 where the characteristic polynomial of A is irreducible

Output: whether or not P_2 is terminating

1: Compute the characteristic polynomial of A and denote it by $D(\lambda)$.
2: Compute the algebraic complement minor of every element in the first (or a fixed) row of $A - \lambda I$ (the characteristic matrix of A) and denote them by A_{1i} $(1 \leq i \leq n)$, respectively.
3: For each row of B, compute $u_j = \sum_{k=1}^{n} b_{jk} A_{1k}$ $(1 \leq j \leq m)$.
4: Construct a semi-algebraic system

$$S : \{D(\lambda) = 0, \ \lambda > 0, \ u_1 u_2 > 0, u_2 u_3 > 0, \ldots, u_{m-1} u_m > 0\}.$$

5: Determine whether S has real solutions. If yes, return "nonterminating". Otherwise, return "terminating".

It is easy to see that S is another representation of formula (5.7) in terms of semi-algebraic system. So, the correctness of Algorithm `TermIrr` is guaranteed by Theorem 5.6 (or Corollary 5.2).

We demonstrate how to use Algorithm `TermIrr` to determine the termination of the loop in Example 5.8. It is easy to compute that

$$D(\lambda) = \lambda^2 - 4\lambda + 1,$$
$$v = (A_{11}, \ A_{12})^{\mathrm{T}} = (2 - \lambda, \ 1)^{\mathrm{T}},$$
$$Bv = (u_1, \ u_2) = (2 - \lambda + b, \ \lambda - 2 + b),$$

and $D(\lambda)$ is irreducible. Then we use `RealRootCounting` to determine whether the following system has real solutions

$$\{D(\lambda) = 0, \ \lambda > 0, \ (2 - \lambda + b)(\lambda - 2 + b) > 0\}.$$

We get that the number of real solutions of the above system is 0. Thus the loop in Example 5.8 terminates.

Example 5.9. Consider the termination of the program

$$\textbf{while} \ \ (Bx > 0) \ \ \{x \leftarrow Ax\},$$

where

$$A = \begin{bmatrix} 3\ 1\ 4\ 1\ 5 \\ 9\ 2\ 6\ 5\ 3 \\ 5\ 8\ 9\ 7\ 9 \\ 3\ 2\ 3\ 8\ 4 \\ 6\ 2\ 6\ 4\ 3 \end{bmatrix}, \quad B = \begin{bmatrix} 3\ -8\ 3\ 2\ -7 \\ 1\ -4\ 1\ 4\ -2 \\ 4\ -2\ 8\ -5\ 7 \end{bmatrix}.$$

The characteristic polynomial of A is irreducible. By Algorithm **TermIrr**, we first compute the algebraic complement minors as follows

$$A_{11}(\lambda) = -48 + 313\lambda + 8\lambda^2 - 22\lambda^3 + \lambda^4,$$
$$A_{12}(\lambda) = 381 + 243\lambda - 117\lambda^2 + 9\lambda^3,$$
$$A_{13}(\lambda) = 74 - 539\lambda + 82\lambda^2 + 5\lambda^3,$$
$$A_{14}(\lambda) = 144 - 60\lambda + 15\lambda^2 + 3\lambda^3,$$
$$A_{15}(\lambda) = -498 + 204\lambda - 54\lambda^2 + 6\lambda^3.$$

Construct $\boldsymbol{v}(\lambda) = (A_{11}(\lambda), \ldots, A_{15}(\lambda))^{\mathrm{T}}$, where λ is an eigenvalue of A. Compute the u_j in Algorithm **TermIrr** as follows

$$\begin{aligned} u_1 &= 3A_{11}(\lambda) - 8A_{12}(\lambda) + 3A_{13}(\lambda) + 2A_{14}(\lambda) - 7A_{15}(\lambda) \\ &= 804 - 4170\lambda + 1614\lambda^2 - 159\lambda^3 + 3\lambda^4, \\ u_2 &= A_{11}(\lambda) - 4A_{12}(\lambda) + A_{13}(\lambda) + 4A_{14}(\lambda) - 2A_{15}(\lambda) \\ &= 74 - 1846\lambda + 726\lambda^2 - 53\lambda^3 + \lambda^4, \\ u_3 &= 4A_{11}(\lambda) - 2A_{12}(\lambda) + 8A_{13}(\lambda) - 5A_{14}(\lambda) + 7A_{15}(\lambda) \\ &= -4568 - 1818\lambda + 469\lambda^2 - 39\lambda^3 + 4\lambda^4. \end{aligned}$$

By Theorem 5.6, the program is nonterminating if and only if the following semi-algebraic system has real solutions

$$\{D(\lambda) = 0,\ \lambda > 0,\ u_1 u_2 > 0,\ u_2 u_3 > 0\},$$

where $D(\lambda)$ is the characteristic polynomial of A.

By calling **RealRootCounting**, we can conclude that the system has no real solutions. Therefore, the program is terminating. If we delete a constraint and set

$$B = \begin{bmatrix} 3\ -8\ 3\ 2\ -7 \\ 1\ -4\ 1\ 4\ -2 \end{bmatrix},$$

by calling **RealRootCounting**, we can conclude that the system $\{D(\lambda) = 0,\ \lambda > 0,\ u_1 u_2 > 0\}$ has 2 distinct real solutions. That is to say, the resulting program is nonterminating.

One may ask whether the assumption in Theorem 5.6 ("the characteristic polynomial of A is irreducible") can be removed or weakened like "the characteristic polynomial of A is square-free". The following example gives a negative answer.

Example 5.10. Given a program as follows

$$Q_4: \quad \textbf{while} \quad (Bx > 0) \quad \{\, x \leftarrow Ax \,\}$$

$$\text{where} \quad A = \begin{bmatrix} 4 & 5 & 2 \\ 9 & -1 & -8 \\ 3 & 2 & 3 \end{bmatrix}, \quad B = \begin{bmatrix} 7 & -7 & -6 \\ 1 & 5 & 1 \end{bmatrix}.$$

The characteristic polynomial of A is

$$\lambda^3 - 6\lambda^2 - 30\lambda + 161 = (\lambda - 7)(\lambda^2 + \lambda - 23)$$

which is square-free with eigenvalues $\lambda_1 = 7, \lambda_2 = -\frac{1}{2} + \frac{\sqrt{93}}{2}$ and $\lambda_3 = -\frac{1}{2} - \frac{\sqrt{93}}{2}$. The corresponding eigenvectors are

$$v_1 = [8,\ 2,\ 7]^{\mathrm{T}}, v_2 = [14,\ -17 + \sqrt{93},\ 11 + \sqrt{93}]^{\mathrm{T}},$$
$$v_3 = [14,\ -17 - \sqrt{93},\ 11 - \sqrt{93}]^{\mathrm{T}},$$

where v_1, v_2 correspond to positive eigenvalues λ_1 and λ_2, respectively. Note that the characteristic polynomial of A is square-free and A is non-singular. It is easy to check that neither $(Bv_1 > 0) \wedge (Bv_2 > 0)$ nor $(-Bv_1 > 0) \wedge (-Bv_2 > 0)$ holds. So we may draw a conclusion that the program terminates. Unfortunately the conclusion is wrong, and we can prove the program is nonterminating by the algorithm given in [Xia *et al.* (2011)].

Chapter 6

Real Root Classification

In this chapter, we discuss the so-called *real root classification* problem of parametric SAS in the form of (3.1), *i.e.* $S : [P, G_1, G_2, H]$,

$$\begin{cases} p_1(\boldsymbol{u}, x_1, \ldots, x_n) = 0, \ldots, p_s(\boldsymbol{u}, x_1, \ldots, x_n) = 0, \\ g_1(\boldsymbol{u}, x_1, \ldots, x_n) \geq 0, \ldots, g_r(\boldsymbol{u}, x_1, \ldots, x_n) \geq 0, \\ g_{r+1}(\boldsymbol{u}, x_1, \ldots, x_n) > 0, \ldots, g_t(\boldsymbol{u}, x_1, \ldots, x_n) > 0, \\ h_1(\boldsymbol{u}, x_1, \ldots, x_n) \neq 0, \ldots, h_m(\boldsymbol{u}, x_1, \ldots, x_n) \neq 0. \end{cases}$$

Herein, $n, s \geq 1$, $r, t, m \geq 0$, p_i, g_j, $h_k \in \mathbb{Q}[\boldsymbol{u}, x_1, \ldots, x_n]$ and $\boldsymbol{u} = (u_1, \ldots, u_d)$ are real parameters.

We are interested in the following questions:

1. What is the condition on $\boldsymbol{u}$ such that S has real solutions?
2. What is the condition on $\boldsymbol{u}$ such that S has positive dimensional real solutions? What is the dimension of the real solutions?
3. What is the condition on $\boldsymbol{u}$ such that S has a prescribed number of (distinct) real solutions?

The answers to the above questions, *i.e.* the conditions we compute, are called the *real root classification* (shortly, RRC) of the system S. It is not hard to see that finding the answers to the above questions 1 and 3 is a special kind of quantifier elimination problems.

Our idea is to make use of the "hierarchical strategy" introduced in Chapter 3 to solve the problem step by step. By Algorithm GRSD in Chapter 3, we first triangularize the system S in $\mathbb{Q}(\boldsymbol{u})[x_1, \ldots, x_n]$. Suppose

$$[\mathcal{TH}, p] = \text{GRSD}(S, \boldsymbol{x}).$$

According to the result of the triangularization, the system S may be in three cases: (A) generic zero-dimensional (every regular set in $\mathcal{TH}$ has n polynomials); (B) generic positive dimensional (some regular sets in $\mathcal{TH}$

133

have less than n polynomials); or (C) generic no solution ($\mathcal{TH}$ is empty). We discuss real root classification in the three cases, respectively. Case (A) is a basic case.

6.1   Border Polynomial and Discrimination Polynomial

Let $Q = \{q_i(\boldsymbol{u}) \in \mathbb{Z}[u_1, \ldots, u_d] \mid 1 \leq i \leq l\}$ be a non-empty set of finitely many nonzero polynomials in parameters. Set $q = \prod_{i=1}^{l} q_i$. For every i ($1 \leq i \leq l$) and every connected component C of $q \neq 0$ in $\mathbb{R}^d$, q_i is sign-invariant in C and the sign is not zero. For any $u \in C$,

$$[\operatorname{sgn}(q_1(u)), \ldots, \operatorname{sgn}(q_l(u))]$$

is called the *sign* of C (with respect to Q or q). Obviously, every connected component of $q \neq 0$ has a unique sign but two different connected components may have the same sign. So, the sign of a component is usually not a defining formula of the component.

Naturally, a sign of a component can be viewed as a first order formula expressed by the polynomials in Q. For example, $[1, -1, 1]$ corresponds to $q_1 > 0 \wedge q_2 < 0 \wedge q_3 > 0$.

Definition 6.1. Suppose S is a parametric SAS and a paremetric polynomial $q(\boldsymbol{u}) = \prod_{i=1}^{l} q_i(\boldsymbol{u})$ satisfies

(a) S has only finitely many real solutions at the parametric values in $\mathbb{R}^d$ such that $q(\boldsymbol{u}) \neq 0$, and
(b) the number of distinct real solutions of S is constant in each connected component of $q(\boldsymbol{u}) \neq 0$ in $\mathbb{R}^d$.

Then $q(\boldsymbol{u})$ is called a *border polynomial* of S. If $q(\boldsymbol{u})$ satisfies an additional condition:

(c) for any two connected components C_1, C_2 of $q(\boldsymbol{u}) \neq 0$ in $\mathbb{R}^d$ with the same sign (with respect to $q(\boldsymbol{u})$), the numbers of real solutions of S in C_1 and C_2 are equal,

then $q(\boldsymbol{u})$ is called a *discrimination polynomial* of S.

The following theorem is clear by the above definition.

Theorem 6.1. *Suppose* $q(\boldsymbol{u}) = \prod_{i=1}^{l} q_i(\boldsymbol{u})$ *and we only consider the parameters in* $\mathbb{R}^d$ *such that* $q(\boldsymbol{u}) \neq 0$.

If $q(\boldsymbol{u})$ is a border polynomial of S, then a necessary condition for S to have N (a prescribed nonnegative integer) distinct real solutions can be expressed by the signs of the factors (i.e. q_i) of $q(\boldsymbol{u})$.

If $q(\boldsymbol{u})$ is a discrimination polynomial of S, then the necessary and sufficient condition for S to have N distinct real solutions can be expressed by the signs of the factors of $q(\boldsymbol{u})$.

Proof.   Suppose $C_1, \ldots, C_k$ are all the connected components of $q(\boldsymbol{u}) \neq 0$ in $\mathbb{R}^d$, in which S has exactly N distinct real solutions. Then

$$\mathrm{sgn}(C_1) \vee \mathrm{sgn}(C_2) \vee \cdots \vee \mathrm{sgn}(C_k)$$

is the condition we want.   □

Theorem 6.1 suggests a possible way for computing real root classification of a parametric SAS, *i.e.* construct border polynomial and discrimination polynomial of the SAS.

6.2   Generic Zero-Dimensional Case

6.2.1   *Regular Zero-Dimensional TSA*

Let us first consider the simplest case: regular zero-dimensional TSAs. Recall that, by Proposition 3.1, any regular zero-dimensional TSA can be transformed equivalently to $[F, [\,], G, [\,]]$. To be concrete, we assume it is of the form

$$\begin{cases} f_1(\boldsymbol{u}, x_1) = 0, \ldots, f_s(\boldsymbol{u}, x_1, \ldots, x_s) = 0, \\ g_1(\boldsymbol{u}, x_1, \ldots, x_s) > 0, \ldots, g_t(\boldsymbol{u}, x_1, \ldots, x_s) > 0. \end{cases} \tag{6.1}$$

By Theorem 3.1, we immediately have the following result.

Theorem 6.2. *Suppose T is a regular zero-dimensional TSA of the form (6.1), then $\mathrm{BP}(\mathrm{T})$ defined by Definition 3.1, i.e.*

$$\mathrm{BP}(\mathrm{T}) = \prod_{1 \leq i \leq s} \mathrm{BP}_{f_i} \cdot \prod_{1 \leq j \leq t} \mathrm{BP}_{g_j}$$

is a border polynomial of T defined by Definition 6.1.

Next, we discuss how to construct the discrimination polynomial for a regular zero-dimensional TSA T.

Let $A = \{A_i | 1 \leq i \leq l\}$ be a non-empty set of finitely many nonzero polynomials. Define

$$\mathrm{mset}(A) = \{1\} \cup \{A_{i_1} A_{i_2} \cdots A_{i_k} | 1 \leq k \leq l, 1 \leq i_1 < i_2 < \cdots < i_k \leq l\}.$$

It is easy to see that

$$\mathrm{mset}(A) = \left\{ \ \prod_{i=1}^{l} A_i^{a_i} \ | \ a_i \in \{0,1\} \ \right\}.$$

So, $\mathrm{mset}(A)$ has 2^l polynomials and every polynomial in $\mathrm{mset}(A)$ can be denoted simply by A^a where $a = (a_1, \ldots, a_l) \in \{0,1\}^l$.

Definition 6.2. Given a regular zero-dimensional TSA T in the form of (6.1), define

$$\mathbf{P}_{s+1} = \{g_1, g_2, \ldots, g_t\};$$
$$\mathbf{P}_i = \bigcup_{q \in \mathrm{mset}(\mathbf{P}_{i+1})} \mathrm{GDL}(f_i, q), \quad i = s, \ldots, 1.$$

Recall that $\mathrm{GDL}(f,g)$ is the discriminant sequence of f with respect to g (see Section 4.2). Sometimes we use $\mathbf{P}_1(g_1, \ldots, g_t)$ instead of $\mathbf{P}_1$ to indicate the inequalities. Denote by $\mathrm{DP}(\mathrm{T})$ or DP the product of all the polynomials in $\mathbf{P}_1$.

Obviously, $\mathrm{BP}(\mathrm{T})$ is a factor of $\mathrm{DP}(\mathrm{T})$ and therefore $\mathrm{DP}(\mathrm{T})$ is also a border polynomial (*i.e.* $\mathrm{DP}(\mathrm{T})$ satisfies the conditions (a) and (b) of Definition 6.1). We prove that it also satisfies the condition (c). That is, $\mathrm{DP}(\mathrm{T})$ is the discrimination polynomial of T.

Suppose $Q = \{q_1(x), \ldots, q_l(x)\}$ is a set of nonzero univariate polynomials with real coefficients and $f(x)$ is a univariate polynomial with real coefficients, which has no common roots with any polynomial in Q. By Theorem 4.6 (Discrimination Theorem I), for any polynomial q, the signs of the elements in $\mathrm{GDL}(f,q)$ can determine the value of

$$n(f,q) = f_{q+} - f_{q-}$$

which is the difference of the number of real roots of f such that $q > 0$ and that of real roots of f such that $q < 0$.

We denote by

$$Q_\sigma \ (\sigma \in \Sigma = \{1, -1\}^l)$$

a list of signs of the polynomials in Q. For example, assume $l = 3$, then $Q_{(1,1,-1)}$ stands for $q_1 > 0, q_2 > 0, q_3 < 0$. Denote by f_σ

$$\mathrm{card}(\{x \in \mathbb{R} \ | \ f(x) = 0 \ \wedge \ Q_\sigma\}),$$

i.e. the number of real roots of $f(x)$ such that the condition Q_σ holds.

Set $1 \prec -1$ and define the lexicographic order on $\Sigma = \{1, -1\}^l$ accordingly. Then

$$(1, 1, \ldots, 1) \prec (-1, 1, \ldots, 1) \prec \cdots \prec (-1, -1, \ldots, -1).$$

Define the following vector according to the lexicographic order on Σ

$$f_\Sigma = (f_{\sigma_1}, \ldots, f_{\sigma_{2^l}})^{\mathrm{T}},$$

where $\sigma_1 \prec \cdots \prec \sigma_{2^l}$.

Similarly, stipulate $q_1 \prec q_2 \prec \cdots \prec q_l$ and define a lexicographic order on $\mathrm{mset}(Q)$. Then

$$1 \prec q_1 \prec q_2 \prec \cdots \prec \prod_{i=1}^{l} q_i,$$

or, equivalently

$$(0, 0, \ldots, 0) \prec (1, 0, \ldots, 0) \prec (0, 1, 0, \ldots, 0) \prec \cdots \prec (1, 1, \ldots, 1).$$

Define the following vector according to the lexicographic order

$$n(f, Q) = (n(f, 1), n(f, q_1), \ldots, n(f, \prod_{i=1}^{l} q_i))^{\mathrm{T}}.$$

Given a condition $\sigma \in \Sigma$, every polynomial $Q^a \in \mathrm{mset}(Q)$ has a fixed sign, denoted by $\mathrm{sgn}(Q^a_\sigma)$. Under the order defined above, we construct a $2^l \times 2^l$ matrix $M^Q = (m_{ij})$ as follows.

$$m_{ij} = \mathrm{sgn}(Q^{a_i}_{\sigma_j}).$$

For example, assume $l = 1$,

$$f_\Sigma = \begin{pmatrix} f_1 \\ f_{-1} \end{pmatrix}, \quad n(f, Q) = \begin{pmatrix} n(f, 1) \\ n(f, q_1) \end{pmatrix}, \quad M^Q = \begin{pmatrix} 1 & 1 \\ 1 & -1 \end{pmatrix}.$$

Note that, if $l = 1$, $M^Q \cdot f_\Sigma = n(f, Q)$ by Theorems 4.6 and 4.7 (Discrimination Theorems I, II). Actually, the fact holds for general cases.

Theorem 6.3. *Let notations be as above. Then $M^Q \cdot f_\Sigma = n(f, Q)$.*

Proof. According to the definitions of M^Q, f_Σ and $n(f, Q)$, the theorem obviously holds. $\square$

The equation in Theorem 6.3 can be viewed as a linear equation where f_Σ are unknowns. Note that every element in $n(f, Q)$ can be computed based on Discrimination Theorems. Therefore, to solve the linear equations, we have to prove that the matrix M^Q is invertible.

Definition 6.3. Let $M = (m_{ij})$ and $M' = (m'_{ij})$ be $n \times m$ and $n' \times m'$ matrices, respectively. The *tensor product* $M \otimes M'$ is defined as an $nn' \times mm'$ matrix $[m'_{ij}M]$.

Set

$$M_1 = \begin{pmatrix} 1 & 1 \\ 1 & -1 \end{pmatrix},$$

then

$$M_1 \otimes M_1 = \begin{pmatrix} M_1 & M_1 \\ M_1 & -M_1 \end{pmatrix} = \begin{pmatrix} 1 & 1 & 1 & 1 \\ 1 & -1 & 1 & -1 \\ 1 & 1 & -1 & -1 \\ 1 & -1 & -1 & 1 \end{pmatrix}.$$

We define inductively that

$$M_{j+1} = M_j \otimes M_1.$$

Proposition 6.1. *Let notations be as above. Then* $M^Q = M_l$.

Proof. By induction on l. First, if $l = 1$, it is clear that $M^Q = M_1$. Assume that the conclusion holds for $l = j$.

If $l = j$, M^Q is a $2^j \times 2^j$ matrix, whose rows are labeled by $Q^{a_1}, \ldots, Q^{a_{2^j}} \in \mathrm{mset}(Q)$ and whose columns are labeled by $\sigma_1, \ldots, \sigma_{2^j} \in \Sigma$. According to the orders on $\mathrm{mset}(Q)$ and Σ, if $l = j + 1$, the matrix M^Q is a $2^{j+1} \times 2^{j+1}$ matrix, whose rows are labeled by

$$Q^{a_1}, \ldots, Q^{a_{2^j}}, Q^{a_1} \cdot q_{j+1}, \ldots, Q^{a_{2^j}} \cdot q_{j+1}$$

and whose columns are labeled by

$$(\sigma_1, 1), \ldots, (\sigma_{2^j}, 1), (\sigma_1, -1), \ldots, (\sigma_{2^j}, -1).$$

Herein, $(\sigma_i, 1)$ (or $(\sigma_i, -1)$) stands for a new vector by appending a new component 1 (or -1) to σ_i.

On the other hand,

$$M_{j+1} = M_j \otimes M_1 = \begin{pmatrix} M_j & M_j \\ M_j & -M_j \end{pmatrix}. \tag{6.2}$$

Therefore, by the definition of M^Q, the above matrix is just M^Q. That completes the proof. $\qquad\square$

Proposition 6.2. M^Q *is invertible.*

Proof. Use induction on l. If $l = 1$, $M^Q = M_1$ is obviously invertible. By (6.2), M_{j+1} is invertible if M_j is invertible. $\qquad\square$

Corollary 6.1. *For any $\sigma \in \Sigma$, the value of f_σ can be determined by the signs of the elements in $\mathrm{GDL}(f, q)$ where $q \in \mathrm{mset}(Q)$.*

Now, we can prove the following theorem.

Theorem 6.4. *Suppose* T *is a regular zero-dimensional TSA of the form (6.1). Then* DP(T) *is the discrimination polynomial of* T.

Proof. First, BP(T) is a factor of DP(T). Thus the conditions (a) and (b) of Definition 6.1 hold. Viewing f_s and each g_i as polynomials in x_s, by Corollary 6.1, we know that the number of real zeros of $f_s = 0$ such that $\{g_i > 0 | 1 \leq i \leq t\}$ can be determined by the signs of polynomials in $\mathbf{P}_s$. Then, viewing f_{s-1} and polynomials in $\mathbf{P}_s$ as polynomials in x_{s-1}, by Corollary 6.1 again, we know that the signs of polynomials in $\mathbf{P}_s$ can be determined by the signs of polynomials in $\mathbf{P}_{s-1}$ under the condition that $f_{s-1} = 0$. In other words, the number of real solutions to $\{f_s = 0, f_{s-1} = 0\}$ such that $\{g_i > 0 | 1 \leq i \leq t\}$ can be determined by the signs of polynomials in $\mathbf{P}_{s-1}$. Continuing similar deduction, we know that the number of real solutions to T is determined by the signs of polynomials in $\mathbf{P}_1$. $\square$

Remark 6.1. Theorem 6.4 was introduced in [Xia (1998); Yang *et al.* (2001)]. The proof here was given by Rong Xiao and the first author of this book. The method used in the proof was first proposed in [Ben-Or *et al.* (1986)]. See also [Basu *et al.* (2003)].

Remark 6.2. Another method for constructing DP of an SAS was given in [Xiao (2009)], based on triangularization and Thom's lemma.

6.2.2 *Generic Zero-Dimensional SAS*

Now, Suppose S is a generic zero-dimensional SAS. First, let

$$[\mathcal{T}, p_1] = \mathtt{GRDforZDSAS}(S, \boldsymbol{x}),$$

i.e. compute a generic regular decomposition of S by Algorithm 3.2. Then, let

$$[\mathcal{T}', p_2] = \mathtt{Disjoint}(\mathcal{T}, \boldsymbol{x}),$$

i.e. make the TSAs in $\mathcal{T}$ pairwise disjoint by Algorithm 3.7. Define

$$\mathrm{BP}(S) = p_1 p_2 \prod_{\mathrm{T} \in \mathcal{T}'} \mathrm{BP}(\mathrm{T}). \tag{6.3}$$

By Theorem 6.2, Theorem 3.3 and the specification of Algorithm 3.7, we have the following

Theorem 6.5. *Suppose* S *is a generic zero-dimensional SAS and* BP(S) *is defined as above by (6.3). Then,* BP(S) *is a border polynomial of* S.

Furthermore, for any $a \in \mathbb{R}^d \setminus V^u(\text{BP}(S))$, the number of distinct real solutions to S at a is the sum of the numbers of distinct real solutions to all the TSAs in $\mathcal{T}'$ at a.

6.2.3 *Algorithm*

By Theorem 6.5, if we can compute the real root classification (RRC) for regular zero-dimensional TSAs, we can certainly compute the RRC for any generic zero-dimensional SAS. So, essentially, we only need to discuss algorithms for computing RRC of regular zero-dimensional TSAs.

Algorithm 6.1 tofind

Input: A regular zero-dimensional TSA T in the form of (6.1) and a nonnegative integer N

Output: $[\Phi, p]$, the necessary and sufficient condition, Φ, for T to have exactly N distinct real solutions provided that $p \neq 0$

1: $Poly \leftarrow \text{BP}(T)$ where $\text{BP}(T)$ is defined by Definition 3.1;

2: **for** i from 1 to t **do**

3:    Compute at least one sample point in every connected component of $Poly \neq 0$;

4:    For each connected component C (sample point s_C), substitute s_C for the parameters in T, denoted by $\text{T}(s_C)$, which is a constant regular zero-dimensional TSA. Compute the number of distinct real solutions to $\text{T}(s_C)$ by Algorithm 5.4 and compute the sign of C with respect to $Poly$, denoted by Φ_C. Set

$$set_1 \leftarrow \{\Phi_C \mid \text{T has exactly } N \text{ distinct real solutions in } C\},$$

$$set_0 \leftarrow \{\Phi_C \mid \text{T does not have exactly } N \text{ distinct real solutions in } C\}.$$

5:    **if** $set_1 \cap set_0 = \emptyset$ **then break; end if**

6:    $Poly \leftarrow Poly \cdot \mathbf{P}_1(g_1, \ldots, g_i)$;

7: **end for**

8: Assume $set_1 = \{\Phi_{C_1}, \ldots, \Phi_{C_m}\}$. Let $\Phi \leftarrow \Phi_{C_1} \vee \cdots \vee \Phi_{C_m}$.

9: **return** $[\Phi, Poly]$

By Theorems 6.2 and 6.4, we may come up with an incremental algorithm: First, test whether the signs of the factors of BP can determine the RRC; if not, then multiply BP by some factors of DP and test again; repeat the procedure until we obtain the RRC. The algorithm is formally described as Algorithm 6.1.

Remark 6.3. Line 3 of Algorithm 6.1 can be accomplished by the algorithm of Cylindrical Algebraic Decomposition (CAD) [Collins (1975)].

Remark 6.4. The termination and correctness of Algorithm 6.1 are guaranteed by Theorems 6.2 and 6.4.

In many applications, we do not care the case when parameters are on a closed set. So, the output of Algorithm `tofind` is enough. However, to obtain a complete RRC, we need to discuss the case when parameters are on the "boundary" (*i.e.* $Poly = 0$).

Suppose $[\Phi, Poly] = \texttt{tofind}(\text{T}, N)$ for a regular zero-dimensional TSA T and $R = R(u_1, \ldots, u_d)$ is a factor of $Poly$. We want to compute the necessary and sufficient condition for T to have exactly N distinct real solutions when $R = 0$.

Remark 6.5. In Algorithm 6.2, without loss of generality, we assume that u_1 appears in $R(\boldsymbol{u})$. For convenience, we also assume that the new system $\{\text{T}, R = 0\}$, viewed as a system in variable $(u_1, x_1, \ldots, x_s)$, is still a regular zero-dimensional TSA. Actually, if it is not generic zero-dimensional, we will discuss how to deal with the case in the next section. Otherwise, it can be decomposed into regular zero-dimensional TSAs by Algorithm 3.2.

Remark 6.6. Suppose $[\Psi, p^*] = \texttt{Tofind}(\text{T}, R(\boldsymbol{u}), N)$. We obtain a new boundary $p^* = 0$ in $\mathbb{R}^{d-1}$. So, if we repeat similar procedure, finally we can obtain a complete RRC for T. From the viewpoint of geometry, the hierarchical strategy works like this: First, we obtain a variety V_d in $\mathbb{R}^d$ and compute the number of real solutions of the system in every (open) connected component of the complement of V_d in $\mathbb{R}^d$; Then, we obtain a new variety $V_{d-1} \subset V_d$ in $\mathbb{R}^{d-1}$ and compute the number of real solutions of the system in every (open) connected component of the complement of V_{d-1} in $\mathbb{R}^{d-1}$ restricted to V_d; and so on. Finally, we get a decomposition of $\mathbb{R}^d$ defined by those open components with different dimensions (from d to 1) and finite points of dimension 0. Moreover, the RRC at all these steps altogether form a complete RRC for the original system.

The following simple example illustrates how to use Algorithms `tofind` and `Tofind` to solve problems. For more examples, see the last several sections of this chapter.

Algorithm 6.2 Tofind

Input: A regular zero-dimensional TSA T in the form of (6.1), a parametric polynomial $R(\boldsymbol{u})$ and a nonnegative integer N

Output: $[\Psi, p^*]$, the necessary and sufficient condition, Ψ, for T to have exactly N distinct real solutions under the condition that $R(\boldsymbol{u}) = 0$ and $p^* \neq 0$

1: Add the equation $R = 0$ into T and denote the new system by TR; (By Remark 6.5, TR is viewed as a regular zero-dimensional TSA with parameters $(u_2, \ldots, u_d)$ and variables $(u_1, \boldsymbol{x})$.)

2: $Poly \leftarrow \mathrm{BP}(TR)$;

3: **for** i from 1 to t **do**

4:      Compute at least one sample point in each connected component of $Poly \neq 0$ in $\mathbb{R}^{d-1}$;

5:      Let $S' \leftarrow \emptyset$. Substitute each sample point s_C for $(u_2, \ldots, u_d)$ in $R = 0$. Assume the real roots of $R(s_C) = 0$ are $a_1 < \cdots < a_{k_C}$. Add every (a_i, s_C) $(1 \leq i \leq k_C)$ to S'.

6:      Substitute every $(a_j, s_C) \in S'$ for $\boldsymbol{u}$ in T, denoted by $\mathrm{T}(a_j, s_C)$. Compute the number of distinct real solutions to $\mathrm{T}(a_j, s_C)$ and the sign of C with respect to $Poly$, denoted by Ψ_C. Replace Ψ_C with (Ψ_C, j) according to every (a_j, s_C). Set

$$set_1 \leftarrow \{(\Psi_C, j)|\ \mathrm{T}\ \text{has exactly}\ N\ \text{distinct real solutions at}\ (a_j, s_C)\},$$

$$set_0 \leftarrow \{(\Psi_C, j)|\ \mathrm{T}\ \text{does not have exactly}\ N\ \text{distinct real solutions at}\ (a_j, s_C)\}.$$

7:      **if** $set_1 \cap set_0 = \emptyset$ **then break; end if**

8:      $Poly \leftarrow Poly \cdot \mathbf{P}_1(g_1, \ldots, g_i)$ where $\mathbf{P}_1(g_1, \ldots, g_i)$ is defined with respect to TR;

9: **end for**

10: Assume $set_1 = \{(\Psi_{C_1}, j_1), \ldots, (\Psi_{C_m}, j_m)\}$. Then let $\Psi \leftarrow (\Psi_{C_1}, j_1) \vee \cdots \vee (\Psi_{C_m}, j_m)$, where (Ψ_{C_i}, j_i) means that $(u_2, \ldots, u_d)$ satisfies Ψ_{C_i} and u_1 is the j_ith real root of $R = 0$ when $(u_2, \ldots, u_d)$ is specified.

11: **return** $[\Psi, Poly]$

Example 6.1. [Brown and McCallum (2005)]

$$(\exists x)(\exists y)[f = g = 0 \wedge y \neq 0 \wedge xy - 1 < 0], \tag{6.4}$$

where $f = x^3 - 3xy^2 + ax + b$, $g = 3x^2 - y^2 + a$.

This is equivalent to finding the condition on a, b such that the system

$$f = g = 0 \wedge y \neq 0 \wedge xy - 1 < 0$$

has real solutions. By Algorithm 3.2, the system is transformed into a regular zero-dimensional TSA:

$$f_1 = 8x^3 + 2ax - b = 0, f_2 = 3x^2 - y^2 + a = 0, y \neq 0, xy - 1 < 0.$$

The RDU for this decomposition is 1.

We then compute the border polynomial

$$r_1 = \mathrm{discrim}(f_1) = 4a^3 + 27b^2, \ \ r_2 = \mathrm{res}(\mathrm{discrim}(f_2), f_1, x) = r_1,$$

and

$$\begin{aligned}
q_1 &= \mathrm{res}(y; f_2, f_1) = r_1, \\
q_2 &= \mathrm{res}(xy - 1; f_2, f_1) \\
&= -4a^3 b^2 - 27b^4 + 16a^4 + 512a^2 + 4096.
\end{aligned}$$

Therefore BP $= r_1 q_2$ (after square-free). The output of Algorithm **tofind** is that the system has real solutions if and only if $r_1 > 0$ provided that $r_1 \neq 0$ and $q_2 \neq 0$.

We use Algorithm **Tofind** to discuss the cases when a, b are on $r_1 = 0$ or $q_2 = 0$ separately. Adding $r_1 = 0$ to the system, by Algorithm **Tofind**, we know that the system has no real solutions if $b \neq 0$. Then adding $b = 0, r_1 = 0$ to the original system and, by Algorithm **Tofind**, we know that the system has no real solutions. In other words, the system has no real solutions if $r_1 = 0$. Similarly, we can discuss the case that $q_2 = 0$. Finally, we get that the system has real solutions if and only if (or (6.4) is equivalent to)

$$(q_2 = 0) \vee (q_2 \neq 0 \wedge r_1 > 0).$$

Moreover, we can easily verify by Algorithm **tofind** that $q_2 = 0$ implies $r_1 > 0$. So the condition can be simplified as $r_1 > 0$.

Remark 6.7. By the definition of DP, it usually contains lots of factors. So, in Algorithms **tofind** and **Tofind**, we may have many different strategies to add "boundaries" to border polynomial. According to our experiments, we use the following strategy in our implementations.

Let $A = \{A_i | 1 \leq i \leq l\}$ be a non-empty set of finitely many nonzero polynomials. Define $A^{(0)} = \{1\}$ and

$$A^{(k)} = A^{(k-1)} \cup \{A_{i_1} A_{i_2} \cdots A_{i_k} | 1 \leq i_1 < i_2 < \cdots < i_k \leq l\}, \ 1 \leq k \leq l.$$

Obviously, $\mathrm{mset}(A) = A^{(l)}$.

Given a regular zero-dimensional TSA T in the form of (6.1) and a positive integer k, define

$$\mathbf{P}_{s+1} = \{g_1, g_2, \ldots, g_t\};$$
$$\mathbf{P}_i^{(k)} = \bigcup_{q \in \mathbf{P}_{i+1}^{(k)}} \mathrm{GDL}(f_i, q), \quad \text{for } i = s, \ldots, 1.$$

Herein $\mathbf{P}_i^{(k)}$ is the set of polynomials in $\mathrm{GDL}(f_i, q)$ where $q \in \mathbf{P}_{i+1}^{(k)}$. It is easy to see that $\mathrm{BP} \subset \mathbf{P}_s^{(1)}$, and $\mathbf{P}_1^{(k)} = \mathbf{P}_1$ if k is large enough.

The initial value of *Poly* in Algorithms `tofind` and `Tofind` is set to BP. Then, if necessary, it is multiplied by the factors of $\mathbf{P}_1^{(1)}, \mathbf{P}_1^{(2)}, \ldots$, successively. From lots of examples we computed, we find that BP is enough to express the necessary and sufficient condition for many practical systems and most of the examples are solved by adding factors of $\mathbf{P}_1^{(1)}$. On the other hand, if we add too many factors to *Poly*, the computation may become very time-consuming. So, how to find as less necessary "boundaries" (polynomials) as possible is an important problem because the necessary and sufficient condition is usually expressed by a few polynomials.

6.3   Positive Dimensional and Over-Determined Cases

As stated at the beginning of this chapter, parametric SASs are divided into three groups: (A) generic zero-dimensional; (B) generic positive dimensional; and (C) generic no solution. We discuss on how to compute RRC in Cases (B) and (C) in this section.

First, we consider Case (B). Let S : $[P, G_1, G_2, H]$ be a generic positive dimensional SAS in the form of (3.1). Step 1, we triangularize the system by Algorithm 3.9 (`GRSD`, generic regular system decomposition), *i.e.* `GRSD`$(P, \boldsymbol{x})$ will lead to a triangular decomposition $\mathcal{TH}$ of S with corresponding RDU p. Because the system S is generic positive dimensional, there exists an ascending chain $F_i : [f_1, \ldots, f_l]$ of some TSA $\mathrm{T}_i : [F_i, G_1, G_2, H]$ in $\mathcal{TH}$ such that $l < n$.

Definition 6.4. Suppose $F = [f_1, \ldots, f_l]$ is a triangular set in $\mathcal{K}[x_1, \ldots, x_n]$. The *dimension* of F is defined to be $n - l$.

Note that the dimension of F is not the dimension of $\mathcal{V}(F)$. For example,

$$F = [x(x - 1), x(y + z)]$$

has dimension $3 - 2 = 1$. However, $V(F) = V([x]) \cup V([x - 1, y + z])$ and thus $V(F)$ has dimension 2.

Because the variety of a regular chain is unmixed (equi-dimensional), we only need to consider the regular chain in $\mathcal{TH}$ with maximal dimension.

Step 2. Each TSA $T_i : [F_i, G_1, G_2, H]$ in $\mathcal{TH}$ is not necessarily regular because F_i may not be coprime with respect to the inequalities. However, it is not hard to modify Algorithms 3.3-3.5 a little so that each T_i can be transformed equivalently into regular TSA with some RDU recording possible "bad" parametric values. Therefore, without loss of generality, we may assume that each T_i is regular and the polynomial p is an RDU for the triangularization.

We want to know the dimension of the real solutions of T_i (and thus S). Suppose a regular set T has s polynomials with dimension $k = n - s > 0$. For general parametric values, the dimension of the complex solutions of T, $V_{\mathbb{C}}(T)$, is k. But, the case of real solution is very different. For example, $V_{\mathbb{C}}(x^2 + y^2 + z^2)$ has dimension 2 while $V_{\mathbb{R}}(x^2 + y^2 + z^2)$ has only one point.

To compute the RRC for a TSA T with dimension $k > 0$, *i.e.* to determine whether T has real solutions and the dimension of real solutions, we take the following "hierarchical strategy": At the top level, we view T as a regular chain in mvar(T) and the other k variables (say $x_1, \ldots, x_k$) as parameters. Then T is a regular zero-dimensional TSA (Case (A)) and we can use Algorithms `tofind` and `Tofind` to compute the condition for T to have (or have no) real solutions.

If we view T as a conjunction of first-order formulas, then we are considering the following QE problem

$$\exists x_1 \cdots \exists x_n(T) \quad \text{or} \quad \forall x_1 \cdots \forall x_n(\neg T),$$

where $\neg T$ means the system T has no real solutions.

The two problems are similar. For convenience, we only take the former for example. Suppose, viewing $x_1, \ldots, x_k$ and $\boldsymbol{u}$ as parameters, we obtain

$$[\Phi, \mathrm{BP}] = \texttt{tofind}(T, 1..k),$$

where $1..k$ means the system has at least one real solution. That is to say, under the condition $\mathrm{BP} \neq 0$, we eliminate the last $n - k$ quantifiers of $(\exists x_1) \cdots (\exists x_n)(T)$ and obtain

$$(\exists x_1) \cdots (\exists x_k)\Phi. \tag{6.5}$$

If Φ is identically true, then the dimension of real solution of T is k for general parameters. If Φ is identically false, then the dimension of real solution of T is generally less than k. Otherwise, we have to perform QE on the formula (6.5) to obtain condition on the parameters $\boldsymbol{u}$. This can be done by CAD [Collins (1975)]. Suppose Ψ is a quantifier-free formula equivalent to (6.5). Note that Φ is actually the signs of some open cells in $\mathbb{R}^{k+d}$, *i.e.* the atom formulas in Φ are of the form $R > 0$ or $R < 0$. So, if there exist $x_1, \ldots, x_k$ satisfying Φ, then $x_1, \ldots, x_k$ in an open set of $\mathbb{R}^k$ also satisfy Φ. Therefore Ψ is the necessary and sufficient condition for T to have k-dimensional real solutions provided that BP $\neq 0$.

For the next level, we consider the case that parameters are on some "boundary". Suppose BP $= p_1 \cdots p_l$ is the irreducible factorization of BP. We may add $p_i = 0$ to the system T one by one. If some $x_1, \ldots, x_k$ appear in p_i, then the dimension of the complex solution to the new system must be less than k. If the dimension of real solution at preceding level is k (Φ is identically true or Ψ is not identically false), then the new system with $p_i = 0$ need no more consideration because we are interested in the maximal dimension of real solutions. If, at preceding level, Φ is identically false or Ψ is identically false, then the new system need to be studied further. If p_i is a polynomial in $\boldsymbol{u}$, then the system may have dimension higher than k. So, we need to study the new system with Algorithm `Tofind` and the hierarchical strategy recursively. We illustrate the procedure by the following simple example.

Example 6.2. Compute the RRC of $f = x^2 + a^2 y^2$, where x, y are variables and a is parameter. That is to say, compute the condition on a such that f has maximal dimensional real solutions.

The result of triangularization on f in $\mathbb{Q}(a)[x, y]$ is still f. This is Case (B). Viewing y as parameter, by Algorithm `tofind`, we obtain that f has no real solutions (*i.e.* Φ is identically false) provided that $y \neq 0$ and $a \neq 0$.

Add the boundary $y = 0$ to the system. The new system is triangularized as $[x = 0, y = 0]$ (Case (A)) in $\mathbb{Q}(a)[x, y]$. Obviously, no matter what value a takes, f has only one real solution if $a \neq 0$.

Add the boundary $a = 0$ to the system. The new system is triangularized as $[a = 0, x = 0]$ (Case (B)) in $\mathbb{Q}[a, x, y]$. By the procedure above, viewing y as parameter and calling Algorithm `Tofind`, we obtain that the dimension of real solution of f is 1 if and only if $a = 0$.

We now consider Case (C), *i.e.* by `GRSD(S, x)`, we obtain an empty set and an RDU q in parameter.

The basic idea here is similar to that we use in Algorithm `Tofind`: add the RDU q to the system S. Without loss of generality, suppose u_d appears in q and denote the new system by S_d. Regard u_d as variable and assume $[\mathcal{T}H_d, q_d] = \text{GRSD}(S_d, [u_d, x_1, \ldots, x_n])$. If $\mathcal{T}H_d$ is also an empty set, then the new system is also in Case (C). We can add q_d to S_d and repeat the above procedure by adding a new parameter, say u_{d-1}, to the variables. If S_d is in Case (A) or (B), we can use Algorithm `Tofind` and the methods proposed in the last section and the beginning of this section. By Remark 6.6, the above hierarchical strategy will finally lead to a complete RRC of the original system S. The details are omitted here.

Example 6.3. [Cox *et al.* (1992); Weispfenning (1998)]

Consider the mapping sending $\mathbb{R}^2$ to $\mathbb{R}^3$ defined by

$$x = uv, \; y = v, \; z = u^2.$$

The so-called "Whiteney Umbrella" is the smallest real variety containing the image of the mapping. By Gröbner basis computation, it is easy to know that the Whiteney Umbrella is implicitly defined by $x^2 - y^2 z = 0$.

Restate the problem as a QE problem as follows:

$$\exists u \exists v (x = uv \wedge y = v \wedge z = u^2).$$

Let $f_1 = x - uv, f_2 = y - v, f_3 = z - u^2$. Then it is equivalent to finding the condition on x, y, z such that the system $F : f_1 = f_2 = f_3 = 0$ has real solutions.

Triangularizing the system F in $\mathbb{Q}(x, y, z)[u, v]$ by $\text{GRSD}(F, [u, v])$, we obtain $[\emptyset, x^2 - y^2 z]$. So, the system is in Case (C). That is to say the system F may have real solutions only when $x^2 - y^2 z = 0$. This is the first level of the hierarchical procedure. Then, we add $x^2 - y^2 z = 0$ to the system F and, by $\text{GRSD}(F \cup \{x^2 - y^2 z = 0\}, [z, u, v])$ (viewing z as variable), we obtain a regular system

$$[[x^2 - y^2 z, yu - x, v - y], y]$$

with an RDU y. By Algorithm `Tofind`, we find that the system F has real solutions if and only if $x^2 - y^2 z = 0$ provided that $y \neq 0$. This is the second level of the hierarchical procedure.

For the next level, we add $y = 0$ to $\tilde{F} = F \cup \{x^2 - y^2 z = 0\}$ and regard y as variable too. The new system is in Case (C) with an RDU x. Again, by adding $x = 0$ to the preceding system and regarding x as variable, we get a regular system

$$[[x, y, u^2 - z, v], 1]$$

with an RDU 1. This is in Case (B) since we have 4 equations but 5 variables. According to the method proposed at the beginning of this section, by Algorithm `Tofind`, we obtain that, under the precondition $x = y = 0$, the original system has real solutions if and only if $z > 0$ while the case $z = 0$ need further discussion. Finally we add $z = 0$ to the preceding system and the system is triangularized to

$$[[x, y, z, u, v], 1].$$

Obviously, the system F has real solution if $x = y = z = 0$. In summary, the system F has real solutions if and only if

$$[\, x^2 - y^2 z = 0 \wedge y \neq 0 \,] \ \vee \ [\, x = y = 0 \wedge z \geq 0 \,].$$

This is coincident with the result in [Weispfenning (1998)].

6.4 DISCOVERER

In 1996, Zhenbing Zeng developed a program INVENTOR using Maple, which aimed at generating necessary conditions for an SAS to have real solutions. DISCOVERER [Xia (2000, 2007)] is a Maple program developed by the first author of this book based on the code of INVENTOR. The theory behind DISCOVERER has been founded by the authors of the book and their collaborators [Yang *et al.* (1999, 2001); Yang and Xia (2004, 2005); Xia *et al.* (2005); Xia and Zhang (2006); Yang and Xia (2008); Tang *et al.* (2014); Chen *et al.* (2015, 2014)] and is clarified in Chapters 2, 3, 4, 5, and 6 of this book. The main functions of DISCOVERER include real root classification, real root isolation and real root counting for SASs.

Since 2009, the main functions of DISCOVERER have been integrated into the `RegularChains` library of Maple. Since then, the implementation has been improved by Chen *et al.* [Chen *et al.* (2012a,b, 2013)].

In this section, we illustrate the usage of `RealRootClassification` (`RRC` for short), the most important function of DISCOVERER, by some examples of inequality proving and discovering.

Example 6.4. Prove that $f \geq 0$ under the constraints that $a \geq 0, b \geq 0, c \geq 0, abc - 1 = 0$, where

$$\begin{aligned}
f = {} & 2b^4 c^4 + 2b^3 c^4 a + 2b^4 c^3 a + 2b^3 c^3 a^2 + 2a^3 c^3 b^2 + 2a^4 c^3 b + 2a^3 c^4 b + 2a^4 c^4 \\
& + 2a^3 b^4 c + 2a^4 b^4 + 2a^3 b^3 c^2 + 2a^4 b^3 c - 3b^5 c^4 a^3 - 6b^4 c^4 a^4 - 3b^5 c^3 a^4 \\
& - 3b^4 c^3 a^5 - 3b^4 c^5 a^3 - 3b^3 c^5 a^4 - 3b^3 c^4 a^5.
\end{aligned}$$

To prove the inequality by Maple, we first start Maple and load two packages of `RegularChains` as follows.

```
> with(RegularChains):
> with(ParametricSystemTools):
> with(SemiAlgebraicSetTools):
```
Then define an order of the unknowns:
```
> R := PolynomialRing([a, b, c]);
```
To get more information from the output of the function directly, we type in:
```
> infolevel[RegularChains] := 1;
```
Now, by calling
```
> RealRootClassification([abc-1], [a, b, c], [-f], [ ], 2, 0, R);
```
we will know at once that the inequality holds.

In general, for an SAS S of the form (3.1), the calling sequence of `RealRootClassification` is

$$\texttt{RealRootClassification}([p_1, \ldots, p_s], [g_1, \ldots, g_r], [g_{r+1}, \ldots, g_t],$$
$$[h_1, \ldots, h_m], k, \lambda, R);$$

where $R := PolynomialRing([x_1, \ldots, x_n, u_1, \ldots, u_d])$ is a predefined order on variables and parameters.

The formal parameter k is a positive integer which indicates the last k elements in R are to be viewed as parameters of the given system.

The formal parameter λ has two possible forms. If λ is a nonnegative integer, then `RealRootClassification` will output the conditions for the system S to have exactly λ distinct real solutions. If λ is a range, *e.g.* 2..3, then `RealRootClassification` will output the conditions for the number of distinct real solutions of the system S falls into the range λ. If the second element of λ is an unassigned name, it means positive infinity.

`RealRootClassification` can handle systems in Case (A), (B) or (C) automatically. Furthermore, `RealRootClassification` integrates the function of `tofind` and `Tofind` and therefore can also handle the case when parameters are on some boundaries.

Before we give more examples, we first give a detailed explanation of Example 6.4.

Example 6.4 (continued). Obviously,

$$a \geq 0 \land b \geq 0 \land c \geq 0 \land abc - 1 = 0 \implies f \geq 0$$

is equivalent to that the following system has no real solutions

$$a \geq 0 \land b \geq 0 \land c \geq 0 \land abc - 1 = 0 \land f < 0.$$

So, in Example 6.4, we call

```
> RealRootClassification([abc-1], [a, b, c], [-f], [ ], 2, 0, R);
```
where the "0" means we want to compute the conditions for the system to have no real solutions.

The output is:

$$\text{There is always given number of real solution(s)!}$$
$$\text{PROVIDED THAT}$$
$$\phi(b, c) \neq 0,$$

where $\phi(b, c)$ is a polynomial of degree 18 in b and c with 19 terms.

The output means that the system always has no real solutions provided that the polynomial $\phi(b, c)$ does not vanish. In other words, RRC proves that the proposition holds for almost all a, b and c except those such that $\phi(b, c) = 0$.

Because the inequality to be proved is a non-strict inequality ($f \geq 0$), by continuity, we know at once that $f \geq 0$ holds for all a, b and c such that $a \geq 0 \wedge b \geq 0 \wedge c \geq 0 \wedge abc - 1 = 0$. Thus, the proposition is proved.

Applying `RealRootClassification` to handle the boundary case is illustrated by the following simple example.

Example 6.5. We want to know the conditions on the coefficients of $f = ax^2 + bx + c$ for f to have real roots if $a \neq 0$.

After loading `RegularChains` library and the two packages, we define the system as follows.
```
> f  := a*x^2+b*x+c;
> P  := [f]; G1 := [ ]; G2 := [ ]; H := [a];
> R  := PolynomialRing([x,a,b,c]);
```
To get more information from the output of the function directly, we type in:
```
> infolevel[RegularChains] := 1;
```
Then, we call
```
> RealRootClassification(P, G1, G2, H, 3, 1..n, R);
```
where the range 1..n means "the polynomial has at least one real roots".

The output is: $R_1 > 0$ where $R_1 = b^2 - 4ac$ provided that $a \neq 0$ and $R_1 \neq 0$. To discuss the case when $R_1 = 0$, we can add this equation into the original system and call `RealRootClassification` again.
```
> RealRootClassification([b^2-4*a*c,op(F)], G1, G2, H, 3, 1..n, R);
```
In this way, we finally know that the condition is $R_1 \geq 0$.

We show by the following two examples how to deal with the situation where no equations appear in the given SAS.

Example 6.6. Prove that

$$a \geq 0 \land b \geq 0 \land c \geq 0 \land d \geq 0 \implies u \geq 0$$

where

$$
\begin{aligned}
u = {} & 1280bd^3c + 624bc^2d^2 + 320ab^4 + 464ac^4 - 112ad^4 - 112a^4b + 464a^4c \\
& -112b^4c + 464b^4d + 208c^3b^2 + 1072d^3b^2 - 224b^3c^2 + 1072b^3d^2 \\
& +320bc^4 + 464bd^4 - 112c^4d + 208d^3c^2 - 224c^3d^2 + 320cd^4 + 128ad^3c \\
& +624ab^2c^2 + 740b^3cd + 1812ab^2d^2 + 516ac^2d^2 + 1812b^2cd^2 \\
& +128bc^3d + 516b^2c^2d + 128a^3bd + 624a^2b^2d + 516a^2bd^2 + 1280a^3cd \\
& +1812a^2c^2d + 624a^2cd^2 + 128ab^3c + 1280ab^3d + 1280ac^3b + 740ac^3d \\
& +740ad^3b + 1812a^2bc^2 + 740a^3bc + 516a^2b^2c + 1896ab^2cd + 1896abc^2d \\
& +1896abcd^2 + 1896a^2bcd + 320a^4d + 208b^3a^2 + 1072c^3a^2 - 224d^3a^2 \\
& -224a^3b^2 + 1072a^3c^2 + 208a^3d^2 + 64a^5 + 64b^5 + 64c^5 + 64d^5.
\end{aligned}
$$

As usual, we want to prove that the following system has no real solutions

$$a \geq 0 \land b \geq 0 \land c \geq 0 \land d \geq 0 \land u < 0.$$

However, the system does not contain equations and thus `RRC` cannot be applied directly.

We introduce a new variable T and the system being inconsistent is equivalent to that the following new system is inconsistent

$$a \geq 0 \land b \geq 0 \land c \geq 0 \land d \geq 0 \land u + T = 0 \land T > 0.$$

For this new problem, we first define
```
> R := PolynomialRing([T, a, b, c, d]);
```
and then call
```
> RealRootClassification([u+T], [a, b, c, d], [T], [ ], 4, 0, R);
```
The problem is solved immediately.

Example 6.7. Prove that

$$a \geq 0 \land b \geq 0 \land c \geq 0 \implies v \geq 0$$

where

$$v = 104976a^{12} + 1679616a^{11}b + 1469664a^{11}c + 10850112a^{10}b^2$$
$$+19046016a^{10}bc + 8076024a^{10}c^2 + 36149760a^9b^3 + 95364864a^9b^2c$$
$$+80561952a^9bc^2 + 22935528a^9c^3 + 65762656a^8b^4 + 228601856a^8b^3c$$
$$+282635520a^8b^2c^2 + 162625040a^8bc^3 + 42710593a^8c^4 + 63474176a^7b^5$$
$$+251921856a^7b^4c + 354740704a^7b^3c^2 + 288770224a^7b^2c^3$$
$$+207550776a^7bc^4 + 83017484a^7c^5 + 29076288a^6b^6 + 60534016a^6b^5c$$
$$-155234320a^6b^4c^2 - 380047056a^6b^3c^3 + 3130676a^6b^2c^4$$
$$+375984436a^6bc^5 + 181119606a^6c^6 + 8313344a^5b^7 - 89738240a^5b^6c$$
$$-760459488a^5b^5c^2 - 1768157568a^5b^4c^3 - 1403613720a^5b^3c^4$$
$$+236428572a^5b^2c^5 + 824797636a^5bc^6 + 291288188a^5c^7$$
$$+13943056a^4b^8 - 3628032a^4b^7c - 514131904a^4b^6c^2 - 1869896304a^4b^5c^3$$
$$-2495402586a^4b^4c^4 - 783163260a^4b^3c^5 + 1171287578a^4b^2c^6$$
$$+1122586500a^4bc^7 + 288706561a^4c^8 + 18028800a^3b^9 + 116005472a^3b^8c$$
$$+171678496a^3b^7c^2 - 347011440a^3b^6c^3 - 1231272792a^3b^5c^4$$
$$-894635820a^3b^4c^5 + 731754984a^3b^3c^6 + 1497257080a^3b^2c^7$$
$$+851454308a^3bc^8 + 170469720a^3c^9 + 10593792a^2b^{10} + 100409472a^2b^9c$$
$$+365510616a^2b^8c^2 + 624203728a^2b^7c^3 + 480156788a^2b^6c^4$$
$$+215762988a^2b^5c^5 + 511667522a^2b^4c^6 + 990571720a^2b^3c^7$$
$$+861820134a^2b^2c^8 + 356931720a^2bc^9 + 58375800a^2c^{10}$$
$$+2985984ab^{11} + 34730496ab^{10}c + 165207744ab^9c^2 + 415788248ab^8c^3$$
$$+606389880ab^7c^4 + 560561092ab^6c^5 + 437187748ab^5c^6 + 422470380ab^4c^7$$
$$+390424292ab^3c^8 + 235263240ab^2c^9 + 77497200abc^{10} + 10692000ac^{11}$$
$$+331776b^{12} + 4478976b^{11}c + 25292160b^{10}c^2 + 778991104b^9c^3$$
$$+1442474489b^8c^4 + 1706066684b^7c^5 + 141892350b^6c^6 + 1020860036b^5c^7$$
$$+767481161b^4c^8 + 52182360b^3c^9 + 247662000b^2c^{10} + 6804000bc^{11}$$
$$+810000c^{12}.$$

Similar to Example 6.6, the inequality is proved by first defining

```
> R := PolynomialRing([T, a, b, c]);
```

and then calling

```
> RealRootClassification([v+T], [a, b, c], [T], [ ], 3, 0, R);
```

6.5 Automated Discovering of Geometric Inequalities

From now on, when we call `RealRootClassification`, it is assumed that those necessary library and packages have been loaded. For simplicity, we use `RRC` to stand for `RealRootClassification`. Moreover, the last argument is given directly by a list of unknowns which should be understood as a predefined data in the same order (see the last section). For example, $[x, y, z, a, b]$ stands for $PolynomialRing([x, y, z, a, b])$.

Example 6.8. Which triangles can be sections of a regular tetrahedron by planes which separate one vertex from the other three?

This was an open problem proposed in [Folke (1994)]. In fact, this is a special case of the "camera calibration" problem called "perspective-three-point (P3P)" problem [Yang (1998); Gao *et al.* (2003)].

If we let $1, a, b$ (assume $b \geq a \geq 1$) be the lengths of three sides of the triangle, and x, y, z the distances from the vertex to the three vertices of the triangle respectively, then, what we need is to find the necessary and sufficient condition that a, b should satisfy for the following system to have real solution(s),

$$\begin{cases} h_1 = x^2 + y^2 - xy - 1 = 0, \\ h_2 = y^2 + z^2 - yz - a^2 = 0, \\ h_3 = z^2 + x^2 - zx - b^2 = 0, \\ x > 0, y > 0, z > 0, a - 1 \geq 0, b - a \geq 0, a + 1 - b > 0. \end{cases}$$

With `RealRootClassification`, we attack this problem by the following two steps. First of all, we type in

$$\texttt{RRC}([h_1, h_2, h_3], [a - 1, b - a], [x, y, z, a + 1 - b], [\,], 2, 1..n, [z, y, x, b, a]);$$

RRC runs 0.874 seconds on a PC (Intel Core i5-3470 CPU @ 3.20GHz, 8G memory, Windows 7 OS) with Maple 17, and outputs

FINAL RESULT :

The system has required real solution(s) IF AND ONLY IF

$$[0 < R_1, 0 < R_2]$$

or

$$[0 < R_1, R_2 < 0, 0 < R_3]$$

where

$$R_1 = a^2 + a + 1 - b^2$$

$$R_2 = a^2 - 1 + b - b^2$$

$$R_3 = 1 - \frac{8}{3}a^2 - \frac{8}{3}b^2 + \frac{16}{9}a^8 - \frac{68}{27}b^6 a^2 + \frac{241}{81}b^4 a^4 - \frac{68}{27}b^2 a^6$$

$$- \frac{68}{27}b^4 a^2 - \frac{68}{27}b^2 a^4 - \frac{2}{9}b^6 + \frac{16}{9}b^8 - \frac{2}{9}a^6 + \frac{46}{9}b^2 a^2$$

$$+ \frac{16}{9}b^4 + \frac{16}{9}a^4 + \frac{46}{9}b^2 a^8 + \frac{46}{9}b^8 a^2 - \frac{68}{27}b^6 a^4 - \frac{68}{27}b^4 a^6$$

$$+ \frac{16}{9}b^4 a^8 - \frac{8}{3}b^{10} a^2 + \frac{16}{9}b^8 a^4 - \frac{2}{9}b^6 a^6 - \frac{8}{3}b^2 a^{10} - \frac{8}{3}b^{10}$$

$$+ b^{12} - \frac{8}{3}a^{10} + a^{12}$$

PROVIDED THAT:

$$-b + a \neq 0$$
$$a - 1 \neq 0$$
$$b - 1 \neq 0$$
$$a^2 - 1 + b - b^2 \neq 0$$
$$a^2 - 1 - b - b^2 \neq 0$$
$$a^2 - a + 1 - b^2 \neq 0$$
$$a^2 + a + 1 - b^2 \neq 0$$
$$a^2 - 1 - ab + b^2 \neq 0$$
$$a^2 - 1 + ab + b^2 \neq 0$$
$$R_3 \neq 0$$

Folke gave a sufficient condition [Folke (1994)] that any triangle with two angles $> 60°$ is a possible section. It is not hard to see that this condition is equivalent to $[R_1 > 0, R_2 > 0]$.

Then we use `Tofind` to handle the case when parametric point (a, b) is on some "boundary" (*i.e.* $R_1 = 0$, $R_2 = 0$, $R_3 = 0$, $a - 1 = 0$, $b - a = 0, \ldots$). If we want to know the result when (a, b) are on a certain boundary, say $R_2 = 0$, we need only to type in

`RRC(`$[h_1, h_2, h_3, R_2], [a - 1, b - a], [x, y, z, a + 1 - b], [\,], 2, 1..n, [z, y, x, b, a])$`;`

`RRC` runs 0.09 seconds on the same machine and outputs that

FINAL RESULT:

The system has required real solution(s) IF AND ONLY IF

$$[S_1 < 0, (2)R_2]$$

where

$$S_1 = 3b^6 + 56b^4 - 122b^3 + 56b^2 + 3$$

PROVIDED THAT :

$$b - 1 \neq 0$$
$$S_1 \neq 0$$

In the above output, $[S_1 < 0, (2)R_2]$ means parameters (a_0, b_0) should satisfy $S_1 < 0$, and a_0 is the second least root of $R_2(a, b_0) = 0$. For the new boundaries (*i.e.* $b - 1 = 0, S_1 = 0$), we may use the same method. For example, suppose (a, b) is on $R_2 = 0 \wedge b - 1 = 0$ or $R_2 = 0 \wedge S_1 = 0$, we may respectively call

`RRC` $([h_1, h_2, h_3, R_2, b - 1], [a - 1, b - a], [x, y, z, a + 1 - b], [\,], 2, 1..n, [z, y, x, b, a])$;

$$\text{RRC}\,([h_1, h_2, h_3, R_2, S_1], [a - 1, b - a], [x, y, z, a + 1 - b], [\,], 2, 1..n, [z, y, x, b, a]);$$

As the equations on parameters determine only a finite number of points, the program automatically call Algorithm 5.4 (**RRI-TSA**, see Chapter 5) to compute a real root isolation of the system. Both systems have one real solution.

This way, we finally get that the original system has real solution if and only if

$$[0 < R_1, 0 < R_2, R_3 \le 0, 0 < a - 1, 0 \le b - a, 0 < a + 1 - b]$$
$$\text{or}$$
$$[0 < R_1, 0 \le R_3, 0 \le a - 1, 0 \le b - a, 0 < a + 1 - b].$$

Actually, by our algorithm and program, we can do more than the request to this problem. If we type in respectively

$$\text{RRC}\,([h_1, h_2, h_3], [a - 1, b - a], [x, y, z, a + 1 - b], [\,], 2, 1, [z, y, x, b, a]);$$

$$\text{RRC}\,([h_1, h_2, h_3], [a - 1, b - a], [x, y, z, a + 1 - b], [\,], 2, 2, [z, y, x, b, a]);$$

$$\text{RRC}\,([h_1, h_2, h_3], [a - 1, b - a], [x, y, z, a + 1 - b], [\,], 2, 3, [z, y, x, b, a]);$$

we will get the condition for the system to have exactly 1 or 2 or 3 real solution(s), respectively. This way, we get a complete real root classification of this problem.

Instead of listing the quantifier-free formulas expressing the RRC, we show the result by Fig. 6.1 which may be more illustrative. It can be seen that the region of $b \ge a \ge 1$ is decomposed into "cells" of different dimensions (open sets in $\mathbb{R}^2$, segments of curves and points) by $R_1 = 0, R_2 = 0, R_3 = 0$. The symbol printed on one cell indicates the number of real solutions to the system over the cell. However, it should be pointed out that rigorous results are those expressed by quantifier-free formulas. We must be very careful when using plots to illustrate results even if we are in $\mathbb{R}^2$ case. For example, the segments of $R_1 = 0$ and $R_3 = 0$ in the region of $b \ge a \ge 1$ do not coincide but it is hard to realize this from the figure. One may need to study the topology of the curves in this case.

Example 6.9. It is well known that, for any triangle, the midpoints of the three sides, the pedals of the three altitudes and the midpoints of the three segments joining the orthocenter and the three vertices are concyclic.

This circle is known as *Feuerbach's circle* or *nine-point circle*. Its radius equals half the radius of the circumcircle.

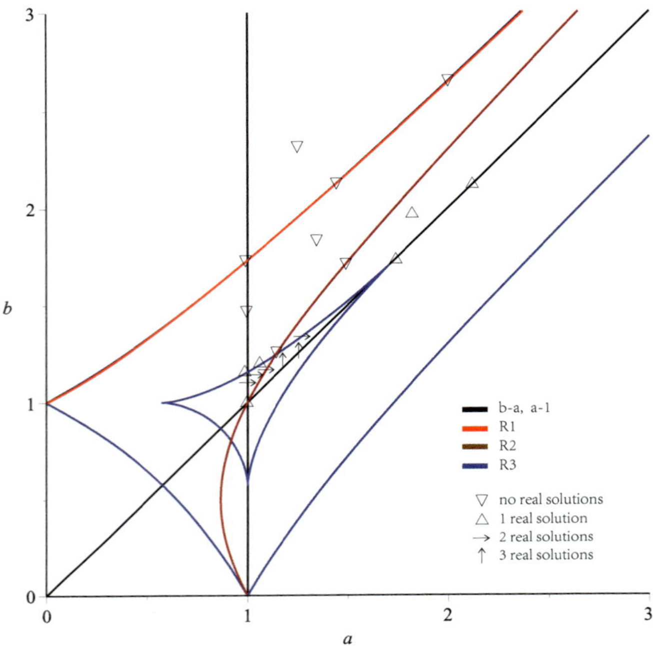

Fig. 6.1   RRC for Example 6.8

Moreover, there are four *tritangent circles* (each tangent to all sides of the triangle), namely, one inscribed circle and three escribed circles. One interesting question is to determine how many tritangent circles have radii smaller than that of the Feuerbach circle.

This problem was studied by Guergueb *et al.* (1994). They obtained a polynomial M in two variables, drew a sketch figure of $M = 0$, chose test points in each connected component of the complements of the curve $M = 0$ and finally worked out their result by computing and comparing the radii at these test points. We solve this problem by our program DISCOVERER and give a complete real root classification to this problem. The case $M = 0$ is also considered.

Given a triangle ABC whose vertices $B(1,0)$ and $C(-1,0)$ are fixed and the vertex $A(u_1, u_2)$ depends on two parameters, we want to find the conditions on u_1 and u_2 such that there are exactly four, three, two, one or none of the tritangent circles whose radii are smaller than that of the Feuerbach circle, respectively. By a routine computation, the system to be dealt with is

$$\begin{cases} f = 16x^2 u_2^2 - (u_1^2 + 2u_1 + 1 + u_2^2)(1 - 2u_1 + u_1^2 + u_2^2) = 0, \\ i = y^4 u_2 + (2 - 2u_2^2 - 2u_1^2)y^3 + u_2(u_1^2 - 5 + u_2^2)y^2 \\ \quad + 4u_2^2 y - u_2^3 = 0, \\ x > 0, \ x^2 - y^2 > 0, \end{cases} \tag{6.6}$$

where x is the radius of the Feuerbach circle and $|y|$ are the radii of the four tritangent circles. For easy reference and comparison, we use the same notations as in [Guergueb *et al.* (1994)] to denote the unknowns and polynomials.

We type in

$$\mathtt{RRC}([f, i], [\,], [x, x^2 - y^2], [\,], 2, 4, [y, x, u_2, u_1]);$$
$$\mathtt{RRC}([f, i], [\,], [x, x^2 - y^2], [\,], 2, 3, [y, x, u_2, u_1]);$$
$$\mathtt{RRC}([f, i], [\,], [x, x^2 - y^2], [\,], 2, 2, [y, x, u_2, u_1]);$$
$$\mathtt{RRC}([f, i], [\,], [x, x^2 - y^2], [\,], 2, 1, [y, x, u_2, u_1]);$$
$$\mathtt{RRC}([f, i], [\,], [x, x^2 - y^2], [\,], 2, 0, [y, x, u_2, u_1]);$$

respectively and get the following outputs (for concision, we combine the outputs of the five instructions):

FINAL RESULT :

The system has 3 (distinct) real solutions IF AND ONLY IF

$$[R_1 < 0, R_2 \geq 0, R_3 \leq 0];$$

The system has 2 (distinct) real solutions IF AND ONLY IF

$$[R_1 > 0];$$

The system has 1 (distinct) real solution IF AND ONLY IF

$$[R_1 < 0, R_2 \leq 0]$$

or

$$[R_1 < 0, R_2 \geq 0, R_3 \geq 0];$$

The system does not have 0 or 4 real solution(s);

PROVIDED THAT :

$$u_1 \neq 0,$$
$$u_2 \neq 0,$$
$$(u_1 + 1)^2 + u_2^2 \neq 0,$$
$$(u_1 - 1)^2 + u_2^2 \neq 0,$$
$$L(u_1, u_2) = 9 + 84u_2^6 u_1^2 + 84u_2^2 - 36u_1^2 - 116u_1^2 u_2^2 + 54u_1^4$$
$$+ 166u_2^4 - 140u_2^6 + 132u_2^4 u_1^2 + 25u_2^8 + 102u_2^4 u_1^4 - 36u_1^6$$
$$+ 9u_1^8 - 20u_1^4 u_2^2 + 52u_2^2 u_1^6 \neq 0,$$
$$R_1 \neq 0,$$

where

$$R_1 = -7 + 20u_2^6u_1^2 + 20u_2^2 + 28u_1^2 - 52u_1^2u_2^2 - 42u_1^4 + 70u_2^4$$
$$-204u_2^6 + 68u_2^4u_1^2 + 9u_2^8 + 6u_2^4u_1^4 + 28u_1^6 - 7u_1^8$$
$$+44u_1^4u_2^2 - 12u_2^2u_1^6,$$

$$R_2 = 189 + 189u_1^{12} + 720u_2^2 - 1134u_1^2 - 1977u_2^8 + 2835u_1^4$$
$$-1235u_2^4 - 3560u_2^6 - 3780u_1^6 + 2835u_1^8 - 8088u_2^6u_1^2$$
$$-1968u_1^2u_2^2 + 2332u_2^4u_1^2 + 558u_2^4u_1^4 + 672u_1^4u_2^2 + 2592u_2^2u_1^6$$
$$+984u_2^6u_1^6 - 1566u_2^8u_1^2 - 40u_2^{10}u_1^2 + 135u_2^8u_1^4 - 2776u_2^6u_1^4$$
$$-3172u_2^4u_1^6 - 2928u_1^8u_2^2 + 1517u_1^8u_2^4 + 912u_2^2u_1^{10} + 15u_2^{12}$$
$$-168u_2^{10} - 1134u_1^{10},$$

$$R_3 = -63 + 225u_2^{14}u_1^2 - 63u_1^{16} + 4284u_1^{12} - 345u_2^2 - 504u_1^2$$
$$+515u_2^8 + 4284u_1^4 + 485u_2^4 + 3347u_2^6 - 11592u_1^6 + 15750u_1^8$$
$$+73991u_2^6u_1^2 - 2851u_1^2u_2^2 + 23658u_2^4u_1^2 - 29957u_2^4u_1^4$$
$$+9791u_1^4u_2^2 - 4163u_2^2u_1^6 + 69174u_2^6u_1^6 - 125788u_2^8u_1^2$$
$$-48997u_2^{10}u_1^2 + 274u_2^8u_1^4 + 89942u_2^6u_1^4 - 22516u_2^4u_1^6$$
$$-12163u_1^8u_2^2 + 36971u_1^8u_2^4 + 13567u_2^2u_1^{10} + 1031u_2^{12}u_1^4$$
$$-1974u2^{12}u_1^2 - 2245u_2^{10}u_1^4 + 1717u_2^{10}u_1^6 - 5609u_2^6u_1^8$$
$$-1052u_2^8u_1^6 + 995u_2^8u_1^8 - 7766u_2^4u_1^{10} - 875u_2^4u_1^{12}$$
$$-3427u_1^{12}u_2^2 - 445u_2^6u_1^{10} - 409u_1^{14}u_2^2 + 407u_2^{12}$$
$$-1643u_2^{10} - 11592u_1^{10} - 15u_2^{14} - 504u_1^{14}.$$

R_1 is just the polynomial $M(u_1, u_2)$ in [Guergueb *et al.* (1994)], but R_2 and R_3 do not appear there.

We remark that the above results are satisfying enough because the only information we do not know is the situation under those "degenerate" conditions, *i.e.* parameters are some "boundary". To get a complete real root classification to the problem, we discuss the "non-degenerate" conditions (*i.e.* the six inequalities) by RRC again as follows. The non-degenerate condition $u_2 \neq 0$ must be satisfied because otherwise the vertices A, B, C are on a line. Thus $(u_1 + 1)^2 + u_2^2 \neq 0$ and $(u_1 - 1)^2 + u_2^2 \neq 0$ are verified. Furthermore, it can be easily shown (by RRC, for example) that $L(u_1, u_2) > 0$ if $u_1 \neq 0$ and $u_2 \neq 0$. So, the "non-degenerate" conditions we need to consider are $u_1 \neq 0$ and $R_1 \neq 0$.

(1) $u_1 \neq 0$ and $R_1 = 0$.

With `RealRootClassification` we type in

$$\text{RRC}([R_1, f, i], [\,], [x, x^2 - y^2], [u_1, u_2], 2, 4, [y, x, u_2, u_1]);$$
$$\dots$$
$$\text{RRC}([R_1, f, i], [\,], [x, x^2 - y^2], [u_1, u_2], 2, 0, [y, x, u_2, u_1]);$$

and obtain that

FINAL RESULT :

The system has 1 real solution IF AND ONLY IF

$$[0 < S_2, S_3 < 0, (3)R_1]$$
or
$$[0 < S_2, S_3 < 0, (2)R_1];$$

The system has 2 real solutions IF AND ONLY IF

$$[S_1 < 0, S_2 < 0, S_3 < 0, (3)R_1]$$
or
$$[S_1 < 0, S_2 < 0, S_3 < 0, (2)R_1]$$
or
$$[0 < S_1, 0 < S_2, S_3 < 0, (4)R_1]$$
or
$$[0 < S_1, 0 < S_2, 0 < S_3, (2)R_1]$$
or
$$[0 < S_1, 0 < S_2, (1)R_1]$$
or
$$[S_1 < 0, S_3 < 0, (4)R_1]$$
or
$$[S_1 < 0, S_3 < 0, (1)R_1];$$

The system does not have 0, 3 or 4 real solution(s);

PROVIDED THAT :

$$S_1 \neq 0, \; S_2 \neq 0, \; S_3 \neq 0,$$

where

$$S_1 = u_2^2 - 3, \; S_2 = u_2^2 - 1/3, \; S_3 = u_2^4 - 22u_2^2 - 7.$$

Let us explain the notations in the above results. A point (a, b) verifying $[0 < S_2, S_3 < 0, (3)R_1]$ means that $S_2(b) > 0 \;\wedge\; S_3(b) < 0$ and a is the third smallest real zero of $R_1(u_1, b)$ in u_1.

The program generates three new "non-degenerate" conditions. Taking $S_2 \neq 0$ for example, we type in

$$\text{RRC}([S_2, R_1, f, i], [\,], [x, x^2 - y^2], [u_1, u_2], 2, 4, [y, x, u_2, u_1]).$$

Because $\{S_2 = 0, R_1 = 0\}$ gives finite points on the curve $R_1 = 0$, RRC invokes a sub-procedure `RealRootIsolate` (*i.e.* Algorithm `RRI-TSA` in Chapter 5) to isolate the real solutions to the system and we can obtain our conclusion from the outputs easily. Our result is as follows

(a) If a point (u_1, u_2) verifies $u_1 \neq 0, R_1 = 0, S_1 = 0$, the system has 1 distinct real solution at the point;

(b) If a point (u_1, u_2) verifies $u_1 \neq 0, R_1 = 0, S_2 = 0$, the system has 2 distinct real solutions at the point;

(c) If a point (u_1, u_2) verifies $u_1 \neq 0, R_1 = 0, S_3 = 0$, the system has 2 distinct real solutions at the point.

(2) $u_1 = 0$.

The system is transformed into the following new system

$$
\begin{cases}
g = 16u_2^2 x^2 - (u_2^2 + 1)^2 = 0, \\
j = u_2 y^4 + (2 - 2u_2^2)y^3 + (u_2^3 - 5u_2)y^2 \\
\qquad + 4u_2^2 y - u_2^3 = 0, \\
x > 0, \ x^2 - y^2 > 0.
\end{cases}
\tag{6.7}
$$

Typing in

$$\text{RRC}([g, j], [\ ], [x, x^2 - y^2], [u_2], 1, 4, [y, x, u_2]);$$
$$\cdots$$
$$\text{RRC}([g, j], [\ ], [x, x^2 - y^2], [u_2], 1, 0, [y, x, u_2]);$$

we obtain that

The system has 2 real solutions IF AND ONLY IF

$$[S_2 > 0, S_3 > 0]$$
$$\text{or}$$
$$[S_2 < 0, S_3 < 0];$$

The system has 1 real solution IF AND ONLY IF

$$[S_2 > 0, S_3 < 0];$$

The system does not have 0, 3 or 4 real solution(s)

PROVIDED THAT :

$$S_1 \neq 0, \ S_2 \neq 0, \ S_3 \neq 0.$$

Because $S_2 < 0 \wedge S_3 > 0$ is impossible, we can write the conditions as $S_2 S_3 > 0$ and $S_2 S_3 < 0$, respectively.

Similarly, for the three "non-degenerate" conditions we type in

$$\text{RRC}([S_1, g, j], [\ ], [x, x^2 - y^2], [u_2], 1, 4, [y, x, u_2]);$$
$$\text{RRC}([S_2, g, j], [\ ], [x, x^2 - y^2], [u_2], 1, 4, [y, x, u_2]);$$
$$\text{RRC}([S_3, g, j], [\ ], [x, x^2 - y^2], [u_2], 1, 4, [y, x, u_2]);$$

DISCOVERER isolates the real solutions of the system, respectively, and we obtain that:

(a) If a point $(0, u_2)$ verifies $S_1 = 0$, the system has no real solutions at this point.

(b) If a point $(0, u_2)$ verifies $S_2 = 0$, the system has 1 distinct real solution at this point.

(c) If a point $(0, u_2)$ verifies $S_3 = 0$, the system has 1 distinct real solution at this point.

The total time for all the computation is 35.9 seconds on a PC (Intel Core i5-3470 CPU @ 3.20GHz, 8G memory, Windows 7 OS) with Maple 17. Finally, we state our result as the following theorem. All the polynomials and notations are defined as before.

Theorem 6.6. *Suppose $u_2 \neq 0$.*

0. The system (6.6) has real solution(s) at all parameters except for two points $(0, \sqrt{3})$ and $(0, -\sqrt{3})$ (i.e. $u_1 = 0 \wedge S_1 = 0$).

1. The system (6.6) has exactly 1 distinct real solution if and only if one of the following conditions holds

(1.1) $u_1 \neq 0 \wedge R_1 < 0 \wedge R_2 \leq 0$,
(1.2) $u_1 \neq 0 \wedge R_1 < 0 \wedge R_2 \geq 0 \wedge R_3 \geq 0$,
(1.3) $u_1 \neq 0 \wedge R_1 = 0 \wedge S_1 = 0$,
(1.4) $u_1 \neq 0 \wedge S_1 \neq 0 \wedge S_2 > 0 \wedge S_3 < 0 \wedge (\,(2)R_1 \vee (3)R_1\,)$,
(1.5) $u_1 = 0 \wedge S_1 \neq 0 \wedge S_2 S_3 \leq 0$.

2. The system (6.6) has exactly 2 distinct real solutions if and only if one of the following conditions holds

(2.1) $u_1 \neq 0 \wedge R_1 > 0$,
(2.2) $u_1 \neq 0 \wedge R_1 = 0 \wedge S_2 S_3 = 0$,
(2.3) $u_1 \neq 0 \wedge S_1 < 0 \wedge S_2 < 0 \wedge S_3 < 0 \wedge (\,(2)R_1 \vee (3)R_1\,)$,
(2.4) $u_1 \neq 0 \wedge S_1 < 0 \wedge S_3 < 0 \wedge (\,(1)R_1 \vee (4)R_1\,)$,
(2.5) $u_1 \neq 0 \wedge S_1 > 0 \wedge S_2 > 0 \wedge (\,(1)R_1 \vee (S_3 < 0 \wedge (4)R_1) \vee (S_3 > 0 \wedge (2)R_1)\,)$,
(2.6) $u_1 = 0 \wedge S_2 S_3 > 0$.

3. The system (6.6) has exactly 3 distinct real solutions if and only if

$$u_1 \neq 0 \wedge R_1 < 0 \wedge R_2 \geq 0 \wedge R_3 \leq 0.$$

4. The system (6.6) does not have four distinct real solutions.

Any parametric point $(u_1, u_2) \in \mathbb{R}^2$ with $u_2 \neq 0$ must verify one of the 13 conditions in the above theorem. Therefore, the theorem gives a complete real root classification of the system (6.6).

Example 6.10. Solving geometric constraints is the central topic in many current work of developing intelligent Computer Aided Design systems and interactive constraint-based graphic systems. This example comes from a group of classical problems on triangles and can also be seen as a problem of solving geometric constraints.

Give the necessary and sufficient condition for the existence of a triangle with elements a, h_a, R, where a, h_a, R means the side-length, altitude, and circumradius, respectively.

By the relations between the quantities of triangles, the problem is transformed to finding the necessary and sufficient condition for the following system to have real solution(s),

$$\begin{cases} f_1 = a^2 h_a^2 - 4s(s-a)(s-b)(s-c) = 0, \\ f_2 = 2Rh_a - bc = 0, \\ f_3 = 2s - a - b - c = 0, \\ a > 0, b > 0, c > 0, a + b - c > 0, b + c - a > 0, \\ c + a - b > 0, R > 0, h_a > 0, \end{cases}$$

where a, b, c are three sides and $s = (a + b + c)/2$.

By similar procedure, we first use Algorithm `tofind` to get

The system has real solution(s) IF AND ONLY IF
$$[\, 0 < R_1, 0 < R_2 \,]$$

or
$$[\, 0 < R_1, R_2 < 0, R_3 \leq 0 \,]$$

PROVIDED THAT
$$R_1 \neq 0, R_2 \neq 0$$

where

$$R_1 = 2R - a,$$

$$R_2 = -4h_a^2 + 8Rh_a - a^2,$$

$$R_3 = -2a^2 h_a - a^2 R + 4h_a^2 R + 8h_a R^2.$$

Then we use Algorithm `Tofind` to deal with boundaries. With `RRC`, it is easy to know that, if $R_1 = 0$, the system has real solutions if and only if $R - h_a \geq 0$; if $R_2 = 0$, the system always has real solution. Finally, we conclude that the system has real solutions if and only if

$$(R_1 > 0 \ \wedge \ R_2 > 0) \ \vee \ (R_1 > 0 \ \wedge \ R_2 < 0 \ \wedge \ R_3 \leq 0)$$
$$\vee \ (R_1 = 0 \ \wedge \ R - h_a \geq 0) \ \vee \ R_2 = 0.$$

The result in [Mitrinović *et al.* (1989)] is: $R_1 \geq 0 \wedge R_2 \geq 0$. Now, we know it is only a sufficient condition.

Our program, `RealRootClassification`, is very efficient for solving this kind of problems as Example 6.10. By `RealRootClassification`, we have discovered or rediscovered about 70 such conditions for the existence of a triangle, and found three mistakes in [Mitrinović *et al.* (1989)].

6.6 Algebraic Analysis of Biological Systems

In this section we apply DISCOVERER to some problems of stability analysis of biological systems. The main part of this section comes from [Wang and Xia (2005b)].

Consider biological networks that may be modeled by autonomous systems of differential equations of the form

$$\begin{cases} \dot{x}_1 = f_1(\boldsymbol{u}, x_1, \ldots, x_n), \\ \dot{x}_2 = f_2(\boldsymbol{u}, x_1, \ldots, x_n), \\ \quad \vdots \\ \dot{x}_n = f_n(\boldsymbol{u}, x_1, \ldots, x_n), \end{cases} \tag{6.8}$$

where $f_1, \ldots, f_n$ are rational functions of $\boldsymbol{u}, x_1, \ldots, x_n$ with real coefficients and $\boldsymbol{u}$ is one or several real parameters. As usual, $x_i = x_i(t)$, $\dot{x}_i = \mathrm{d}\, x_i/\mathrm{d}\, t$, and the parameters $\boldsymbol{u}$ are independent of the derivation variable t.

The class of biological networks we consider is large enough to cover many complex systems, including biological positive-feedback loops for cell and protein signaling, such as the well-known Cdc2-cyclin B/Wee1 system [Novák and Tyson (1993); Pomerening *et al.* (2003)] and the Mos/MEK/p42 MAPK cascade [Angeli *et al.* (2004); Ferrell and Machleder (1998)], which have been studied extensively and experimentally in the literature. The detection of bistability or multistability of such systems is an essential step for understanding how the systems function. Systems are bistable when they toggle between two discrete, alternative stable steady states without being able to rest in intermediate states. We refer to [Angeli *et al.* (2004);

Pomerening *et al.* (2003)] for technical discussions on the behavior and importance of bistability and multistability in the biological context.

A classical and widely used method for analyzing the stability of biological systems is based on phase plane or space diagrams, which plot the trajectories of the differential system around equilibria by numerical computation [Angeli *et al.* (2004); Novák and Tyson (1993)]. This method is limited to plane and spatial differential systems. A more powerful and theoretical approach for analyzing stability behaviors together with a simple graphical method for deducing bifurcation diagrams for biological positive-feedback systems is described in [Angeli *et al.* (2004)]. The visualization technique is very useful in practice, but its theoretical rigor cannot be easily guaranteed. Our symbolic approach based on real root classification provides a mathematically rigorous framework for the stability analysis of a large class of biological systems of arbitrary dimension.

Given a differential system (6.8). If it does not have parameter $\boldsymbol{u}$, the problem is to compute exactly the real *steady states* (*equilibria*) of the system and detect the stability of the steady states. If the system has parameter $\boldsymbol{u}$, the problem is to determine the condition on $\boldsymbol{u}$ such that the system has prescribed number of stable (unstable) steady states. The two problems correspond exactly to real root isolation and real root classification of the system, respectively.

For an arbitrary but fixed real value $\bar{\boldsymbol{u}}$ of $\boldsymbol{u}$, let $\bar{\boldsymbol{x}}$ be an equilibrium of (6.8). We want to analyze the stability of $\bar{\boldsymbol{x}}$. To do so, we use Lyapunov's first method with the technique of linearization, that is, by considering the Jacobian matrix

$$
\mathbf{J} = \begin{pmatrix}
\dfrac{\partial f_1}{\partial x_1} & \dfrac{\partial f_1}{\partial x_2} & \cdots & \dfrac{\partial f_1}{\partial x_n} \\[2ex]
\dfrac{\partial f_2}{\partial x_1} & \dfrac{\partial f_2}{\partial x_2} & \cdots & \dfrac{\partial f_2}{\partial x_n} \\[2ex]
\vdots & \vdots & & \vdots \\[2ex]
\dfrac{\partial f_n}{\partial x_1} & \dfrac{\partial f_n}{\partial x_2} & \cdots & \dfrac{\partial f_n}{\partial x_n}
\end{pmatrix}.
$$

Then system (6.8) may be written in the following matrix form:

$$
\dot{\boldsymbol{x}}^{\mathrm{T}} = \mathbf{J}(\bar{\boldsymbol{u}}, \bar{\boldsymbol{x}})(\boldsymbol{x} - \bar{\boldsymbol{x}})^{\mathrm{T}} + \mathcal{G},
$$

where the superscript T denotes matrix transpose and

$$
\mathcal{G} = [f_1(\bar{\boldsymbol{u}}, \boldsymbol{x}), \ldots, f_n(\bar{\boldsymbol{u}}, \boldsymbol{x})]^{\mathrm{T}} - \mathbf{J}(\bar{\boldsymbol{u}}, \bar{\boldsymbol{x}})(\boldsymbol{x} - \bar{\boldsymbol{x}})^{\mathrm{T}}
$$

is $o(|\boldsymbol{x} - \bar{\boldsymbol{x}}|)$ as $\boldsymbol{x} \to \bar{\boldsymbol{x}}$. The following well-known theorem serves to determine the stability of the equilibrium $\bar{\boldsymbol{x}}$.

Theorem 6.7. *(a) If all the eigenvalues of the matrix $\mathbf{J}(\bar{\boldsymbol{u}}, \bar{\boldsymbol{x}})$ have negative real parts, then $\bar{\boldsymbol{x}}$ is asymptotically stable.*

(b) If the matrix $\mathbf{J}(\bar{\boldsymbol{u}}, \bar{\boldsymbol{x}})$ has at least one eigenvalue with positive real part, then $\bar{\boldsymbol{x}}$ is unstable.

Remark 6.8. It is more difficult to determine the stability of $\bar{\boldsymbol{x}}$ when some of the eigenvalues of $\mathbf{J}(\bar{\boldsymbol{u}}, \bar{\boldsymbol{x}})$ have zero real parts, but none of them has positive real part. In this case, if the eigenvalues with zero real parts correspond to a simple zero of the characteristic polynomial of $\mathbf{J}(\bar{\boldsymbol{u}}, \bar{\boldsymbol{x}})$, then $\bar{\boldsymbol{x}}$ is stable; otherwise, it may be unstable.

A standard method for detect the stability of polynomials is the so-called *Routh–Hurwitz's criterion* (pp. 184–186 of [Miller and Michel (1982)]). Let

$$p = a_0 \lambda^m + b_0 \lambda^{m-1} + a_1 \lambda^{m-2} + b_1 \lambda^{m-3} + \cdots \quad (a_0 \neq 0)$$

be a polynomial in λ with real coefficients. Construct an $m \times m$ matrix

$$\mathbf{P} = \begin{pmatrix} b_0 & b_1 & b_2 & \cdots & b_{m-1} \\ a_0 & a_1 & a_2 & \cdots & a_{m-1} \\ 0 & b_0 & b_1 & \cdots & b_{m-2} \\ 0 & a_0 & a_1 & \cdots & a_{m-2} \\ 0 & 0 & b_0 & \cdots & b_{m-3} \\ \vdots & \vdots & \vdots & & \vdots \end{pmatrix},$$

where $a_i = 0$ if $i > m/2$ and $b_j = 0$ if $j \geq m/2$. All the principal minors, $\Gamma_1, \ldots, \Gamma_m$, of $\mathbf{P}$ are called *Hurwitz's determinants* of p.

Theorem 6.8 (Routh–Hurwitz's criterion). *All roots of p have negative real parts if and only if $V(a_0, \Gamma_1, \Gamma_3, \ldots) = V(1, \Gamma_2, \Gamma_4, \ldots) = 0$. Herein $V(\cdots)$ means the sign changes of a number sequence.*

For an autonomous system of differential equations of the form (6.8), the main steps of our algorithm are described below:

1. Set the numerators of the right-hand sides of (6.8) to 0. This gives a set of equations

$$p_1(\boldsymbol{u}, \boldsymbol{x}) = 0, p_2(\boldsymbol{u}, \boldsymbol{x}) = 0, \ldots, p_n(\boldsymbol{u}, \boldsymbol{x}) = 0,$$

where $p_1, \ldots, p_n$ are polynomials in $\boldsymbol{u}$ and $\boldsymbol{x}$ with rational coefficients. In practical problems, the equations are subject to some constraints, such as the denominators of the right-hand sides of (6.8) should be nonzero, some variables must be positive etc. The equations and the constraints form an SAS S'.

2. Compute Jacobian matrix $\mathbf{J}(\mathbf{u}, \mathbf{x})$ and its characteristic polynomial $h(\mathbf{u}, \mathbf{x}, \lambda)$. Compute the Hurwitz determinants of $h(\mathbf{u}, \mathbf{x}, \lambda)$ and, according to Routh–Hurwitz's criterion, obtain a set of inequalities such that $h(\mathbf{u}, \mathbf{x}, \lambda)$ is stable. Form a new SAS S by adding the inequalities to S'.

3. If S is a constant SAS (without $\mathbf{u}$), isolate the real solutions of S. The real solutions are all the steady states of (6.8). If S is a parametric SAS, compute the RRC of it, which is the necessary and sufficient condition for (6.8) to have prescribed number of stable steady states.

Example 6.11. We analyze the bistability of the Cdc2-cyclin B/Wee1 system [Angeli *et al.* (2004); Novák and Tyson (1993); Pomerening *et al.* (2003)]. For the biological meaning and background of the system, refer to [Angeli *et al.* (2004)]. Its stability behavior may be determined numerically by the classical phase plane analysis [Novák and Tyson (1993)] and another graphical method proposed in [Angeli *et al.* (2004)].

We refer to [Angeli *et al.* (2004)] for the setting details of this example. Under certain assumptions, the system of differential equations that model the Cdc2-cyclin B/Wee1 system is reduced to the following form

$$
\begin{cases}
\dot{x}_1 = \alpha_1(1 - x_1) - \dfrac{\beta_1 x_1 (vy_1)^{\gamma_1}}{K_1 + (vy_1)^{\gamma_1}}, \\[2ex]
\dot{y}_1 = \alpha_2(1 - y_1) - \dfrac{\beta_2 y_1 x_1^{\gamma_2}}{K_2 + x_1^{\gamma_2}},
\end{cases}
\tag{6.9}
$$

where $\alpha_1, \alpha_2, \beta_1, \beta_2$ are rate constants, K_1, K_2 are Michaelis (saturation) constants, γ_1, γ_2 are Hill coefficients, and v is a coefficient (feedback) that reflects the strength of the influence of Wee1 on Cdc2-cyclin B. For easy reference and comparison, we take the same numerical values for the biological constants as in [Angeli *et al.* (2004)]:

$$
\gamma_1 = \gamma_2 = 4, \quad \alpha_1 = \alpha_2 = 1,
$$
$$
\beta_1 = 200, \ \beta_2 = 10, \quad K_1 = 30, \ K_2 = 1.
$$

For simplicity of notation, let $x = x_1$ and $y = y_1$. Then system (6.9) becomes

$$
\dot{x} = \frac{P}{30 + v^4 y^4}, \quad \dot{y} = \frac{Q}{1 + x^4},
\tag{6.10}
$$

where

$$P = 30 - 30\,x + v^4(1 - 201\,x)\,y^4,$$
$$Q = 1 + x^4 - (1 + 11\,x^4)\,y,$$

and $v \geq 0$ is a real parameter. Our problem is to detect the stability of (6.10). In particular, we want to know for what parametric value of v bistability may arise in this system, *i.e.* for what value of v system (6.10) may have two stable steady states.

First, we compute the number of steady states. RRC may automatically find a polynomial $R = v\bar{R}$, where $\bar{R}$ is a polynomial in v with degree 32, 9 terms and 4 real roots. Denote the 5 real roots of R by $\bar{v}_2 < \bar{v}_1 < v_0 = 0 < v_1 < v_2$ ($v_1 = -\bar{v}_1 \approx 0.83157, v_2 = -\bar{v}_2 \approx 1.79687$), which may be isolated as follows:

$$\left[-2, -\frac{3}{2}\right], \quad \left[-1, -\frac{1}{2}\right], \quad [0, 0], \quad \left[\frac{1}{2}, 1\right], \quad \left[\frac{3}{2}, 2\right].$$

From the output of the program, we get directly the following results:

(1) when $0 < v < v_1$ or $v_2 < v < +\infty$, system (6.10) has only one steady state (or equilibrium);
(2) when $v_1 < v < v_2$, system (6.10) has three steady states;
(3) when $v = 0$, system (6.10) has a unique steady state;
(4) when $v = v_1$ or $v = v_2$, system (6.10) has two steady states.

We now determine the stability of the steady states, *i.e.* to determine whether the steady states are stable or not. To this end, we consider the Jacobian matrix of (6.10), whose entries are the partial derivatives of

$$F = \frac{P}{30 + v^4 y^4}, \quad G = \frac{Q}{1 + x^4}$$

with respect to x and y, *i.e.*

$$a = \frac{\partial F}{\partial x} = -\frac{3\,(10 + 67\,v^4 y^4)}{30 + v^4 y^4}, \quad b = \frac{\partial F}{\partial y} = -\frac{24000\,v^4 x y^3}{(30 + v^4 y^4)^2},$$

$$c = \frac{\partial G}{\partial x} = -\frac{40\,x^3 y}{(1 + x^4)^2}, \quad d = \frac{\partial G}{\partial y} = -\frac{1 + 11 x^4}{1 + x^4}.$$

Let

$$p = -(a + d) = \frac{2\,\bar{p}}{(30 + v^4 y^4)\,(1 + x^4)},$$

$$q = ad - bc = \frac{3\,\bar{q}}{(30 + v^4 y^4)^2\,(1 + x^4)^2},$$

where

$$\bar{p} = 30 + 180\,x^4 + 101\,v^4 y^4 + 106\,v^4 x^4 y^4,$$
$$\bar{q} = 67\,y^8\,(1 + 11\,x^4)\,(1 + x^4)\,v^8$$
$$\qquad + 20\,y^4\,(101 - 14788\,x^4 + 1111\,x^8)\,v^4$$
$$\qquad + 300\,(1 + 11\,x^4)\,(1 + x^4).$$

It is easy to see that $a < 0$, $d < 0$, $p > 0, p^2 - 4q \geq 0$ always hold. So, $q > 0$ is the condition for stability we should add to the system. As output, our program gives the following results:

(1) when $0 < v < v_1$ or $v_2 < v < +\infty$, the only steady state is stable;
(2) when $v_1 < v < v_2$, two of the three steady states are stable and the other one is not;
(3) when $v = 0$, the only steady state is stable;
(4) when $v = v_1$ or $v = v_2$, one of the two steady states is stable. Because $q = 0$ (*i.e.* the Jacobian matrix of (6.10) is singular) at the other steady state, the method of linearization is inapplicable, but it is not difficult to see that the steady state in this case is unstable.

Therefore, it is rigorously proved that the system exhibits bistability when $v_1 < v < v_2$. This completes our analysis of the stability of (6.10).

The condition on v derived above for the Cdc2-cyclin B/Wee1 system to exhibit bistability is for the given values of the biological constants $\alpha_1, \alpha_2, \beta_1, \beta_2, K_1, K_2, \gamma_1, \gamma_2$. Estimation of the constant values are very difficult: some of the values may be determined experimentally and others may be chosen so that the model can simulate the type of biological behavior that is observed or expected. Our symbolic approach allows us to establish conditions on some constant parameters for the system to exhibit certain desired behavior such as bistability or multistability.

To fix the idea, let us consider the Cdc2-cyclin B/Wee1 system again, but without taking values for the Michaelis constants K_1, K_2. We want to know for what values of K_1, K_2 and v the system exhibits bistability.

From the meanings of the biological constants and variables, we know that $K_1 > 0$, $K_2 > 0$ and v, x_1, y_1 are nonnegative. Our program may compute a polynomial R_1 of degree 32 in v and degree 8 in either of K_1 and K_2 with 81 terms. Under the above assumption, we have $a < 0, d < 0, p > 0, r \geq 0, b \leq 0, c \leq 0$. Then we can conclude that

(1) when $R_1 < 0$, the system has three steady states, of which two are stable (in this case $q > 0$) and the other is unstable (in this case $q < 0$);

(2) when $R_1 > 0$, the system has only one steady state which is stable.

It follows that the system exhibits bistability if and only if $R_1 < 0$. This generalizes the result given in [Angeli *et al.* (2004)]. The computation in the case $R_1 = 0$ is too heavy and could not be completed within one hour in Maple 17 on a PC (Intel Core i5-3470 CPU @ 3.20GHz, 8G memory, Windows 7 OS). From the results in the cases with specialized values of K_1, K_2, we guess that the system has two steady states, of which one is stable and the other is unstable, when $R_1 = 0$.

To determine the range of K_1, K_2 for the system to exhibit bistability, we may compute a polynomial R_2 of K_2:

$$R_2 = 1123963607439473175421875\, K_2^4 - 9244704652117591783090536\, K_2^3$$
$$- 5088828365064957511326382\, K_2^2 - 62301929415679096\, K_2 + 51046875.$$

Let the two positive real roots of R_2 be $k_1 \approx 0.77 \cdot 10^{-9}$ and $k_2 \approx 8.74$. Our computation shows that the system exhibits bistability when $K_1 > 0$ and $k_1 < K_2 < k_2$, or no bistability otherwise. It follows that the system always exhibits bistability for some range of v, no matter what value K_1 takes. This conclusion is related to a question in [Angeli *et al.* (2004)].

For more examples and advances along this direction, please refer to [Wang and Xia (2005b,a); Niu and Wang (2008); Li *et al.* (2011); Hong *et al.* (2015)].

6.7 Program Verification Through SASs Solving

Termination analysis and reachability computation are very important topics in program verification. We show in this section how to solve the problems with DISCOVERER by reducing them to RRC of SASs [Yang *et al.* (2010b)].

An *atomic polynomial formula* over $\mathcal{K}[\boldsymbol{x}] = \mathcal{K}[x_1, ..., x_n]$ is of the form $p(\boldsymbol{x}) \rhd 0$, where $p(\boldsymbol{x}) \in \mathcal{K}[\boldsymbol{x}]$ and $\rhd \in \{=, >, \geq, \neq\}$, while a *polynomial formula* over $\mathcal{K}[\boldsymbol{x}]$ is a Boolean combination of atomic polynomial formulas over $\mathcal{K}[\boldsymbol{x}]$. *Conjunctive polynomial formulas* are those that are built from atomic polynomial formulas only with the logical operator $\wedge$. We denote by $PF(\mathcal{K}[\boldsymbol{x}])$ the set of polynomial formulas over $\mathcal{K}[\boldsymbol{x}]$ and by $CPF(\mathcal{K}[\boldsymbol{x}])$ the set of conjunctive polynomial formulas over $\mathcal{K}[\boldsymbol{x}]$, respectively.

Programs can be represented as transition systems.

Definition 6.5. A *transition system* is a quintuple $\langle V, L, T, \ell_0, \Theta \rangle$, where V is a set of program variables, L is a set of locations, and T is a set of transitions. Each transition $\tau \in T$ is a quadruple $\langle \ell_1, \ell_2, \rho_\tau, \theta_\tau \rangle$, where ℓ_1 and ℓ_2 are the pre- and post-locations of the transition, the transition relation ρ_τ is a first-order formula over $V \cup V'$, and θ_τ is a first-order formula over V, which is the *guard* of the transition. Only if θ_τ holds, the transition can take place. Here, we use V' (variables with prime) to denote the next-state variables. The location ℓ_0 is the initial location, and the initial condition Θ is a first-order formula over V.

If all formulas of a transition system are from $CPF(\mathcal{K}[\boldsymbol{x}])$, the system is also called *semi-algebraic transition system* (SATS). Similarly, a system is called *polynomial transition system* (PTS), if all its formulas are in $PF(\mathcal{K}[\boldsymbol{x}])$.

A *state* is an evaluation of the variables in V and all states are denoted by $Val(V)$. Without confusion we use V to denote both the variable set and an arbitrary state, and use $F(V)$ to mean the (truth) value of function (formula) F under the state V. The semantics of transition systems can be explained through state transitions as usual.

For convenience, we denote the transition $\tau = (l_1, l_2, \rho_\tau, \theta_\tau)$ by $l_1 \overset{\rho_\tau, \theta_\tau}{\to} l_2$, or simply by $l_1 \overset{\tau}{\to} l_2$. A sequence of transitions

$$l_{11} \overset{\tau_1}{\to} l_{12}, \ldots, l_{n1} \overset{\tau_n}{\to} l_{n2}$$

is called *composable* if $l_{i2} = l_{(i+1)1}$ for $i = 1, \ldots, n-1$, and written as

$$l_{11} \overset{\tau_1}{\to} l_{12}(l_{21}) \overset{\tau_2}{\to} \cdots \overset{\tau_n}{\to} l_{n2}.$$

A composable sequence is called *transition circle* at l_{11}, if $l_{11} = l_{n2}$. For any composable sequence $l_0 \overset{\tau_1}{\to} l_1 \overset{\tau_2}{\to} \cdots \overset{\tau_n}{\to} l_n$, it is easy to show that there is a transition of the form $l_0 \overset{\tau_1;\tau_2;\ldots;\tau_n}{\to} l_n$ such that the composable sequence is equivalent to the transition, where $\tau_1; \tau_2 \ldots; \tau_n$, $\rho_{\tau_1;\tau_2;\ldots;\tau_n}$ and $\theta_{\tau_1;\tau_2;\ldots;\tau_n}$ are the compositions of $\tau_1, \tau_2, \ldots, \tau_n$, $\rho_{\tau_1}, \ldots, \rho_{\tau_n}$ and $\theta_{\tau_1}, \ldots, \theta_{\tau_n}$, respectively. The composition of transition relations is defined in the standard way, for example, $x' = x^4 + 3; x' = x^2 + 2$ is $x' = (x^4 + 3)^2 + 2$; while the composition of transition guards have to be given as a conjunction of the guards, each of which takes into account the past state transitions. In the above example, if we assume the first transition with the guard $x + 7 = x^5$, and the second with the guard $x^4 = x + 3$, then the composition of the two guards is $x + 7 = x^5 \wedge (x^4 + 3)^4 = (x^4 + 3) + 3$.

6.7.1 *Non-linear Ranking Function Discovering*

Constructing ranking functions is one of the methods to prove termination of programs. It is easy to reduce the problem of constructing non-linear ranking functions to that of SAS solving. We show in this section how to solve the resulting problem with DISCOVERER.

Definition 6.6 (Ranking Function). *Assume* $P = \langle V, L, T, l_0, \Theta \rangle$ *is a transition system. A* ranking function *is a function* $\gamma : Val(V) \to \mathbb{R}^+$ *such that the following conditions are satisfied:*

Initiation: $\Theta(V_0) \models \gamma(V_0) \geq 0.$
Decreasing: *There exists a constant* $C \in \mathbb{R}^+$ *such that* $C > 0$ *and, for any transition circle,* $l_0 \xrightarrow{\tau_1} l_1 \xrightarrow{\tau_2} \cdots \xrightarrow{\tau_{n-1}} l_{n-1} \xrightarrow{\tau_n} l_0$, *at* l_0,

$$\rho_{\tau_1;\tau_2;\dots;\tau_n}(V, V') \wedge \theta_{\tau_1;\tau_2;\dots;\tau_n}(V) \models \gamma(V) - \gamma(V') \geq C \wedge \gamma(V') \geq 0.$$

In Definition 6.6, if γ is a polynomial, it is called a *polynomial ranking function*.

Remark 6.9. According to Definition 6.6, for any transition system, if we can find such a ranking function, the system will not go through l_0 infinitely often.

The approach to discover nonlinear ranking functions mainly contains the following four steps:

Step 1–Predefining a Ranking Function Template Predetermine a template of ranking functions.

Step 2–Encoding Initial Condition According to the initial condition of ranking function, we have $\Theta \models \gamma \geq 0$ which means that each real solution of Θ must satisfy $\gamma \geq 0$. In other words, $\Theta \wedge \gamma < 0$ has no real solutions. It is easy to see that $\Theta \wedge \gamma < 0$ is an SAS. Therefore, applying the tool DISCOVERER, we get a necessary and sufficient condition of the derived SAS such that it has no real solutions. The condition may contain some program variables. In this case, the condition should hold for any instantiations of the variables. Thus, by introducing universal quantifications of these variables (we usually add a scope to each of these variables according to different situations) and then applying QEPCAD, we can get a necessary and sufficient condition only on the presumed parameters.

Step 3–Encoding Decreasing Condition From Definition 6.6, there exists a positive constant C such that for any transition circle $l_0 \xrightarrow{\tau_1} l_1 \xrightarrow{\tau_2} \cdots \xrightarrow{\tau_n} l_0$,

$$\rho_{\tau_1;\tau_2;\dots;\tau_n} \wedge \theta_{\tau_1;\tau_2;\dots;\tau_n} \models \gamma(V) - \gamma(V') \geq C \wedge \gamma(V') \geq 0, \quad (6.11)$$

is equivalent to

$$\rho_{\tau_1;\tau_2;\ldots;\tau_n} \wedge \theta_{\tau_1;\tau_2;\ldots;\tau_n} \wedge \gamma(V') < 0 \qquad \text{and} \qquad (6.12)$$

$$\rho_{\tau_1;\tau_2;\ldots;\tau_n} \wedge \theta_{\tau_1;\tau_2;\ldots;\tau_n} \wedge \gamma(V) - \gamma(V') < C \qquad\qquad (6.13)$$

both have no real solutions. So applying the tool DISCOVERER, we obtain some conditions on the parameters. Subsequently, similar to Step 2, we may need to exploit QEPCAD or DISCOVERER to simplify the resulting condition in order to get a necessary and sufficient condition only on the presumed parameters.

Step 4–Solving Final Constraints According to the results obtained from Steps 1-3, we can get the final necessary and sufficient condition only on the parameters of the ranking function template. Then, by utilizing DISCOVERER or QEPCAD, we check whether or not the condition is satisfied and produce the instantiations of these parameters such that the condition holds. Thus, we can get a ranking function of the predetermined form by replacing the parameters with the instantiations, respectively.

Example 6.12. Consider the polynomial program given in the following figure.

$$
\boxed{
\begin{array}{ll}
& \textbf{Real } x = A \\
& \textbf{where } 1 < A < 10 \\
l_0 : & \textbf{while } x > 1 \wedge x < 10 \textbf{ do} \\
& \quad \textbf{if } x > 1 \wedge x < 3 \textbf{ then} \\
& \qquad x := x(5 - x) \\
& \quad \textbf{else} \\
& \qquad x := x + 1 \\
& \quad \textbf{end if} \\
& \textbf{end while}
\end{array}
}
$$

The program is represented by an SATS in the following figure.

$$
\boxed{
\begin{array}{l}
P = \{ \\
\quad V = \{x\} \\
\quad L = \{l_0\} \\
\quad T = \{\tau_1, \tau_2\} \\
\quad \Theta = x = A \wedge A > 1 \wedge A < 10 \\
\quad \text{where} \\
\qquad \tau_1 : \langle l_0, l_0, x' - 5x + x^2 = 0, x > 1 \wedge x < 3 \rangle \\
\qquad \tau_2 : \langle l_0, l_0, x' - x - 1 = 0, x \geq 3 \wedge x < 10 \rangle \\
\}
\end{array}
}
$$

First, we assume a ranking function template with degree 1 in the form $\gamma(\{x\}) = ax + b$.

After encoding the initial condition and then applying DISCOVERER and QEPCAD, we get a condition on a and b:

$$b + 10\,a \geq 0 \wedge b + a \geq 0. \tag{6.14}$$

Afterwards, encoding the decreasing condition with respect to the transition circle $l_0 \xrightarrow{\tau_1} l_0$ and then applying DISCOVERER and QEPCAD, we obtain

$$b + 4\,a \geq 0 \wedge 4\,b + 25\,a \geq 0 \wedge C + 4\,a \leq 0 \wedge C + 3\,a \leq 0. \tag{6.15}$$

Similarly, encoding the decreasing condition with respect to the transition circle $l_0 \xrightarrow{\tau_2} l_0$ and then applying DISCOVERER and QEPCAD, we get a condition

$$b + 11\,a \geq 0 \wedge b + 4\,a \geq 0 \wedge C + a \leq 0. \tag{6.16}$$

Thus, a necessary and sufficient condition on these parameters is

$$C > 0 \wedge a + C \leq 0 \wedge b + 11\,a \geq 0.$$

So, if we assume $C = 1$, we can get a linear ranking function $11 - x$.

If we assume a ranking function template with degree 2 in the form $\gamma(\{x\}) = ax^2 + bx + c$, and let $C = 1$, we get a necessary and sufficient condition on a, b, c as

$$
\begin{aligned}
&c + 10\,b + 100\,a \geq 0 \wedge c + b + a \geq 0 \wedge b + 9\,a + 1 \leq 0 \wedge b + 21\,a + 1 \\
&\leq 0 \wedge (b + 2\,a \geq 0 \vee b + 20\,a \leq 0 \vee 4\,ac - b^2 \geq 0) \wedge 16\,c + 100\,b + 625\,a \\
&\geq 0 \wedge c + 4\,b + 16\,a \geq 0 \wedge (b + 8\,a \geq 0 \vee 2\,b + 25\,a \leq 0 \vee 4\,ac - b^2 \geq 0) \\
&\wedge 3\,b + 15\,a + 1 \leq 0 \wedge c + 11\,b + 121\,a \geq 0 \wedge c + 4\,b + 16\,a \\
&\geq 0 \wedge (b + 8\,a \geq 0 \vee b + 22\,a \leq 0 \vee 4\,ac - b^2 \geq 0) \wedge b + 7\,a + 1 \leq 0.
\end{aligned}
\tag{6.17}
$$

For (6.17), applying `PartialCylindricalAlgebraicDecomposition` of `RegularChains` in Maple (or the function **PCAD** of DISCOVERER), we get a sample point $(1, -22, 150)$ and therefore obtain a nonlinear ranking function $x^2 - 22x + 150$.

6.7.2 *Reachability Computation*

Applications of *hybrid systems*, which are known as Cyber-Physical Systems nowadays, span over multiple domains, including communication, healthcare, manufacturing, aerospace, transportation, etc., many of which are safety-critical. In the field of control theory, stability and controllability of hybrid systems are of main concerns. In the field of computer science, one of the most important topics on hybrid systems is to guarantee the correctness of these systems. One main method is reachability computation.

As hybrid systems consist of deep interaction between continuous evolutions and discrete transitions, the reachability problem of most of hybrid systems is undecidable [Henzinger *et al.* (1998)]. Lafferrierre *et al.* (2001) investigated linear hybrid systems defined by the following differential equations:

$$\dot{\xi} = A\xi + Bu \tag{6.18}$$

where $\xi(t) \in \mathbb{R}^n$ is the state of the system at time t, $A \in \mathbb{R}^{n \times n}, B \in \mathbb{R}^{n \times m}$ are the system matrices, and $u : \mathbb{R} \to \mathbb{R}^m$ is a piecewise continuous function which is called the *control input*.

Given an initial state $x = \xi(0)$ at time 0 and a control input u, the solution of the above differential equations at time $t \geq 0$ is

$$\xi(t) = \Phi(x, u, t) = e^{At}x + \int_0^t e^{A(t-\tau)}Bu(\tau)d\tau$$

where e^{At} is defined as

$$e^{At} = \sum_{k=0}^{\infty} \frac{t^k}{k!} A^k.$$

State y is said to be *reachable* from state x, if there exist control input u and time $t \geq 0$ such that $y = \Phi(x, u, t)$. Given $Y \subseteq \mathbb{R}^n$, define $Pre(Y)$ to be

$$Pre(Y) = \{x \in \mathbb{R}^n \mid \exists y \exists u \exists t (y \in Y \wedge u \in \mathcal{U} \wedge t \geq 0 \wedge \Phi(x, u, t) = y)\},$$

where $\mathcal{U}$ is a set of control inputs. Dually, define $Post(Y)$ to be

$$Post(Y) = \{x \in \mathbb{R}^n \mid \exists y \exists u \exists t (y \in Y \wedge u \in \mathcal{U} \wedge t \geq 0 \wedge \Phi(y, u, t) = x)\}.$$

For this system, *reachability computation* means computing $Pre(Y)$ or $Post(Y)$ for a given Y defined by first-order formulas.

Lafferrierre *et al.* (2001) proved that the reachability problem of the following three families of systems are decidable.

(1) A is *nilpotent, i.e.* $A^n = 0$, and each component of u is a polynomial;
(2) A is *diagonalizable* with rational eigenvalues, and each component of u is of the form $\sum_{i=1}^{m} c_i e^{\lambda_i t}$, where λ_is are rationals and c_is are subject to some semi-algebraic constraints;

(3) A is *diagonalizable* with purely imaginary eigenvalues, whose imaginary parts are rationals, and each component of u is of the form $\sum_{i=1}^{m} c_i \sin(\lambda_i t) + d_i \cos(\lambda_i t)$, where λ_is are rationals and c_is and d_is are subject to some semi-algebraic constraints.

Their method is to reduce the reachability problem of those families of systems to quantifier elimination problem first (and thus the decidability of the problem follows). Then, they use some famous tools for QE, such as REDLOG [Dolzman and Sturm (1997)] and QEPCAD [Collins and Hong (1991)], to solve the resulting QE problems. However, according to their report, some of the examples in their paper cannot be solved well by the tools.

We use our tool DISCOVERER instead of REDLOG or QEPCAD, and find that the result for some examples in [Lafferrierre *et al.* (2001)] can be improved greatly [Yang *et al.* (2005)]. The matrix B in all the following examples is unit matrix.

Example 6.13. (Example 3.5 in [Lafferrierre *et al.* (2001)])
Consider the following system: $A \in \mathbb{Q}^{2\times 2}$ and $\mathcal{U} = \{u\}$ are defined as

$$A = \begin{bmatrix} 2 & 0 \\ 0 & -1 \end{bmatrix}, \quad u(t) = \begin{bmatrix} u_1(t) \\ u_2(t) \end{bmatrix} = \begin{bmatrix} -ae^{\frac{1}{2}t} \\ ae^t \end{bmatrix}, a \geq 0.$$

So

$$\Phi(x_1, x_2, u, t) = \begin{bmatrix} x_1 e^{2t} + \frac{2}{3}a(-e^{2t} + e^{\frac{1}{2}t}) \\ x_2 e^{-t} + \frac{1}{2}a(e^t - e^{-t}) \end{bmatrix}.$$

Suppose the initial set is $X = \{(0,0)\}$. Then $Post(X)$ is

$$\{(y_1, y_2) \in \mathbb{R}^2 \mid \exists a \exists t : 0 \leq a \wedge t \geq 0$$
$$\wedge \ y_1 = \frac{2}{3}a(-e^{2t} + e^{\frac{1}{2}t})$$
$$\wedge \ y_2 = \frac{1}{2}a(e^t - e^{-t})\}.$$

Set $z = e^{\frac{1}{2}t}$, the reachability problem is reduced to

$$\exists a \exists z (0 \leq a \wedge z \geq 1 \wedge p_1 = 0 \wedge p_2 = 0) \tag{6.19}$$

where

$$p_1 = y_1 - \frac{2}{3}a(-z^4 + z),$$
$$p_2 = y_2 z^2 - \frac{1}{2}a(z^4 - 1).$$

According to [Lafferrierre *et al.* (2001)], the quantifiers in (6.19) cannot be eliminated if one uses REDLOG or QEPCAD alone. So, they use REDLOG to eliminate a first, and then use QEPCAD to eliminate z. The $Post(X)$ obtained in [Lafferrierre *et al.* (2001)] is

$$\{(y_1, y_2) \in \mathbb{R}^2 \mid (y_2 > 0 \wedge y_1 + y_2 \leq 0) \vee (y_2 < 0 \wedge y_1 + y_2 \geq 0)$$
$$\vee (4y_2 + 3y_1 = 0)\}. \tag{6.20}$$

We use DISCOVERER to solve the problem. First, calling

$$\texttt{RRC}([p_1, p_2], [a, z - 1], [\ ], [\ ], 2, 1..n, [a, z, y_1, y_2], 1..n);$$

we find that (6.19) has real solutions if and only if

$$y_2 > 0 \wedge y_1 + y_2 < 0,$$

provided that $y_1 \neq 0, y_2 \neq 0, y_1 + y_2 \neq 0$ and $R \neq 0$, where

$$R = 192y_2^3 y_1^2 - 63y_1^3 y_2^2 + 112y_1 y_2^4 - 6y_1^4 y_2 + 3y_1^5 + 16y_2^5.$$

Then, we use $\texttt{Tofind}$ to discuss the situation when (y_1, y_2) is on those boundaries. Finally we get that (6.19) is equivalent to

$$(y_2 > 0 \wedge y_1 + y_2 < 0) \vee (y_1 = y_2 = 0) \vee$$
$$(y_2 > 0 \wedge (y_1 \text{ is the smallest real root of } R = 0 \text{ when } y_2 \text{ is fixed})). \tag{6.21}$$

The total time for all the computation is no more than 1.5 seconds in Maple 17 on a PC (Intel Core i5-3470 CPU @ 3.20GHz, 8G memory, Windows 7 OS).

Remark 6.10. Calling

$$\texttt{RRC}([R], [y_1 + y_2], [y_2], [\ ], 1, 1..n, [y_1, y_2]);$$

the output indicates the input system has no real solutions. That means we have proven that $y_2 > 0 \wedge R = 0$ implies $y_1 + y_2 < 0$ by DISCOVERER. Therefore (6.21) can be further simplified as

$$(y_2 > 0 \wedge y_1 + y_2 < 0) \vee (y_1 = y_2 = 0). \tag{6.22}$$

Note that our result (6.22) is different from (6.20) which is the result given in [Lafferrierre *et al.* (2001)]. We may use DISCOVERER again to demonstrate that their result is incorrect. One counterexample found by DISCOVERER is as follows. Set $(y_1, y_2) = (4, -3)$, then $y_2 < 0 \wedge y_1 + y_2 \geq 0$ $(4y_2 + 3y_1 = 0$ holds in this case). However, equations $p_1 = p_2 = 0$ are triangularized as

$$\{a^5 + 24a^4 + 216a^3 + 1080a^2 + 2592a + 7776 = 0, 6z + a + 6 = 0\}.$$

It is obvious that the equations cannot have real solutions such that $a \geq 0, z \geq 1$.

Example 6.14. (Example 3.6 in [Lafferrierre *et al.* (2001)])

Suppose

$$A = \begin{bmatrix} 0 & 1 \\ -4 & 0 \end{bmatrix}, \quad u(t) = \begin{bmatrix} u_1(t) \\ u_2(t) \end{bmatrix} = \begin{bmatrix} \cos(t) \\ -\sin(t) \end{bmatrix}.$$

We want to know the condition such that a point (y_1, y_2) (where $y_2 > 0$) can be reached from an initial point (x_1, x_2), *i.e.*

$$\{(y_1, y_2) \mid y_2 > 0\} \cap Post(\{(x_1, x_2)\}) \neq \emptyset.$$

This problem is much more general than Example 3.6 in [Lafferrierre *et al.* (2001)], where the points are fixed as $x_1 = 1$, $x_2 = -5/3$, $y_1 = 0$ and $y_2 > 0$.

Set $w = \sin(t)$, $z = \cos(t)$. The problem is transformed to finding the condition on x_1, x_2, y_1, y_2 such that the following system has real solutions:

$$\begin{cases} f_1 = w^2 + z^2 - 1 = 0, \\ f_2 = x_1(z^2 - w^2) + \frac{1}{3}(3x_2 + 5)zw - 2/3w - y_1 = 0, \\ f_3 = 1/3(3x_2 + 5)(z^2 - w^2) - 4x_1zw - 5/3z - y_2 = 0, \\ y_2 > 0. \end{cases}$$

The result obtained by RRC is

$$R_1 = 0 \wedge [\, S_2 \neq 0 \vee (\, S_2 = 0 \wedge x_1 = 0\,)\,],$$

where

$S_2 = 36x_1^2 + 9x_2^2 + 30x_2 + 25,$

$$\begin{aligned} R_1 = &(432x_2^2 + 1440x_2 + 1728x_1^2 + 1200)y_1^4 + (720y_2^2x_2 + 72y_2x_2 - 4440x_2 \\ &-3456x_1^4 - 2025 - 216x_2^4 + 216x_2^2y_2^2 - 5760x_1^2x_2 + 120y_2 + 864x_1^2y_2^2 \\ &-3732x_2^2 + 600y_2^2 - 1440x_2^3 - 1728x_1^2x_2^2 - 5328x_1^2)y_1^2 + 18x_1(72x_1^2 \\ &+45 + 18x_2^2 + 60x_2 + 2y_2^2)y_1 + 4896x_1^2x_2^2 - 810y_2x_2^2 - 1386y_2x_2 \\ &-1080x_1^2y_2 + 810x_2 - 810y_2 + 1215x_1^2 + 1062x_2^4 + 27x_2^6 + 2592x_1^4 \\ &-360x_2^3y_2^2 + 2043x_2^2 - 657y_2^2 + 4320x_1^2x_2 - 648x_1^2y_2x_2 - 432x_2^2y_2^2x_1^2 \\ &-1440x_1^2y_2^2x_2 + 324x_2^4x_1^2 - 54x_2^4y_2^2 - 864x_1^4y_2^2 + 1296x_1^4x_2^2 \\ &+2160x_1^2x_2^3 + 12y_2^3x_2 - 162x_2^3y_2 - 987x_2^2y_2^2 + 240x_1y_1^3 + 4320x_1^4x_2 \\ &-1548x_1^2y_2^2 - 1290y_2^2x_2 + 27y_2^4x_2^2 + 90y_2^4x_2 + 108y_2^4x_1^2 + 20y_2^3 \\ &+1728x_1^6 + 2080x_2^3 + 270x_2^5 + 75y_2^4. \end{aligned}$$

If we evaluate R_1 at $x_1 = 1, x_2 = -5/3, y_1 = 0$, that will give the result $2916y_2^4 - 32688y_2^2 + 22445 = 0$ in [Lafferrierre *et al.* (2001)].

Example 6.15. (Example 3.7 in [Lafferrierre *et al.* (2001)])

Suppose

$$A = \begin{bmatrix} 0 & -1 \\ 1 & 0 \end{bmatrix}, \quad u(t) = \begin{bmatrix} u_1(t) \\ u_2(t) \end{bmatrix} = \begin{bmatrix} a\cos(2t) \\ -a^{-1}\sin(2t) \end{bmatrix}, a > 0.$$

We consider whether $Y = \{(-1,1)\}$ is reachable from initial set $X = \{(0,0)\}$. Set $w = \sin(2t)$, $z = \cos(2t)$. Then, Y is reachable from X if and only if

$$\exists w \exists z \exists a : a > 0 \wedge g_1 = 0 \wedge g_2 = 0 \wedge g_3 = 0, \tag{6.23}$$

where

$$g_1 = w^2 + z^2 - 1,$$
$$g_2 = w((4a^2 - 2)z + 2 - a^2) + 3a,$$
$$g_3 = (a^2 - 2)(w^2 - z^2 + z) - 3a.$$

In [Lafferrierre *et al.* (2001)], they use REDLOG to eliminate w first. Because the resulting formula is too complicated to be solved by REDLOG or QEPCAD, they then set $z = 0$ and simplify the formula as

$$\exists a \, (\, a > 0 \wedge a^2 \neq 2 \wedge 3a^4 - 9a^3 - 12a^2 + 18a + 12 = 0$$
$$\wedge -a^4 + 13a^2 - 4 = 0). \tag{6.24}$$

Finally, QEPCAD is applied to verify that the above formula is true.

Actually, (6.23) is a problem that asks whether a constant SAS has real solutions. So, we may use `RealRootCounting` (see Section 5.3) to answer this kind of questions. However, to compare with the result in [Lafferrierre *et al.* (2001)], we use the real root isolation function

`RealRootIsolate`$([g_1, g_2, g_3], [\,], [a], [\,], [a, w, z], method =' Discoverer');$

The output is

$$\left[\left[a = \left[\frac{7}{2}, \frac{15}{4} \right], w = 1, z = 0 \right], \right.$$
$$\left[a = \left[\frac{125}{256}, \frac{143}{256} \right], w = \left[\frac{-21}{32}, \frac{-167}{256} \right], z = \left[\frac{-777}{1024}, \frac{-97}{128} \right] \right],$$
$$\left. \left[a = \left[\frac{27}{8}, \frac{7}{2} \right], w = \left[\frac{-231105}{262144}, \frac{-231091}{262144} \right], z = \left[\frac{30919}{65536}, \frac{3865}{8192} \right] \right] \right].$$

It is easy to see that $z = 0$ is only one of the three solutions. It is worth to point out that, by DISCOVERER, we find that any point can be reached from $\{(0,0)\}$ for this system.

Remark 6.11. We have generalized the decidable results in [Lafferrierre *et al.* (2001)] to the following two families of systems of the form (6.18):

(1) A is diagonalizable with *real* eigenvalues, and each component of u is of the form $\sum_{i=1}^{m} c_i e^{\lambda_i t}$, where λ_is are *reals* and c_is are subject to some semi-algebraic constraints;
(2) A is diagonalizable with purely imaginary eigenvalues, whose imaginary parts are *reals*, and each component of u is of the form $\sum_{i=1}^{m} c_i \sin(\lambda_i t) + d_i \cos(\lambda_i t)$, where λ_is are *reals* and c_is and d_is are subject to some semi-algebraic constraints.

See [Gan *et al.* (2015, 2016)] for details.

Chapter 7

Open Weak CAD

Open CAD (see Definition 7.8) can be applied to many problems involving only non-strict polynomial inequalities. Combined with "hierarchical strategy", it is also applied to the real root classification problem (see Chapter 6). In this chapter, we consider the following three problems and introduce two new projection operators which may produce fewer sample points than the usual open CAD and thus lead to faster algorithms and tools for solving the three problems.

The first problem concerns the global infimum (or supremum) of a polynomial.

Problem 1. For $f \in \mathbb{R}[x, k]$, find all $r \in \mathbb{R}$ such that $f(x, r) \geq 0$ on $\mathbb{R}^n$, where $x = (x_1, \ldots, x_n)$ are ordered variables.

The problem can be expressed as:

$$\forall x (f(x, k) \geq 0).$$

If we apply some QE algorithm, *e.g.* cylindrical algebraic decomposition (CAD), to the problem, we will obtain a quantifier free formula in k, say $\phi(k)$, which defines a semi-algebraic set of $\mathbb{R}$. By the obvious fact that a semi-algebraic set of $\mathbb{R}$ is either empty or union of finitely many points and (or) intervals, $\phi(k)$ indeed gives an answer to Problem 1.

If $f(x, k) = g(x) - k$, the supremum of the semi-algebraic set defined by $\phi(k)$ is obviously the global infimum $\inf g(\mathbb{R}^n)$. From this viewpoint, the following problem can be viewed as a special case of Problem 1.

Problem 2. For $f \in \mathbb{R}[x]$, find the global infimum $\inf f(\mathbb{R}^n)$.

Depending on the global infimum, we may answer

Problem 3. For $f \in \mathbb{R}[x]$, prove or disprove $f(x) \geq 0$ on $\mathbb{R}^n$.

Of course, to answer Problem 3, there is no need to compute the global infimum $\inf f(\mathbb{R}^n)$. It may be solved by directly applying QE tools to the quantified formula $\forall x (f(x) \geq 0)$.

A common characteristic of the three problems is that the inequalities involved are non-strict. So, if we use CAD-based method to solve these problems, we only need "open CAD" instead of complete CAD.

The main content of this chapter is from [Han *et al.* (2016, 2014); Dai *et al.* (2015)].[a]

Although most of the theorems of this chapter are valid for $\mathbb{R}[\boldsymbol{x}]$, we restrict ourselves to $\mathbb{Z}[\boldsymbol{x}]$ when we design algorithms because they need effective (square-free) factorization and real root isolation. Actually, suppose $\mathcal{R}$ is a subring of $\mathbb{R}$ and takes $\mathbb{Z}$ as a subring. If $\mathcal{R}[\boldsymbol{x}]$ admits effective (square-free) factorization and $\mathcal{R}[x]$ admits effective real root isolation, all the algorithms in this chapter are effective. Some examples of such rings are $\mathbb{Z}$, $\mathbb{Q}$ and the field of real algebraic numbers. In this chapter, we use $\mathcal{R}$ to denote such a ring.

7.1 Quantifier Elimination and Cylindrical Algebraic Decomposition

Quantifier elimination can be interpreted in any suitable theory. For our purpose, we state it in the field of real numbers.

Definition 7.1. An *atom formula* is $f \triangleright 0$ where $f \in \mathbb{Z}[\boldsymbol{x}]$ and $\triangleright \in \{=, >\}$. A *formula* is constructed by atom formulas with the logical connectives $\wedge$, $\vee$ and $\neg$ and the quantifiers $\forall$ and $\exists$. A *quantifier free formula* is a formula with no quantifiers. If $\forall x_i$ or $\exists x_i$ appears in a formula, x_i is a *quantified variable*. All the variables of a formula which are not quantified are the *free variables* of the formula. A formula with all its variables quantified is called a *sentence*. A formula

$$Q_1 x_{i_1} \cdots Q_k x_{i_k}(\phi(\boldsymbol{x})),$$

where $Q_i \in \{\forall, \exists\}$, $1 \leq i_j \leq n$ and $\phi(\boldsymbol{x})$ is a quantifier free formula, is called a *prenex normal form*. Two formulas are *equivalent* if they define the same set over $\mathbb{R}$.

Now the quantifier elimination problem in the theory of $\mathbb{R}$ can be stated as:

Given a prenex normal form Φ whose variables are interpreted in $\mathbb{R}$, compute a quantifier free formula Ψ in the free variables of Φ such that Ψ is equivalent to Φ.

The famous result of [Tarski (1951)] told us that the above QE problem is decidable. Since then, many algorithms for QE based on different theories

---

[a]Reprinted from Han, J., Jin, Z. and Xia, B. (2016). Proving inequalities and solving global optimization problems via simplified CAD projection, *J. Symbolic Computation* **72**, pp. 206-230, with permission from Elsevier.

with different complexity have appeared. In this chapter, we focus on using CAD-based methods. So, in the following, we recall briefly some basic concepts and results of CAD. The reader is referred to [Collins (1975)], [Hong (1990)], [McCallum (1988, 1998)], [Brown (2001)] and [Strzeboński (2000)] for a detailed discussion on the properties of CAD and open CAD. For the extension of CAD to the decision problems involving exponential-polynomials, see for example [Achatz *et al.* (2008); Strzeboński (2011); Xu *et al.* (2015); Gan *et al.* (2015)].

Definition 7.2. A non-empty connected subset of $\mathbb{R}^n$ is called a *region*. For a subset $S \subseteq \mathbb{R}^n$, a group of pairwise disjoint regions $(\mathcal{D}_1, \ldots, \mathcal{D}_m)$ is called a *decomposition* of S if $\mathcal{D}_1 \cup \cdots \cup \mathcal{D}_m = S$. Every region in the decomposition is a *cell*. Any point in a cell is a *sample point*. If $s_i \in \mathcal{D}_i$ $(1 \leq i \leq m)$, then $(s_1, \ldots, s_m)$ is said to be a *sample* of the decomposition.

Definition 7.3. Suppose $P \subset \mathbb{Z}[\boldsymbol{x}]$. A decomposition $\mathcal{D}$ of $\mathbb{R}^n$ is called *P-sign invariant* if every polynomial $p \in P$ has a constant sign in every cell of $\mathcal{D}$.

Given a set P of polynomials in $\mathbb{R}[\boldsymbol{x}]$, $V_{\mathbb{R}}(P)$ partitions $\mathbb{R}^n$ into finitely many connected regions (semi-algebraic sets) and each polynomial in P keeps constant *sign* (either $+$, $-$ or 0) on each region. The goal of CAD is to compute at least one sample point in each of the regions. For example, suppose $f_1 = y - x$, $f_2 = y + x$. The graphs of $f_1 = 0$ and $f_2 = 0$ decompose $\mathbb{R}^2$ into 9 regions with different dimensions: four of which are 2-dimensional (open) regions (*i.e.* $f_1 \sim 0 \wedge f_2 \sim 0$, where $\sim \in \{>, <\}$); four of which are 1-dimensional regions (*i.e.* $f_1 \sim 0 \wedge f_2 = 0$ or $f_1 = 0 \wedge f_2 \sim 0$, where $\sim \in \{>, <\}$); and one of which is 0-dimensional region (*i.e.* $f_1 = 0 \wedge f_2 = 0$). Complete CAD takes at least one sample point from each of the 9 regions, while GCAD or open CAD takes at least one sample point only from each of the four 2-dimensional (open) regions.

Definition 7.4. For a region $S \subseteq \mathbb{R}^i$, $S \times \mathbb{R}$ is the *cylinder* over S, denoted by $\mathcal{C}(S)$. Suppose continuous real-valued function $f(x_1, \ldots, x_n)$ is defined over the region S. The set
$$\{(a_1, \ldots, a_n, f(a_1, \ldots, a_n)) \mid (a_1, \ldots, a_n) \in S\}$$
is called a *section* (or *f-section*) of the cylinder $\mathcal{C}(S)$. Let
$$f_1 = f_1(x_1, \ldots, x_n), \quad f_2 = f_2(x_1, \ldots, x_n)$$
be two continuous real-valued functions over S with $f_1 < f_2$. The set
$$\{(a_1, \ldots, a_n, b) \mid (a_1, \ldots, a_n) \in S, \ f_1(a_1, \ldots, a_n) < b < f_2(a_1, \ldots, a_n)\}$$

is called the *sector* (or (f_1, f_2)-*sector*) of $\mathcal{C}(S)$. Herein, it is allowed that $f_1 = -\infty$, $f_2 = +\infty$.

Suppose $f_1 < \cdots < f_k$ $(k \geq 0)$ are continuous real-valued functions over S and let $f_0 = -\infty$, $f_{k+1} = +\infty$. Then, f_i-sections $(1 \leq i \leq k)$ and (f_i, f_{i+1})-sectors $(0 \leq i \leq k)$ form a decomposition of $\mathcal{C}(S)$, which is called a *stack* over S defined by $f_1, \ldots, f_k$.

Notation 7.1. Suppose $\boldsymbol{x} = (x_1, ..., x_n)$. We denote $(x_1, ..., x_j)$ by $\boldsymbol{x}^{[j]}$ for $j = 1, ..., n - 1$.

Definition 7.5. An n-variate polynomial $f(\boldsymbol{x}^{[n-1]}, x_n)$ over the reals is said to be *delineable* on a subset S (usually a region) of $\mathbb{R}^{n-1}$ if (1) the portion of the real variety of f that lies in the cylinder $S \times \mathbb{R}$ over S consists of the union of the graphs of some $t \geq 0$ continuous functions $\theta_1 < \cdots < \theta_t$ from S to $\mathbb{R}$; and (2) there exist integers $m_1, \ldots, m_t \geq 1$ such that for every $a \in S$, the multiplicity of the root $\theta_i(a)$ of $f(a, x_n)$ (considered as a polynomial in x_n alone) is m_i. If additionally S is an algebraic submanifold and the θ_is are analytic functions, f is said to be *analytic delineable* on S.

In the above definition, the θ_is are called the *real root functions* of f on S. Let f be an analytic function defined in some open region U. We say that f has *order* k at $p \in U$ if k is the least nonnegative integer such that some partial derivative of f of order k does not vanish at p.

Definition 7.6. A *cylindrical decomposition* $\mathcal{D}$ of $\mathbb{R}^n$ is defined inductively as follows.

If $n = 1$, $\mathcal{D} = \mathbb{R}$ or there exist some real numbers $t_1 < \cdots < t_m$ $(m \geq 1)$ such that $\mathcal{D} = (\mathcal{D}_1, \ldots, \mathcal{D}_{2m+1})$, where

$$\mathcal{D}_1 = (-\infty, t_1),$$
$$\mathcal{D}_{2i} = \{t_i\}, \quad 1 \leq i \leq m,$$
$$\mathcal{D}_{2i+1} = (t_i, t_{i+1}), \quad 1 \leq i < m,$$
$$\mathcal{D}_{2m+1} = (t_m, +\infty).$$

If $n > 1$, there exists a cylindrical decomposition $\mathcal{D}' = (\mathcal{D}_1, \ldots, \mathcal{D}_l)$ of $\mathbb{R}^{n-1}$ such that, for any i $(1 \leq i \leq l)$, $(\mathcal{D}_{i,1}, \ldots, \mathcal{D}_{i,2m_i+1})$ is a stack over $\mathcal{D}_i$ and

$$\mathcal{D} = (\mathcal{D}_{1,1}, \ldots, \mathcal{D}_{1,2m_1+1}, \ldots, \mathcal{D}_{l,1}, \ldots, \mathcal{D}_{l,2m_l+1}).$$

If $\mathcal{D}$ is determined by polynomials, it is called a *cylindrical algebraic decomposition*.

A *cylindrical sample* $s = (s_1, \ldots, s_m)$ of a cylindrical algebraic decomposition $\mathcal{D} = (\mathcal{D}_1, \ldots, \mathcal{D}_m)$ can be defined similarly.

If $n = 1$, s is always a cylindrical sample. If $n > 1$, there exist a cylindrical sample $s' = (s_1, \ldots, s_l)$ of a cylindrical decomposition $\mathcal{D}'$ of $\mathbb{R}^{n-1}$ such that

$$s = (s_{1,1}, \ldots, s_{1,2m_1+1}, \ldots, s_{l,1}, \ldots, s_{l,2m_l+1}).$$

Herein, for any $s_{i,j}$ ($1 \leq i \leq l$, $1 \leq j \leq 2\,m_i + 1$), its first $n - 1$ coordinates are exactly those of s_i.

If every point in s is algebraic, s is called a *cylindrical algebraic sample*.

Theorem 7.1. *[McCallum (1988, 1998)] Let $f(\boldsymbol{x}, x_{n+1})$ be a polynomial in $\mathbb{R}[\boldsymbol{x}, x_{n+1}]$ of positive degree and $\mathrm{discrim}(f, x_{n+1})$ is a nonzero polynomial. Let S be a connected submanifold of $\mathbb{R}^n$ on which f is degree-invariant and does not vanish identically, and in which $\mathrm{discrim}(f, x_{n+1})$ is order-invariant. Then f is analytic delineable on S and is order-invariant in each f-section over S.*

Based on this theorem, McCallum (1988) proposed the projection operator MCproj, which consists of the discriminant of f and all coefficients of f.

Theorem 7.2. *[Brown (2001)] Let $f(\boldsymbol{x}, x_{n+1})$ be an $(n + 1)$-variate polynomial of positive degree m in the variable x_{n+1} with $\mathrm{discrim}(f, x_{n+1}) \neq 0$. Let S be a connected submanifold of $\mathbb{R}^n$ where $\mathrm{discrim}(f, x_{n+1})$ is order-invariant, the leading coefficient of f is sign-invariant, and such that f vanishes identically at no point in S. f is degree-invariant on S.*

Based on this theorem, Brown (2001) obtained a reduced McCallum projection in which only leading coefficients and discriminants appear.

Recall that the *class* of a polynomial f is the largest i such that $\deg(f, x_i) > 0$ (see Chapter 2) and $\mathrm{sqrfree}(f)$ is the square free part of f (see Definition 7.9). Note that $\mathrm{lc}(g)\mathrm{discrim}(g)$ divides $\mathrm{res}(g, \frac{\partial g}{\partial x_n}, x_n)$ for any polynomial g of class n. So, we have

Definition 7.7. [Brown (2001)] Given a square-free polynomial $f \in \mathcal{R}[\boldsymbol{x}]$ of class n, the Brown-McCallum projection operator for f is

$$\mathrm{BMproj}(f, x_n) = \mathrm{res}(\mathrm{sqrfree}(f), \frac{\partial(\mathrm{sqrfree}(f))}{\partial x_n}, x_n).$$

If L is a polynomial set and the class of any polynomial in L is n, then

$$\mathrm{BMproj}(L, x_n) = \cup_{f \in L} \{\mathrm{res}(\mathrm{sqrfree}(f), \frac{\partial(\mathrm{sqrfree}(f))}{\partial x_n}, x_n)\} \bigcup$$

$$\cup_{f,g \in L, f \neq g} \{\mathrm{res}(\mathrm{sqrfree}(f), \mathrm{sqrfree}(g), x_n)\}.$$

Define

$$\mathrm{BMproj}(f, [x_n]) = \mathrm{BMproj}(f, x_n),$$
$$\mathrm{BMproj}(f, [x_n, x_{n-1}, \ldots, x_i])$$
$$= \mathrm{BMproj}(\mathrm{BMproj}(f, [x_n, x_{n-1}, \ldots, x_{i+1}]), x_i).$$

Algorithm 7.1 `BMprojection` [Brown (2001)]

Input: A set A of polynomials in $\mathbb{Z}[x]$ and the ordered variables x.
Output: A projection factor set F.
  1: $f(x) \leftarrow \mathrm{sqrfree}(\prod_{q \in A} q)$;
  2: $F \leftarrow \{f(x)\}$;
  3: **for** i from n downto 2 **do**
  4:     $F \leftarrow F \bigcup \{\mathrm{BMproj}(F^{[i]}, x_i)\}$; ($F^{[i]}$ is the set of polynomials in F of class i).
  5: **end for**
  6: **return** F

7.2   Open CAD

The following definition of open CAD [Xiao (2009)] is essentially the GCAD introduced in [Strzeboński (2000)]. For convenience, we use the terminology of open CAD in this book.

Definition 7.8 (Open CAD). For a polynomial $f(x) \in \mathcal{R}[x]$, an *open CAD* defined by $f(x)$ is a set of sample points in $\mathbb{R}^n$ obtained through the following three phases:

(1) Projection. Use the Brown-McCallum projection operator on $f(x)$, *i.e.* let $F = \mathtt{BMprojection}(\{f\}, x)$ (See Algorithm 7.1).

(2) Base. Assume $F^{[1]}$ has $k(k \geq 0)$ distinct real roots, then choose one rational point in each of the $k + 1$ open intervals on $\mathbb{R}$ defined by the k real roots.

(3) Lifting. Substitute each sample point, $(a_1, \ldots, a_{i-1})$, of $\mathbb{R}^{i-1}$ for $x^{[i-1]}$ in $F^{[i]}$ to get a univariate polynomial $F_i = F_i(a_1, \ldots, a_{i-1}, x_i)$ and then, by the same method as Base phase, choose sample points for F_i. If $b_1, \ldots, b_{m+1}$ are the sample points for F_i, all $(a_1, \ldots, a_{i-1}, b_j)$ $(1 \leq j \leq m+1)$ are sample points of $\mathbb{R}^i$. Repeat the process for i from 2 to n.

Remark 7.1. By Theorem 7.1 and Theorem 7.2, we can deduce easily that at least one sample point can be taken from every highest dimensional cell via the lifting phase of open CAD. This is an important property of open CAD.

Suppose $\mathcal{S}$ is an open CAD defined by $f(\boldsymbol{x})$. Then, by checking the sign of $f(\boldsymbol{x})$ at every sample point in $\mathcal{S}$, we can immediately determine whether or not $\forall \boldsymbol{x}(f(\boldsymbol{x}) \geq 0)$. Based on this procedure, [Yang and Xia (2000); Yang (2001)] designed an algorithm (`SRes`) for solving polynomial optimization problem, which is described below.

Algorithm 7.2 `SRes` (Successive Resultant Method)

Input: A square-free polynomial $f \in \mathbb{Z}[\boldsymbol{x}]$.

Output: The supremum of $k \in \mathbb{R}$, such that $\forall \boldsymbol{a}_n \in \mathbb{R}^n(f(\boldsymbol{a}_n) \geq k)$. If there is no such k, then return $-\infty$.

1: $g \leftarrow f - k$; $//g$ is viewed as a polynomial in $k \prec x_1 \prec \cdots \prec x_n$

2: $F \leftarrow \texttt{BMprojection}(\{g\}, (k, \boldsymbol{x}))$; $// F^{[i]}$ is the set of polynomials in F of class i. Here $F^{[i]}$ has no more than one polynomial, we denote this polynomial by F_i.

3: $C_0 \leftarrow$ an open CAD of $\mathbb{R}$ defined by $F_0(k)$; $//$ Suppose $C_0 = \bigcup_{i=0}^{m}\{p_i\}$, $p_i \in (k_i, k_{i+1})$, where k_i $(1 \leq i \leq m)$ are the real roots of F_0 and $k_0 = -\infty$, $k_{m+1} = +\infty$.

4: **for** l from 0 to m **do**

5:     $C_{ln} \leftarrow$ an open CAD of $\mathbb{R}^n$ defined by $F_n(\boldsymbol{x}, p_l)$; $// F_n = g$

6:     **if** there exists $\boldsymbol{a}_n$ in C_{ln} such that $F_n(\boldsymbol{a}_n, p_l) < 0$ **then**

7:         **return** k_l

8:     **end if**

9: **end for**

10: **return** $-\infty$

For a polynomial $f(\boldsymbol{x})$, the `SRes` method first applies Algorithm 7.1 on polynomial $f(\boldsymbol{x}) - K$ to get a polynomial $g(K)$. Suppose $g(K)$ has m distinct real roots $k_i(1 \leq i \leq m)$. Then computes $m + 1$ rational numbers $p_i(0 \leq i \leq m)$ such that $k_i \in (p_{i-1}, p_i)$. Finally, substitutes each p_i in turn for K in $f(\boldsymbol{x}) - K$ to check if $f(\boldsymbol{x}) - p_i \geq 0$ holds for all $\boldsymbol{x}$. If p_j is the first such that $f(\boldsymbol{x}) - p_j \geq 0$ does not hold, then k_j is the infimum (let $k_0 = -\infty$). To check if $f(\boldsymbol{x}) - p_i \geq 0$ holds for all $\boldsymbol{x}$, the `SRes` method applies Brown-McCallum's projection on $f(\boldsymbol{x}) - p_i$ and choose sample points by open CAD in the lifting phase.

Remark 7.2. If $h(x) \geq 0$ for all $x \in \mathbb{R}^n$, Algorithm 7.2 can also be applied to compute $\inf\{\frac{f(x)}{h(x)} \mid x \in \mathbb{R}^n\}$. We just need to replace $g \leftarrow f - k$ in Line 1 by $g \leftarrow f - kh$. The proof of the correctness is the same.

We introduce some notations before we prove the correctness of Algorithm 7.2.

For a positive integer n, a_n, b_n and 0_n denote the points $(a_1, \ldots, a_n) \in \mathbb{R}^n$, $(b_1, \ldots, b_n) \in \mathbb{R}^n$, and $(0, \ldots, 0) \in \mathbb{R}^n$, respectively. For $a_n, b_n \in \mathbb{R}^n$, the Euclidean distance of a_n and b_n is defined by $\rho(a_n, b_n) = \sqrt{\sum_{i=1}^n (a_i - b_i)^2}$. For $a_n \in \mathbb{R}^n$, let $B_{a_n}(r)$ be the open ball which centered in a_n with radius r, that is

$$B_{a_n}(r) = \{b_n \in \mathbb{R}^n \mid \rho(a_n, b_n) < r\}.$$

For $a_n, b_n \in \mathbb{R}^n$, denote by $a_n b_n$ the segment $a_n \to b_n$. For m points $a_n^1, \ldots, a_n^m$, denote by $a_n^1 \to a_n^2 \to \cdots \to a_n^m$ the broken line through $a_n^1, \ldots, a_n^m$ in turn.

The following lemma can be inferred from the results of [McCallum (1998)] and [Brown (2001)], *i.e.* f is delineable over the maximal connected regions defined by $\mathrm{BMproj}(f, x_n) \neq 0$.

Lemma 7.1. *Let F_i, F_{i-1} be as in Algorithm 7.2. Let U be a connected component of $F_{i-1} \neq 0$ in $\mathbb{R}^i$ and $y_i^1(\gamma) < y_i^2(\gamma) < \cdots < y_i^m(\gamma)$ be all real roots of $F_i(\gamma, x_i) = 0$ for any given $\gamma \in U$. Then for all $\alpha, \beta \in U$ and $j = 2, \ldots, m$, $\alpha \times (y_i^{j-1}(\alpha), y_i^j(\alpha))$ and $\beta \times (y_i^{j-1}(\beta), y_i^j(\beta))$ are in the same connected component of $F_i \neq 0$ in $\mathbb{R}^{i+1}$.*

Proof. For $\alpha \in U$, let $\varepsilon = \min_{2 \leq i \leq m} \mid y_i(\alpha) - y_{i-1}(\alpha) \mid$, by Lemma 3.1, $\exists \delta > 0$, such that $\forall \alpha' \in B_\alpha(\delta)$, $\max_{1 \leq i \leq m} \mid y_i(\alpha) - y_i(\alpha') \mid < \frac{\varepsilon}{6}$.

Consider segment $(\alpha, \frac{y_{j-1}(\alpha) + y_j(\alpha)}{2}) \to (\alpha', \frac{y_{j-1}(\alpha') + y_j(\alpha')}{2})$ where $\alpha' \in B_\alpha(\delta)$. For any point (α'', y) on the segment, we have

$$
\begin{aligned}
&\mid y - y_s(\alpha'') \mid \\
&= \mid (y_s(\alpha'') - y_s(\alpha)) + (y_s(\alpha) - \tfrac{y_{j-1}(\alpha) + y_j(\alpha)}{2}) + (\tfrac{y_{j-1}(\alpha) + y_j(\alpha)}{2} - y) \mid \\
&\geq \mid y_s(\alpha) - \tfrac{y_{j-1}(\alpha) + y_j(\alpha)}{2} \mid - \mid y_s(\alpha'') - y_s(\alpha) \mid - \mid \tfrac{y_{j-1}(\alpha) + y_j(\alpha)}{2} - y \mid \\
&\geq \tfrac{\varepsilon}{2} - \tfrac{\varepsilon}{6} - \mid \tfrac{y_{j-1}(\alpha) + y_j(\alpha)}{2} - \tfrac{y_{j-1}(\alpha') + y_j(\alpha')}{2} \mid \\
&\geq \tfrac{\varepsilon}{2} - \tfrac{\varepsilon}{6} - \tfrac{\varepsilon}{6} \\
&> 0
\end{aligned}
$$

So the points satisfying $F_i = 0$ are not on the segment.

Therefore, for any points $r_1 \in (y_{j-1}(\alpha), y_j(\alpha))$, $r_2 \in (y_{j-1}(\alpha'), y_j(\alpha'))$, $(\alpha' \in B_\alpha(\delta))$, the points satisfying $F_i = 0$ are not on the broken line $(\alpha, r_1) \to (\alpha, \frac{y_{j-1}(\alpha) + y_j(\alpha)}{2}) \to (\alpha', \frac{y_{j-1}(\alpha') + y_j(\alpha')}{2}) \to (\alpha', r_2)$.

Hence we know that for any $\alpha \in U$, there exists $\delta > 0$ such that for any point $\alpha' \in B_\alpha(\delta)$ and $2 \leq s \leq m$, $\alpha \times (y_{s-1}(\alpha), y_s(\alpha))$ and $\alpha' \times (y_{s-1}(\alpha'), y_s(\alpha'))$ are in the same connected component of $F_i \neq 0$ in $\mathbb{R}^{i+1}$. For all $\alpha, \beta \in U$, there exists a path $\gamma : [0,1] \to U$ that connects α and β. Due to the compactness of the path, there are finitely many open sets $B_{\alpha_t}(\delta_t)$ covering $\gamma([0,1])$ with $\alpha_t \in \gamma([0,1])$ such that, for all $\alpha' \in B_{\alpha_t}(\delta_t)$ and $2 \leq j \leq m$, $\alpha \times (y_{j-1}(\alpha), y_j(\alpha))$ and $\alpha' \times (y_{j-1}(\alpha'), y_j(\alpha'))$ are in the same connected component of $F_i \neq 0$. Since the union of these open sets are connected, the lemma is proved. $\square$

Remark 7.3. By the above lemma, in Algorithm 7.2, for any two points p_l, $p_l' \in (k_l, k_{l+1})$, their corresponding sample points obtained through the open CAD lifting phase are in the same connected component of $F_n \neq 0$ in $\mathbb{R}^{n+1}$. Since at least one sample point can be taken from every highest dimensional cell via the open CAD lifting phase, the set of the corresponding sample points of p_l obtained through the open CAD lifting phase, *i.e.* C_{ln} in Algorithm 7.2, contains at least one point from every connected component U of $F_n(\boldsymbol{x}, k) \neq 0$, in which $U \bigcap (\mathbb{R}^n \times (k_{l-1}, k_l)) \neq \emptyset$.

Theorem 7.3. *The Successive Resultant Method is correct.*

Proof. Let notations be as in Algorithm 7.2. If there exists a $k' \in (k_i, k_{i+1})$, such that $F_n(\boldsymbol{x}, k') \geq 0$ for all $\boldsymbol{x} \in \mathbb{R}^n$, then by Lemma 7.1, for any $k \in (k_i, k_{i+1})$, $F_n(\boldsymbol{x}, k) \geq 0$ for all $\boldsymbol{x} \in \mathbb{R}^n$ (since their corresponding sample points obtained through the open CAD lifting phase are in the same connected component of $F_n(\boldsymbol{x}, k) \neq 0$ in $\mathbb{R}^{n+1}$). Therefore, for any $k \in [k_i, k_{i+1}]$, $F_n(\boldsymbol{x}, k) \geq 0$ for all $\boldsymbol{x} \in \mathbb{R}^n$. The global optimum k will be found by checking whether $\forall \boldsymbol{a}_n \in \mathbb{R}^n (F_n(\boldsymbol{a}_n, p_i) \geq 0)$ holds where p_i is the sample point of (k_i, k_{i+1}). Since Algorithm 7.2 ensures that at least one point is chosen from every connected component of $F_n(\boldsymbol{x}, p_i) \neq 0$ in $\mathbb{R}^n$, the theorem is proved. $\square$

7.3 Projection Operator Np

7.3.1 *An Illustrative Example*

We first show the comparison of Brown-McCallum's projection operator and the projection operator Np on a simple example. Formal description and proofs of new algorithms are given subsequently.

Example 7.1. Prove or disprove

$$\forall (x, y, z) \in \mathbb{R}^3 \, (f(x, y, z) \geq 0)$$

where

$$f(x, y, z) = 4z^4 - 4z^2 y^2 - 4z^2 + 4y^2 x^4 + 4x^2 y^4 + 8x^2 y^2 + 5y^4 + 6y^2 + 4x^4 + 4x^2 + 1.$$

We solve this example by constructing an open CAD for f. First we apply Brown-McCallum's projection operator and take the following steps:

Step 1. (First projection: apply Brown-McCallum's projection operator to f)

$$f_1 := \mathrm{res}(\mathrm{sqrfree}(f), \frac{\partial}{\partial z} \mathrm{sqrfree}(f), z) = 1048576 g_1^3 g_2 h_1^2 h_2^2,$$

where

$$g_1 = y^2 + 1, \ g_2 = 4x^4 + 4x^2 y^2 + 4x^2 + 5y^2 + 1, \ h_1 = x^2 + 1, \ h_2 = x^2 + y^2,$$

Step 2. (Second projection: apply Brown-McCallum's projection operator to f_1)

$$f_2 := \mathrm{res}(\mathrm{sqrfree}(f_1), \frac{\partial}{\partial y} \mathrm{sqrfree}(f_1), y)$$
$$= \mathrm{res}(g_1 g_2 h_1 h_2, \frac{\partial (g_1 g_2 h_1 h_2)}{\partial y}, y)$$
$$= 16384(x^2 + 1)^{15}(x - 1)^{12}(x + 1)^{12}(2x^2 + 1)^2(4x^2 + 5)^2 x^2.$$

Actually, computing f_2 is equivalent to computing the following 6 resultants.

(a) $\mathrm{res}(g_i, \frac{\partial}{\partial y} g_i, y)$ $(i = 1, 2)$,
(b) $\mathrm{res}(h_2, \frac{\partial}{\partial y} h_2, y)$,
(c) $\mathrm{res}(g_1, g_2, y), \mathrm{res}(g_i, h_2, y)$ $(i = 1, 2)$.

Step 3. (Base and lifting phases)
By real root isolation of $f_2 = 0$, choose 4 sample points of x: $x_1 = -2$, $x_2 = -\frac{1}{2}$, $x_3 = \frac{1}{2}$, $x_4 = 2$. At the lifting phase, we first get 4 sample points of (x, y) for $f_1(x_i, y) \neq 0$: $(-2, 0), (-\frac{1}{2}, 0), (\frac{1}{2}, 0), (2, 0)$. Then get 4 sample points of (x, y, z) for $f(x_i, y_i, z) \neq 0$: $(-2, 0, 0), (-\frac{1}{2}, 0, 0), (\frac{1}{2}, 0, 0), (2, 0, 0)$.

Step 4. (Consistency determination)

Finally we should check that whether or not $f(x, y, z) \geq 0$ at all the 4 sample points. Because $f(x, y, z) \geq 0$ at all the sample points, the answer is that the given sentence is true.

Now, we apply the new algorithm (see Algorithm DPS) to the problem and take the following steps:

Step 1. (First projection: apply new projection operator to f)

We first apply Brown-McCallum's projection operator to f, collect the odd and even factors of f_1, and let

$$L_1 = \{g_1, g_2\}, L_2 = \{h_1, h_2\}.$$

According to Theorem 7.7, $\forall (x, y, z) \in \mathbb{R}^3 (f(x, y, z) \geq 0)$ if and only if (1) both of g_1 and g_2 are positive semi-definite on $\mathbb{R}^2$, and (2) there exists a set of points $\mathcal{A} \subset \mathbb{R}^2$ satisfying that (i) the intersection of $\mathcal{A}$ and each connected component of $h_1 h_2 \neq 0$ in $\mathbb{R}^2$ is nonempty, (ii) the intersection of $\mathcal{A}$ and the real zeros of $g_1 g_2$ is empty, and for all $\alpha \in \mathcal{A}$, $f(\alpha, z)$ is positive semi-definite on $\mathbb{R}$. Thus, we need to check the nonnegativity of g_1 and g_2 in the next step.

Step 2. (Check the nonnegativity of g_1 and g_2)

Step 2(i). (Decide the nonnegativity of g_1)

By real root isolation of $g_1 = 0$, we choose $y_1 = 0$ as a sample point for g_1, and verify that $g_1 \geq 0$ at this point.

Step 2(ii). (Decide the nonnegativity of g_2)

We apply Brown-McCallum's projection operator to g_2.

$$\mathrm{res}(g_2, \frac{\partial}{\partial y} g_2, y) = 4(2x^2 + 1)^2 (4x^2 + 5)^2.$$

By real root isolation of $(2x^2 + 1)^2 (4x^2 + 5)^2$, we choose $x_1 = 0$ as a sample point of x. At lifting phase, compute a sample point $(0, 0)$ of (x, y) and verify that $g_2 \geq 0$ at this point.

Step 3. (Second projection: apply Brown-McCallum's projection operator to $h_1 h_2$)

$$\mathrm{res}(h_1 h_2, \frac{\partial}{\partial y} h_1 h_2, y) = h_1 \mathrm{res}(h_2, \frac{\partial}{\partial y} h_2, y) = 4x^2 (x^2 + 1).$$

Step 4. (Base and lifting phases)

By real root isolation of $x^2 (x^2 + 1) = 0$, we choose $x_1 = -1$ and $x_2 = 1$ as sample points of x. At lifting phase, compute 2 sample points $(-1, 0), (1, 0)$ of (x, y) and verify that $g_1 g_2 \neq 0$ at these two points. Then compute 2 sample points $(-1, 0, 0), (1, 0, 0)$ of (x, y, z).

Step 5. (Consistency determination)

Check whether or not $f(x, y, z) \geq 0$ at all the 2 sample points. Because $f(x, y, z) \geq 0$ at all the sample points, the answer is that the given sentence is true.

For this example, the projection operator $\mathtt{Np}$ avoids computing 4 resultants compared to Brown-McCallum's operator. In general, for a polynomial $f(x_1, \ldots, x_n) \in \mathbb{Z}[x_1, \ldots, x_n]$, $\mathtt{Np}$ first computes $f_1 = \mathrm{res}(\mathrm{sqrfree}(f), \frac{\partial}{\partial x_n}\mathrm{sqrfree}(f), x_n)$ as other CAD based methods do. Then, divides the irreducible factors of f_1 into two groups: L_1 and L_2, where L_1 contains all factors with odd multiplicities and L_2 contains all factors with even multiplicities. Compared to Brown-McCallum's projection, at the next level of projection, neither the resultants of those polynomial pairs of which one is from L_1 and the other from L_2 nor the resultants of the polynomial pairs in L_1 are to be computed. Therefore, the scale of $\mathtt{Np}$ is no larger than that of Brown-McCallum's. For a wide class of problems (see for example Remark 7.6), especially when $n \geq 3$, the scale of $\mathtt{Np}$ is much smaller than that of Brown-McCallum's. Based on the new operator, we obtain a new algorithm $\mathtt{DPS}$ to determining the nonnegativity of a polynomial.

7.3.2 *Notations*

Let $f(\boldsymbol{x}) \in \mathcal{R}[\boldsymbol{x}]$, say $f(\boldsymbol{x}) = \sum_{i=0}^{l} c_i x_n^i, c_l \neq 0$, where $c_i(i = 0, \ldots, l)$ is a polynomial in $\boldsymbol{x}^{[n-1]}$. Then, the *leading base coefficient* of f in $\mathcal{R}$, $\mathrm{lbcf}(f)$, is defined by induction on n: $\mathrm{lbcf}(f) = \mathrm{lc}(f, x_1)$ if $n = 1$ and $\mathrm{lbcf}(f) = \mathrm{lbcf}(\mathrm{lc}(f, x_n))$ if $n > 1$.

Definition 7.9. Suppose $h \in \mathcal{R}[\boldsymbol{x}]$ can be factorized in $\mathcal{R}[\boldsymbol{x}]$ as:

$$h = a l_1^{2j_1-1} \ldots l_t^{2j_t-1} h_1^{2i_1} \ldots h_m^{2i_m},$$

where $a \in \mathcal{R}$, $l_i(i = 1, \ldots, t)$ and $h_j(j = 1, \ldots, m)$ are pairwise different irreducible primitive nonconstant polynomials with positive leading base coefficient in $\mathcal{R}[\boldsymbol{x}]$. Define

$$\mathrm{sqrfree}(h) = \prod_{i=1}^{t} l_i \prod_{i=1}^{m} h_i,$$

$$\mathrm{sqrfree}_1(h) = \{l_i, i = 1, 2, \ldots, t\},$$

$$\mathrm{sqrfree}_2(h) = \{h_i, i = 1, 2, \ldots, m\}.$$

If h is a constant, let $\mathrm{sqrfree}(h) = 1$ and $\mathrm{sqrfree}_1(h) = \mathrm{sqrfree}_2(h) = \emptyset$.

Definition 7.10. Suppose $f \in \mathcal{R}[x]$ is a polynomial of class n. Define

$$\mathrm{Oc}(f, x_n) = \mathrm{sqrfree}_1(\mathrm{lc}(f, x_n)), \ \mathrm{Od}(f, x_n) = \mathrm{sqrfree}_1(\mathrm{discrim}(f, x_n)),$$
$$\mathrm{Ec}(f, x_n) = \mathrm{sqrfree}_2(\mathrm{lc}(f, x_n)), \ \mathrm{Ed}(f, x_n) = \mathrm{sqrfree}_2(\mathrm{discrim}(f, x_n)),$$
$$\mathrm{Ocd}(f, x_n) = \mathrm{Oc}(f, x_n) \cup \mathrm{Od}(f, x_n), \ \mathrm{Ecd}(f, x_n) = \mathrm{Ec}(f, x_n) \cup \mathrm{Ed}(f, x_n).$$

The *secondary* and *principal parts* of the projection $\mathtt{Np}$ are defined as

$$\mathtt{Np}_1(f, x_n) = \mathrm{Ocd}(f, x_n),$$

$$\mathtt{Np}_2(f, x_n) = \{ \prod_{g \in \mathrm{Ecd}(f,x_n) \backslash \mathrm{Ocd}(f,x_n)} g \}.$$

If L is a set of polynomials of class n, define

$$\mathtt{Np}_1(L, x_n) = \cup_{g \in L} \mathrm{Ocd}(g, x_n),$$

$$\mathtt{Np}_2(L, x_n) = \bigcup_{g \in L} \{ \prod_{h \in \mathrm{Ecd}(g,x_n) \backslash \mathtt{Np}_1(L,x_n)} h \}.$$

Remark 7.4. In case $n = 1$, both sets $\mathtt{Np}_1(f, x_n)$ and $\mathtt{Np}_2(f, x_n)$ are empty.

Remark 7.5. We use irreducible factorization instead of square-free factorization in definitions, theorems and algorithms, since the implementation of our algorithms based on irreducible factorization performs better in practice. It should be pointed out that all those definitions, theorems and algorithms are still valid if we use square-free factorization instead of irreducible factorization.

7.3.3 *Algorithm* DPS

Algorithm 7.3 DPS (Decide Positive Semi-definiteness)

Input: A polynomial $f(x) \in \mathbb{Z}[x]$, of class n, with positive leading base coefficient

Output: Whether or not $f(x) \geq 0$ on $\mathbb{R}^n$

1: $L_1 \leftarrow \mathrm{sqrfree}_1(f)$;
2: **for** each g in L_1 **do**
3:    **if** not $\mathrm{DPSIP}(g)$ **then return** false **end if**
4: **end for**
5: **return** true

Algorithm 7.4 DPSIP (Decide Positive Semi-definiteness for Irreducible Polynomials)

Input: An irreducible polynomial $f(\boldsymbol{x}) \in \mathbb{Z}[\boldsymbol{x}]$, of class n, with positive leading base coefficient

Output: Whether or not $f(\boldsymbol{x}) \geq 0$ on $\mathbb{R}^n$

1: **if** f is a univariate polynomial **then**
2:    **if** f has real root **then**
3:      **return** `false`
4:    **else**
5:      **return** `true`
6:    **end if**
7: **end if**
8: $L_1 \leftarrow \mathrm{Np}_1(f, x_n)$;
9: **for** each g in L_1 **do**
10:    **if** not DPSIP(g) **then return** `false` **end if**
11: **end for**
12: $L_2 \leftarrow \mathrm{Np}_2(f, x_n)$;
13: Construct an open CAD C_{n-1} of $\mathbb{R}^{n-1}$ w.r.t. L_2 such that for all $\boldsymbol{a}_{n-1}$ in C_{n-1}, $\boldsymbol{a}_{n-1}$ is not a zero of an element of L_1; (Use a slight modification of open CAD.)
14: Lift from C_{n-1} to C_n using f;
15: **if** there exists $\boldsymbol{a}_n$ in C_n such that $f(\boldsymbol{a}_n) < 0$ **then**
16:    **return** `false`
17: **end if**
18: **return** `true`

We first describe the basic idea of Algorithm DPSIP (Decide Positive Semi-definiteness of Irreducible Polynomials) which is fully described and validated below. Let f be an irreducible polynomial of class $n \geq 2$ with positive leading base coefficient. By Theorem 7.7, the task of proving $f(\boldsymbol{x}) \geq 0$ on $\mathbb{R}^n$ can be accomplished by (1) proving that all the polynomials in $\mathrm{Np}_1(f, x_n)$ are positive semi-definite on $\mathbb{R}^{n-1}$; and (2) computing sample points of $\mathrm{Np}_2(f, x_n) \neq 0$ in $\mathbb{R}^{n-1}$ and checking $f(\alpha, x_n) \geq 0$ on $\mathbb{R}$ for all sample points α. To ensure (1) holds, we apply Algorithm DPSIP recursively to each element $g \in \mathrm{Np}_1(f, x_n)$. In case we find that some $g \in \mathrm{Np}_1(f, x_n)$ is not positive semi-definite on $\mathbb{R}^{n-1}$ we can immediately return a negative answer for f (by the necessity part of Theorem 7.7). In case no such g is found, this means that (1) is satisfied. To ensure that (2) holds, we construct an open CAD of $\mathbb{R}^{n-1}$ with respect to $\mathrm{Np}_2(f, x_n)$, with a slight modification. In case $f(\alpha, x_n)$ is not positive semi-definite, for some sample point α, we return a negative answer for f. In case $f(\alpha, x_n) \geq 0$ for all sample points α, (2) holds and hence we can return an affirmative answer for f (by the sufficiency part of Theorem 7.7). Algorithm 7.4 is a formal description of this algorithm.

We describe the basic idea of Algorithm DPS (Decide Positive Semi-definiteness) next. Let f be a polynomial $f(x) \in \mathbb{Z}[x]$, of class n, with positive leading base coefficient. According to Proposition 7.2, in order to check the positive semi-definiteness of f on $\mathbb{R}^n$, it suffices to check the positive semi-definiteness of every polynomial $g \in \mathrm{sqrfree}_1(f)$. Since g is an irreducible polynomial of class $k \leq n$ with positive leading base coefficient, we can call Algorithm DPSIP for this step. Algorithm 7.3 is a formal description of this algorithm.

To give the readers a picture of how the projection operator is different from existing CAD projection operators, we give Algorithm 7.5 here, which returns all possible polynomials that may appear in the projection phase of Algorithm 7.3 and Algorithm 7.4.

Algorithm 7.5 Np

Input: A polynomial $f(x) \in \mathbb{Z}[x]$

Output: Two projection factor sets containing all possible polynomials that may appear in the projection phase of Algorithm 7.3

1: $L_1 \leftarrow \mathrm{sqrfree}_1(f)$;
2: $L_2 \leftarrow \{\}$;
3: **for** i from n downto 2 **do**
4: $\quad L_2 \leftarrow L_2 \bigcup \mathrm{Np}_2(L_1^{[i]}, x_i) \bigcup \cup_{g \in L_2^{[i]}} \mathrm{BMproj}(g, x_i)$; (Recall that $L^{[i]}$ is the set of polynomials in L of class i.)
5: $\quad L_1 \leftarrow L_1 \bigcup \mathrm{Np}_1(L_1^{[i]}, x_i)$;
6: **end for**
7: **return** (L_1, L_2)

Remark 7.6. For polynomial $\tilde{p}(x_1, \ldots, x_{n-1}, x_n) = p(x_1, \ldots, x_{n-1}, x_n^2)$ $(\deg(\tilde{p}, x_n) \geq 2, n \geq 2)$, the resultant of $\tilde{p}$ and $\tilde{p}'_{x_n}$ with respect to x_n is (may differ from a constant)

$$p(x_1, \ldots, x_{n-1}, 0)\mathrm{res}(p, p'_{x_n}, x_n)^2.$$

If $p(x_1, \ldots, x_{n-1}, 0)$ is not a square, the set $\mathrm{Np}_1(\tilde{p}, x_n)$ is not empty and thus the scale of $\mathrm{Np}(\tilde{p})$ is smaller than that of $\mathrm{BMprojection}(\tilde{p})$.

If for any polynomial $f \in \mathbb{Z}[x_1, \ldots, x_n]$, the iterated discriminants of f always have odd factors and are reducible (for generic f or for most polynomials, it is quite likely), then for $n \geq 3$, the scale of $\mathrm{Np}(f)$ is always strictly smaller than that of $\mathrm{BMprojection}(f)$.

7.3.4 *The Correctness of Algorithm* DPS

The following lemma is a well-known result.

Lemma 7.2. *Let $f(\boldsymbol{x}) \in \mathbb{R}[\boldsymbol{x}]$ and r be a positive real number. If $f(\boldsymbol{a}_n) = 0$ for all $\boldsymbol{a}_n \in B_{\mathbf{0}_n}(r)$, then $f(\boldsymbol{x}) \equiv 0$.*

Lemma 7.3. *For $f, g \in \mathbb{R}[\boldsymbol{x}]$, if f and g are coprime in $\mathbb{R}[\boldsymbol{x}]$, then after any linear invertible transform, f and g are still coprime in $\mathbb{R}[\boldsymbol{x}]$, namely for $A \in GL_n(\mathbb{R})$, $B_n \in \mathbb{R}^n$, $\boldsymbol{x}_n^{*T} = A\boldsymbol{x}_n^T + B_n^T$, then $\gcd(f(\boldsymbol{x}_n^*), g(\boldsymbol{x}_n^*)) = 1$ in $\mathbb{R}[\boldsymbol{x}]$.*

Proof. If $\gcd(f(\boldsymbol{x}_n^*), g(\boldsymbol{x}_n^*)) = h(\boldsymbol{x})$ and h is not a constant, then $h(A^{-1}(\boldsymbol{x} - B_n^T))$ is a non-trivial common divisor of f and g in $\mathbb{R}[\boldsymbol{x}]$, which is a contradiction. $\qquad\square$

Lemma 7.4. *Suppose $f, g \in \mathbb{R}[\boldsymbol{x}]$ and $\gcd(f, g) = 1$ in $\mathbb{R}[\boldsymbol{x}]$. For any $\boldsymbol{a}_{n-1} \in \mathbb{R}^{n-1}$ and $r > 0$, there exists $\boldsymbol{a}'_{n-1} \in \mathbb{R}^{n-1}$ such that $\rho(\boldsymbol{a}_{n-1}, \boldsymbol{a}'_{n-1}) < r$ and for all $a'_n \in \mathbb{R}$, $(\boldsymbol{a}'_{n-1}, a'_n) \notin V_{\mathbb{R}}(f, g)$.*

Proof. Otherwise, there exist $\boldsymbol{a}_{n-1}^0 = (a_1^0, \ldots, a_{n-1}^0) \in \mathbb{R}^{n-1}$ and $r_0 > 0$, such that for any $\boldsymbol{a}_{n-1}^1 = (a_1^1, \ldots, a_{n-1}^1)$ satisfying $\rho(\boldsymbol{a}_{n-1}^0, \boldsymbol{a}_{n-1}^1) < r_0$, there exists an $a_n^1 \in \mathbb{R}$ such that $f(\boldsymbol{a}_{n-1}^1, a_n^1) = g(\boldsymbol{a}_{n-1}^1, a_n^1) = 0$. Thus $\mathrm{res}(f, g, x_n) = 0$ at every point of $B_{\boldsymbol{a}_{n-1}^0}(r_0)$. From Lemma 7.2, we get that $\mathrm{res}(f, g, x_n) \equiv 0$, meaning $\gcd(f, g)$ is non-trivial, which is impossible. $\qquad\square$

Let $f(x_{n+1}) = c_l x_{n+1}^l + \cdots + c_0$. Suppose the coefficients of f are given parametrically as polynomials in $\boldsymbol{x}$. If the leading coefficient $\mathrm{lc}(f, x_{n+1}) =$

$c_l \not\equiv 0$, the discriminant of $f(\boldsymbol{x}, x_{n+1})$ can be written as

$$\mathrm{discrim}(f, x_{n+1}) = (-1)^{\frac{l(l-1)}{2}} \begin{vmatrix} 1 & c_{l-1} & c_{l-2} & \cdots & c_j & \cdots \\ 0 & c_l & c_{l-1} & \cdots & c_{j+1} & \cdots \\ 0 & 0 & c_l & \cdots & c_{j+2} & \cdots \\ \vdots & \vdots & \vdots & \ddots & \vdots & \ddots \\ l & (l-1)c_{l-1} & (l-2)c_{l-2} & \cdots & jc_j & \cdots \\ 0 & lc_l & (l-1)c_{l-1} & \cdots & (j+1)c_{j+1} & \cdots \\ 0 & 0 & lc_l & \cdots & (j+2)c_{j+2} & \cdots \\ \vdots & \vdots & \vdots & \ddots & \vdots & \ddots \end{vmatrix}.$$

If $c_l = c_{l-1} = 0$ at point $\boldsymbol{a}_n$, from the above expression, $\mathrm{discrim}(f, x_{n+1}) = 0$ at this point.

Lemma 7.5. *Given a polynomial $f(\boldsymbol{x}, x_{n+1}) \in \mathbb{R}[\boldsymbol{x}, x_{n+1}]$, say*

$$f(\boldsymbol{x}, x_{n+1}) = \sum_{i=0}^{l} c_i x_{n+1}^i, c_l \not\equiv 0,$$

where $c_i (i = 0, \ldots, l)$ is a polynomial in $\boldsymbol{x}$. Let U be an open set in $\mathbb{R}^n$. If $f(\boldsymbol{x}, x_{n+1}) \geq 0$ on $U \times \mathbb{R}$, then l is even and

$$(-1)^{\frac{l}{2}}\mathrm{discrim}(f, x_{n+1}) \geq 0 \quad and \quad \mathrm{lc}(f, x_{n+1}) \geq 0 \quad for\ all \quad \boldsymbol{a}_n \in U.$$

Proof. Since f is positive semi-definite for any given $\boldsymbol{a}_n \in U$, $\mathrm{lc}(f, x_{n+1})$ is positive semi-definite on U and l is even. If $c_l > 0$ at $\boldsymbol{a}_n$ and $f(\boldsymbol{a}_n, x_{n+1})$ is square-free, then

$$(-1)^{\frac{l}{2}}\mathrm{discrim}(f(\boldsymbol{x}, x_{n+1}), x_{n+1}) \mid_{\boldsymbol{x}=\boldsymbol{a}_n} = (-1)^{\frac{l}{2}}\mathrm{discrim}(f(\boldsymbol{a}_n, x_{n+1}), x_{n+1}) > 0$$

by Corollary 4.3. Otherwise, either $c_l = 0$ at $\boldsymbol{a}_n$ which suggests $c_{l-1} = 0$ at $\boldsymbol{a}_n$, or $c_l > 0$ at $\boldsymbol{a}_n$ and $f(\boldsymbol{a}_n, x_{n+1})$ is not square-free. In both cases we can deduce

$$(-1)^{\frac{l}{2}}\mathrm{discrim}(f(\boldsymbol{x}, x_{n+1}), x_{n+1}) = 0$$

at $\boldsymbol{a}_n$. That completes the proof. $\qquad\square$

Theorem 7.4. *Let $f(\boldsymbol{x})$ and $g(\boldsymbol{x})$ be coprime in $\mathbb{R}[\boldsymbol{x}]$. For any connected open set U in $\mathbb{R}^n$, the open set $V = U \backslash V_{\mathbb{R}}(f, g)$ is also connected.*

This theorem plays an important role in our proof. It can be proved by the fact that closed and bounded semi-algebraic set is semi-algebraically triangulable (Theorem 9.2.1 of [Bochnak *et al.* (1998)]) and Alexander duality. Here we give an elementary proof.

Proof. For any two points α, β in V, we only need to prove that there exists a path $\gamma(t) : [0,1] \to V$ such that $\gamma(0) = \alpha, \gamma(1) = \beta$. Choose a path γ_U that connects α and β in U. Notice that U is an open set, so for any $X_n \in \gamma_U$, there exists $\delta_{X_n} > 0$ such that $U \supset B_{X_n}(\delta_{X_n})$. Since γ_U is compact and $\bigcup B_{X_n}(\delta_{X_n})$ is an open covering of γ_U, there exists an $m \in \mathbb{N}$, such that $\bigcup_{k=1}^{m} B_{X_n^k}(\delta_{X_n^k}) \supset \gamma_U$ and $\alpha \in B_{X_n^1}(\delta_{X_n^1})$, $\beta \in B_{X_n^m}(\delta_{X_n^m})$, $B_{X_n^i}(\delta_{X_n^i}) \bigcap B_{X_n^{i+1}}(\delta_{X_n^{i+1}}) \neq \emptyset$ $(i = 1, \ldots, m-1)$.

Now we only need to prove that for every k, $B_{X_n^k}(\delta_{X_n^k}) \backslash V_{\mathbb{R}}(f, g)$ is connected. If this is the case, we can find k paths γ_1, γ_2, $\ldots$, γ_k with $\gamma_1(0) = \alpha$, $\gamma_1(1) \in B_{X_n^1}(\delta_{X_n^1}) \bigcap B_{X_n^2}(\delta_{X_n^2})$, $\gamma_{i+1}(0) = \gamma_i(1)$, $\gamma_{i+1}(1) \in B_{X_n^i}(\delta_{X_n^i}) \bigcap B_{X_n^{i+1}}(\delta_{X_n^{i+1}})$ $(i = 1, 2, \ldots, m-1)$, $\gamma_m(1) = \beta$. Let γ be the path: $[0,1] \to U$ which satisfies $\gamma([\frac{j-1}{m}, \frac{j}{m}]) = \gamma_j([0,1])$ $(j = 1, \ldots, m)$, then γ is the path as desired.

Choose a, $b \in B_{X_n^k}(\delta_{X_n^k}) \backslash V_{\mathbb{R}}(f, g)$. There exists an affine coordinate transformation T such that $T(B_{X_n^k}(\delta_{X_n^k})) = B_{\mathbf{0}_n}(1)$ and $\overrightarrow{T(a)T(b)}$ and $\overrightarrow{(\mathbf{0}_{n-1}, 1)}$ are parallel. Thus the first $n-1$ coordinates of $T(a)$ and $T(b)$ are the same. Let $T(a) = (Y_{n-1}, a')$, $T(b) = (Y_{n-1}, b')$. Without loss of generality, we assume that $a' > b'$.

In the new coordinate, f and g become $T(f)$ and $T(g)$, respectively. $B_{\mathbf{0}_n}(1)$ is an open set and $T(a), T(b) \notin V_{\mathbb{R}}(T(f), T(g))$, so there exists $r > 0$ such that the cylinder $B_{Y_{n-1}}(r) \times [b', a'] \subseteq B_{\mathbf{0}_n}(1)$, $B_{T(a)}(r) \bigcap V_{\mathbb{R}}(T(f), T(g)) = \emptyset$ and $B_{T(b)}(r) \bigcap \mathrm{Zero} \, (T(f), T(g)) = \emptyset$. By Lemma 7.3, $T(f)$ and $T(g)$ are coprime in $\mathbb{R}[\boldsymbol{x}]$. So by Lemma 7.4, there exists $X'_{n-1} \in B_{Y_{n-1}}(r)$, such that for any $x_n \in \mathbb{R}$, $(X'_{n-1}, x_n) \notin V_{\mathbb{R}}(T(f), T(g))$. Thus the broken line $T(a) \to (X'_{n-1}, a') \to (X'_{n-1}, b') \to T(b)$ is a path that connects $T(a)$ and $T(b)$ in $B_{\mathbf{0}_n}(1) \backslash V_{\mathbb{R}}(T(f), T(g))$. The theorem is proved. $\qquad\square$

Proposition 7.1. *Suppose $U \subseteq \mathbb{R}^n$ is a connected open set, $f, g \in \mathbb{R}[\boldsymbol{x}]$, $\gcd(f, g) = 1$ in $\mathbb{R}[\boldsymbol{x}]$ and for all $\alpha \in U$, $f(\alpha)g(\alpha) \geq 0$. Then either $f(\alpha) \geq 0, g(\alpha) \geq 0$ for all $\alpha \in U$ or $f(\alpha) \leq 0, g(\alpha) \leq 0$ for all $\alpha \in U$. Similarly, if for all $\alpha \in U$, $f(\alpha)g(\alpha) \leq 0$, then either $f(\alpha) \geq 0, g(\alpha) \leq 0$ for all $\alpha \in U$ or $f(\alpha) \leq 0, g(\alpha) \geq 0$ for all $\alpha \in U$.*

Proof. If not, there exist α_1, $\alpha_2 \in U$, such that $f(\alpha_1) \leq 0$, $g(\alpha_1) \leq 0$ and at least one inequality is strict, and $f(\alpha_2) \geq 0$, $g(\alpha_2) \geq 0$ and at least one inequality is strict. By Theorem 7.4, $U \backslash V_{\mathbb{R}}(f, g)$ is connected. So we can choose a path γ in U that connects α_1 with α_2 and $\gamma \bigcap V_{\mathbb{R}}(f, g) = \emptyset$.

Consider the sign of $f + g$ on γ. Since the sign is different at α_1 and α_2, by intermediate value theorem we know there exists α_3 on γ such that $f(\alpha_3) + g(\alpha_3) = 0$. From an assumption we know that $f(\alpha_3)g(\alpha_3) \geq 0$, hence $\alpha_3 \in V_{\mathbb{R}}(f, g)$, which contradicts the choice of γ.

The second part of the proposition can be proved similarly. $\qquad\square$

The following proposition is an easy corollary of Proposition 7.1.

Proposition 7.2. *Let $f \in \mathcal{R}[\boldsymbol{x}]$ be a polynomial with positive leading base coefficient and* $\mathrm{cls}(f) = n$. *Then the necessary and sufficient condition for $f(\boldsymbol{x})$ to be positive semi-definite on $\mathbb{R}^n$ is, for every polynomial $g \in$ $\mathrm{sqrfree}_1(f)$, g is positive semi-definite on $\mathbb{R}^n$.*

Proof. Let

$$f = al_1^{2j_1-1} \ldots l_t^{2j_t-1} h_1^{2i_1} \ldots h_m^{2i_m},$$

where $a \geq 0$, $h_i (i = 1, \ldots, m)$ and $l_j (i = 1, \ldots, t)$ are pairwise different irreducible primitive nonconstant polynomials with positive leading base coefficient in $\mathcal{R}[\boldsymbol{x}]$.

Since $f \geq 0$, $l_1^{2j_1-1} \ldots l_t^{2j_t-1} \geq 0$. By Proposition 7.1, $l_{j_s} \geq 0$ for all $t \geq s \geq 1$. This proves the necessity.

If $l_{j_s} \geq 0$ for all $t \geq s \geq 1$, then $l_1^{2j_1-1} \ldots l_t^{2j_t-1} \geq 0$, thus

$$f = al_1^{2j_1-1} \ldots l_t^{2j_t-1} h_1^{2i_1} \ldots h_m^{2i_m} \geq 0.$$

This proves the sufficiency. $\qquad\square$

Proposition 7.3. *Suppose $f \in \mathbb{R}[\boldsymbol{x}]$ is a nonzero square-free polynomial and U is a connected open set of $\mathbb{R}^n$. If $f(\boldsymbol{x})$ is semi-definite on U, then $U \backslash V_{\mathbb{R}}(f)$ is also a connected open set.*

Proof. Without loss of generality, we assume $f(\boldsymbol{x}) \geq 0$ on U. Since f is nonzero, we only need to consider the case that the class of f is nonzero. Let $i > 0$ be the class of f and consider f as a polynomial of x_i. Because $f(\boldsymbol{x}) \geq 0$ on U, we claim that

$$V_{\mathbb{R}}(f) \bigcap U = V_{\mathbb{R}}(f, f'_{x_i}) \bigcap U.$$

Otherwise, we may assume there exists a point $X_n^0 = (x_1^0, \ldots, x_n^0) \in U$ such that $f(X_n^0) = 0$ and $f'_{x_i}(X_n^0) > 0$. Thus, there exists r such that $\forall X_n \in B_{X_n^0}(r) \subset U$, $f'_{x_i}(X_n) > 0$. Let $F(x_i) = f(x_1^0, \ldots, x_{i-1}^0, x_i, x_{i+1}^0, \ldots, x_n^0)$. The Taylor series of F at point x_i^0 is

$$F(x_i) = F(x_i^0) + (x_i - x_i^0)F'_{x_i}(x_i^0 + \theta(x_i - x_i^0)),$$

where $\theta \in (0,1)$. Let $x_i^0 > x_i^1 > x_i^0 - r$, then $F(x_i^1) < 0$, which contradicts the definition of F.

If f is irreducible in $\mathbb{R}[\boldsymbol{x}]$, f and f'_{x_i} are coprime in $\mathbb{R}[\boldsymbol{x}]$. Thus $U\backslash V_\mathbb{R}(f, f'_{x_i})$ is connected by Theorem 7.4. So $U\backslash V_\mathbb{R}(f)$ is a connected open set.

If f is reducible in $\mathbb{R}[\boldsymbol{x}]$, let $f = a\prod_{t=1}^{j} f_t$, where $a \in \mathbb{R}$ and $f_t(t = 1,\ldots,j)$ are pairwise different irreducible nonconstant primitive polynomials with positive leading base coefficient in $\mathcal{R}[\boldsymbol{x}]$, then $U\backslash V_\mathbb{R}(f) = U\backslash \bigcup_{t=1}^{j} V_\mathbb{R}(f_t)$ is a connected open set. The proposition is proved. $\square$

Theorem 7.5. *Given a positive integer $n \geq 2$. Let $f \in \mathcal{R}[\boldsymbol{x}]$ be a nonzero square-free polynomial and $U \subseteq \mathbb{R}^{n-1}$ be a (maximal) connected region in which the elements of $\mathrm{Np}_2(f, x_n)$ have constant, nonzero sign. If the polynomials in $\mathrm{Np}_1(f, x_n)$ are semi-definite on U, then f is delineable on $V = U\backslash \bigcup_{h\in \mathrm{Np}_1(f,x_n)} V_\mathbb{R}(h)$.*

Proof. According to Theorem 7.1 and Theorem 7.2, f is delineable over every (maximal) connected component of $\mathrm{res}(f, f'_{x_n}, x_n) \neq 0$. By Proposition 7.3, $V = U\backslash \bigcup_{h\in \mathrm{Np}_1(f,x_n)} V_\mathbb{R}(h)$ is a connected open set. Thus, f is delineable on V. $\square$

Suppose the conditions of Theorem 7.5 are satisfied. Since f is delineable on V, we can determine semi-definiteness of f on $V \times \mathbb{R}$ by testing semi-definiteness of f at sample points over one sample point of V. Therefore we have the following theorem.

Theorem 7.6. *Let $f \in \mathcal{R}[\boldsymbol{x}]$ be a square-free polynomial of class $n(n \geq 2)$ and $U \subseteq \mathbb{R}^{n-1}$ be a (maximal) connected region in which the elements of $\mathrm{Np}_2(f, x_n)$ have constant, nonzero sign. The necessary and sufficient condition for $f(\boldsymbol{x})$ to be semi-definite on $U \times \mathbb{R}$ is the following two conditions hold.*
(1) The polynomials in $\mathrm{Np}_1(f, x_n)$ are semi-definite on U;
(2) There exists a point $\alpha \in U\backslash \bigcup_{h\in \mathrm{Np}_1(f,x_n)} V_\mathbb{R}(h)$, $f(\alpha, x_n)$ is semi-definite on $\mathbb{R}$.

Proof. $\Longrightarrow$: By Lemma 7.5, $\mathrm{discrim}(f, x_n)$ is semi-definite on U. Thus by Proposition 7.1, the polynomials in $\mathrm{Np}_1(f, x_n)$ are semi-definite on U. It is obvious that $f(\alpha, x_n)$ is semi-definite on $\mathbb{R}$.

$\Longleftarrow$: If the polynomials in $\mathrm{Np}_1(f, x_n)$ are semi-definite on U, by Theorem 7.5, f is delineable on the connected open set $V = U\backslash \bigcup_{h\in \mathrm{Np}_1(f,x_n)} V_\mathbb{R}(h)$.

From that $f(\alpha, x_n)$ is semi-definite on $\mathbb{R}$, we know that $f(\boldsymbol{x})$ is semi-definite on $U \times \mathbb{R}$. $\qquad\square$

The following theorem is an easy corollary of the above theorem.

Theorem 7.7. *Let $f \in \mathcal{R}[\boldsymbol{x}]$ be a square-free polynomial of class $n(n \geq 2)$ with positive leading base coefficient, the necessary and sufficient conditions for $f(\boldsymbol{x})$ to be positive semi-definite on $\mathbb{R}^n$ are*
(1) The polynomials in $\mathrm{Np}_1(f, x_n)$ are positive semi-definite on $\mathbb{R}^{n-1}$;
(2) For every (maximal) connected region $U \subseteq \mathbb{R}^{n-1}$ in which the elements of $\mathrm{Np}_2(f, x_n)$ have constant, nonzero sign, there exists a point $\alpha \in U$, and α is not a zero of any polynomial in $\mathrm{Np}_1(f, x_n)$, such that $f(\alpha, x_n) \geq 0$ on $\mathbb{R}$.

Proof. $\impliedby$: By Theorem 7.6, $f(\boldsymbol{x})$ is semi-definite on $\mathbb{R}^n$. Because f has positive leading base coefficient, it must be positive semi-definite on $\mathbb{R}^n$.

$\implies$: By Theorem 7.6, the polynomials in $\mathrm{Np}_1(f, x_n)$ are semi-definite on $\mathbb{R}^{n-1}$. Note that, by the definition of $\mathrm{Np}_1(f, x_n)$, all polynomials in $\mathrm{Np}_1(f, x_n)$ have positive leading base coefficients. Therefore, they must be positive semi-definite on $\mathbb{R}^{n-1}$. The other statement of the Theorem is obvious by Theorem 7.6. $\qquad\square$

Theorem 7.8. *Algorithm 7.4 is correct.*

Proof. It is easy to see that the algorithm is correct for univariate polynomials.

By the necessity part of Theorem 7.7, if one of the polynomials in $\mathrm{Np}_1(f, x_n)$ is not positive semi-definite on $\mathbb{R}^{n-1}$, f is not positive semi-definite on $\mathbb{R}^n$. If we do not get an early negative answer, then this means that the first hypothesis of the sufficiency part of Theorem 7.7 is satisfied.

So next we construct an open CAD C_{n-1} of $\mathbb{R}^{n-1}$ with respect to $\mathrm{Np}_2(f, x_n)$, with a slight modification, such that for all $\boldsymbol{a}_{n-1}$ in C_{n-1}, $\boldsymbol{a}_{n-1}$ is not a zero of any element of $\mathrm{Np}_1(f, x_n)$. By the second hypothesis of the sufficiency part of Theorem 7.7, we can determine positive semi-definiteness of f on $\mathbb{R}^n$ by testing positive semi-definiteness of f at all sample points of C_n. $\qquad\square$

Theorem 7.9. *Algorithm 7.3 is correct.*

Proof. The correctness of Algorithm 7.3 follows directly from Proposition 7.2 and the correctness of Algorithm 7.4. $\qquad\square$

7.3.5￼ *Examples of Proving Polynomial Inequalities*

A careful and detailed worst case of computing time analysis of the original general CAD algorithm appears in [Collins (1975)]. An upper bound on the time complexity of CAD, comprising an expression in which the number of variables in the input polynomials appears doubly exponentially, is derived. However that analysis makes the simplifying assumption that no square-free basis computation is made prior to each projection step. Since square-free decomposition is essential in algorithm DPS, the analysis of Collins is not directly applicable to DPS.

If we replace irreducible factorization with square-free factorization in Algorithm DPS and Algorithm DPSIP, the two algorithms are still valid (see Remark 7.5). In this case, the modified DPS does no more work than Mc-Callum's algorithm. So clearly McCallum's upper bound is also valid for the modified DPS. Now a similar kind of careful computing time analysis might yield a still further improved upper bound for the modified DPS, but such a bound is still likely to be doubly exponential in the number of variables. Note that the complexity of irreducible factorization of polynomials over $\mathbb{Q}$ is polynomial [Lenstra *et al.* (1982)], the complexity of Algorithm DPS is also likely to be doubly exponential in the number of variables.

On the other hand, if the worst case time complexity of open CAD could be proved to be codominant with some expression doubly exponential in the number of variables, then it would follow that the worst case time complexity of DPS is codominant with the same expression. Consider for example the polynomial

$$f(x, y) = y^2 + (g(x) + 2)y + g(x) + 1.$$

The discriminant of f with respect to y is $g(x)^2$. So, the computation of our method is the same as that of open CAD.

Since our main contribution is an improvement on the CAD projection, we choose to do some comparison with other CAD based tools on several non-trivial examples in this subsection. The program DPS we implemented using Maple will be compared with the function PartialCylindricalAlgebraicDecomposition (PCAD) of RegularChains package in Maple 15, function FindInstance in Mathematica 9, and QEPCAD B.

As we do not have Mathematica and QEPCAD B installed in our computer, the computations were performed on different computers. FindInstance (FI) was performed on a laptop with Inter Core(TM) i5-3317U 1.70GHz CPU, 4GB RAM, Windows 8 and Mathematica 9. QEPCAD B was performed on a PC with Intel(R) Core(TM) i5 3.20GHz CPU, 4GB

RAM and ubuntu. The other computations were performed on a laptop with Inter Core2 2.10GHz CPU, 2GB RAM, Windows XP and Maple 15.

Example 7.2. [Han (2011)] Prove

$$f(\boldsymbol{x}, n) = \prod_{i=1}^{n}(x_i^2 + n - 1) - n^{n-2}(\sum_{i=1}^{n} x_i)^2 \geq 0 \text{ on } \mathbb{R}^n.$$

When $n = 3, 4, 5, 6, 7$, we compared DPS, FI, PCAD, QEPCAD in the following table. Hereafter >3000 means either the running time is over 3000 seconds or the software fails to obtain an answer.

n/Time(s)	DPS	FI	PCAD	QEPCAD
3	0.063	0.015	0.078	0.020
4	0.422	0.062	0.250	0.024
5	0.875	2.312	2.282	0.372
6	4.188	>3000	>3000	>3000
7	>3000	>3000	>3000	>3000

When $n = 3, 4, 5, 6$, we compared the number of polynomials in the projection sets of BMprojection with Np (under the same ordering) as well as the number of sample points needed to be chosen through the lifting phase under these two projection operators.

n	BMprojection		Np	
	# polys	# points	# polys	# points
3	11	4	8	3
4	22	10	12	3
5	88	36	18	5
6	Unknown	Unknown	32	15

Example 7.3. Decide the nonnegativity of $g(n, k)$

$$g(n, k) = (\sum_{i=1}^{n} x_i^2)^2 - k \sum_{i=1}^{n} x_i^3 x_{i+1},$$

where $x_{n+1} = x_1$.

In the following table, (T) means that the corresponding program outputs $g(n, k) \geq 0$ on $\mathbb{R}^n$. (F) means the converse.

(n, k)/Time(s)	DPS	FI	PCAD	QEPCAD
$(3, 3)$	0.047(T)	0.031(T)	0.078(T)	0.032(T)
$(4, 3)$	0.171(T)	284.484(T)	0.891(T)	196.996(T)
$(5, 3)$	244.188(T)	>3000	>3000	>3000
$(6, 3)$	>3000	>3000	>3000	>3000
$(4, k_1)$	13.782(F)	5638.656(F)	24.656(F)	>3000

$k_1 = \dfrac{22791210893985024517609}{75557863725914323419136}.$

n	BMprojection		DPS	
	# polys	# points	# polys	# points
3	5	10	4	5
4	6	4	5	2
5	Unknown	Unknown	16	20

The following example was once studied by [Parrilo (2000)].

Example 7.4.

$$\forall X_{3m+2} \in \mathbb{R}^{3m+2}, B(x) = (\sum_{i=1}^{3m+2} x_i^2)^2 - 2 \sum_{i=1}^{3m+2} x_i^2 \sum_{j=1}^{m} x_{i+3j+1}^2 \geq 0.$$

$3m+2$/Time(s)	DPS	FI	PCAD	QEPCAD
5	0.297	0.109	0.265	0.104
8	27.218	>3000	>3000	>3000
11	>3000	>3000	>3000	>3000

$3m+2$	BMprojection		DPS	
	# polys	# points	# polys	# points
5	13	96	10	88
8	Unknown	Unknown	27	6720

The above examples demonstrate that in terms of proving non-strict inequalities, among CAD based methods, Algorithm DPS is faster and can work out some examples which could not be solved by other existing (open) CAD tools.

7.3.6   *Polynomial Optimization via* Np

Recall the problem proposed at the beginning of this chapter: For $f \in \mathbb{R}[\boldsymbol{x}, k]$, find all $r \in \mathbb{R}$ such that $f(\boldsymbol{x}, r) \geq 0$ on $\mathbb{R}^n$. Since this is a typical QE problem, any CAD based QE algorithms can be applied. Under a suitable ordering on variables, *e.g.*, $k \prec x_1 \prec \cdots \prec x_n$, by CAD projection, one can obtain a polynomial in k, say $g(k)$. Assume $k_1 < \cdots < k_m$ are the real roots of $g(k)$ and $p_j \in (k_{j-1}, k_j)(1 \leq j \leq m+1)$ are rational sample points in the $m+1$ intervals where $k_0 = -\infty, k_{m+1} = +\infty$. Then checking whether or not $f(\boldsymbol{x}, k_i) \geq 0(1 \leq i \leq m)$ and $f(\boldsymbol{x}, p_j) \geq 0(1 \leq j \leq m+1)$ on $\mathbb{R}^n$ will give the answer. Namely, if there exist p_j such that $f(\boldsymbol{x}, p_j) \geq 0$

then (k_{j-1}, k_j) is an output. If $f(\boldsymbol{x}, k_i) \geq 0$ holds for some k_i, $\{k_i\}$ is an output.

Thus, a natural idea is to apply the projection operator $\mathtt{Np}$ instead of Brown-McCallum's projection in the above procedure. In the following, we first show by an example why $\mathtt{Np}$ cannot be applied directly to the problem. Then we propose an algorithm based on $\mathtt{Np}$ for solving the problem and prove its correctness.

Example 7.5. Find all $k \in \mathbb{R}$ such that

$$\forall x, y \in \mathbb{R}(f(x, y, k) = x^2 + y^2 - k^2 \geq 0).$$

If we apply $\mathtt{Np}$ directly (with an ordering $k \prec x \prec y$), we will get

$$\mathtt{Np}(f) = (\{f(x, y, k), x - k, x + k, 1\}, \{1\}).$$

Because $L_2 = \{1\}$, there is only one sample point with respect to k, say $k_0 = 0$. Substituting k_0 for k in $f(x, y, k)$, we check whether $\forall x, y \in \mathbb{R}(x^2 + y^2 \geq 0)$. This is obviously true. So, it leads to a wrong result: $\forall k, x, y \in \mathbb{R}(x^2 + y^2 - k^2 \geq 0)$.

The reason for the error is that $(x - k)(x + k)$ will be a square if $k = 0$. The point $k = 0$ should have been found by computing the resultant $\mathrm{res}(x - k, x + k, x)$ which is avoided by $\mathtt{Np}$ since $x - k \in L_1$ and $x + k \in L_1$.

This example indicates that, if we use $\mathtt{Np}$ to solve the problem, we have to consider some "bad" values of k at which some odd factors of $\mathrm{sqrfree}_1(f)$ or $\mathtt{Np}_1(f)$ may become some new even factors. In the following, we first show that such "bad" values of k are finite and propose an algorithm for computing all possible "bad" values. Then we give an algorithm for solving the problem, which handles the "bad" values and the "good" values of k obtained by $\mathtt{Np}$ separately.

Definition 7.11. Let $f(\boldsymbol{x}, k) \in \mathbb{Z}[\boldsymbol{x}, k]$ and $(L_1, L_2) = \mathtt{Np}(f(\boldsymbol{x}, k))$ with the ordering $k \prec x_1 \prec \cdots \prec x_n$. If $r \in \mathbb{R}$ satisfying that

(1) there exist two different polynomials $g_1, g_2 \in \mathrm{sqrfree}_1(f(\boldsymbol{x}, k))$ such that $g_1|_{k=r}$ and $g_2|_{k=r}$ have non-trivial common factors in $\mathbb{R}[\boldsymbol{x}]$; or
(2) there exist $i(2 \leq i \leq n)$, a polynomial $g \in L_1^{[i]}$ and two different polynomials $g_1, g_2 \in \mathtt{Np}_1(g, x_i)$ such that $g_1|_{k=r}$ and $g_2|_{k=r}$ have non-trivial common factors in $\mathbb{R}[\boldsymbol{x}]$; or
(3) there exists a polynomial $g \in L_1$ such that $g|_{k=r}$ has non-trivial square factors in $\mathbb{R}[\boldsymbol{x}]$,

then r is called a *bad value* of k. The set of all the bad values is denoted by $\mathrm{Bad}(f, k)$.

For two coprime multivariate polynomials with parametric coefficients, the problem of finding all parameter values such that the two polynomials have non-trivial common factors at those parameter values is very interesting. We believe that there should have existed some work on this problem. However, we do not find such work in the literature. So, we use an algorithm in [Qian (2013)]. The details of the correctness and improvements on the algorithm are omitted. Note that $\mathrm{indets}(E)$ in the algorithm means the set of unknowns in E.

Algorithm 7.6 BK

Input: Two coprime polynomials $f(\boldsymbol{x}, k), g(\boldsymbol{x}, k) \in \mathbb{Z}[\boldsymbol{x}, k]$ and k
Output: B, a finite set of polynomials in k, such that

$$V_{\mathbb{R}}(B) \supseteq \{\alpha \in \mathbb{R} |\ \gcd(f(\boldsymbol{x}, \alpha), g(\boldsymbol{x}, \alpha)) \text{ is non-trivial}\}$$

1: $B \leftarrow \emptyset$;
2: $r \leftarrow \mathrm{res}(f, g, k)$; Let S be the set of all irreducible factors of r.
3: Let $X = \mathrm{indets}(S)$; (X is the set of variables appearing in S)
4: **while** $X \neq \emptyset$ **do**
5:     Choose a variable $x \in X$ such that the cardinal number of
    $T = \{p \in S |\ x \text{ appears in } p\}$ is the biggest;
6:     $h \leftarrow \mathrm{res}(f, g, x)$;
7:     $B \leftarrow B \cup \{q(k) |\ q(k) \text{ is irreducible and divides } h\}$;
8:     $S \leftarrow S \setminus T$; $X \leftarrow \mathrm{indets}(S)$;
9: **end while**
10: **return** B

It is not hard to prove the following lemmas.

Lemma 7.6. *Algorithm* BK *is correct, i.e.*
    $V_{\mathbb{R}}(\mathrm{BK}(f, g, k)) \supseteq \{\alpha \in \mathbb{R} |\ \gcd(f(\boldsymbol{x}, \alpha), g(\boldsymbol{x}, \alpha)) \text{ is non-trivial}\}.$

Lemma 7.7. *Let notations be as in Algorithm 7.7.*

(1) The first two outputs, L_1 and L_2, are the same as $\mathrm{Np}(f(\boldsymbol{x}, k))$ with the ordering $k \prec x_1 \prec \cdots \prec x_n$.
(2) $\bigcup_{h \in B} V_{\mathbb{R}}(h) \supseteq \mathrm{Bad}(f, k)$. Thus, $\mathrm{Bad}(f, k)$ is finite.
(3) If k_0 is not a bad value and $f(\boldsymbol{x}, k_0) \geq 0$ on $\mathbb{R}^n$, then for any $h \in \mathrm{sqrfree}_1(f)$, $h(\boldsymbol{x}, k_0)$ is semi-definite on $\mathbb{R}^n$.

Algorithm 7.7 NKproj

Input: A polynomial $f(\boldsymbol{x}, k) \in \mathbb{Z}[\boldsymbol{x}]$ and an ordering $k \prec x_1 \prec \cdots \prec x_n$.

Output: Two projection factor sets as in Algorithm 7.5 and a set of polynomials in k.

1: $L_1 \leftarrow \mathrm{sqrfree}_1(f)$; $L_2 \leftarrow \{\}$; $B \leftarrow \emptyset$;

2: **for** i from n downto 1 **do**

3: $\quad L_2 \leftarrow L_2 \bigcup \mathrm{Np}_2(L_1^{[i]}, x_i) \bigcup \cup_{g \in L_2^{[i]}} \mathrm{BMproj}(g, x_i)$;

4: $\quad$ **for** $h \in L_1^{[i]}$ **do**

5: $\quad\quad L_{1h} \leftarrow \mathrm{Np}_1(h, x_i)$;

6: $\quad\quad B \leftarrow B \cup \mathrm{BK}(h, \frac{\partial}{\partial x_i} h, k)$;

7: $\quad\quad B \leftarrow B \bigcup \cup_{h_1 \neq h_2 \in L_{1h}} \mathrm{BK}(h_1, h_2, k)$;

8: $\quad\quad L_1 \leftarrow L_1 \cup L_{1h}$;

9: $\quad$ **end for**

10: **end for**

11: **return** (L_1, L_2, B)

Proof. (1) and (2) are obvious. For (3), because $k_0 \notin \mathrm{Bad}(f)$, $g_1(\boldsymbol{x}, k_0)$ and $g_2(\boldsymbol{x}, k_0)$ are coprime in $\mathbb{Z}[\boldsymbol{x}]$ for any $g_1 \neq g_2 \in \mathrm{sqrfree}_1(f)$. Since $f(\boldsymbol{x}, k_0) \geq 0$, by Proposition 7.1, for any $h \in \mathrm{sqrfree}_1(f(\boldsymbol{x}, k))$, $h(\boldsymbol{x}, k_0)$ is semi-definite on $\mathbb{R}^n$. $\qquad\square$

Lemma 7.8. *Let* $f(\boldsymbol{x}, k) \in \mathbb{Z}[\boldsymbol{x}, k]$ *and* $(L_1, L_2, B) = \mathtt{NKproj}(f)$. *Suppose* $\mathrm{V}_{\mathbb{R}}(L_1^{[0]} \cup L_2^{[0]}) = \cup_{i=1}^m \{k_i\}$ *with* $k_1 < \cdots < k_m$, $k_0 = -\infty$, $k_{m+1} = +\infty$ *and, for every* $l(1 \leq l \leq m+1)$, $p_l \in (k_{l-1}, k_l) \setminus \bigcup_{h \in B} \mathrm{V}_{\mathbb{R}}(h)$. *Denote by* C_{li} *an open CAD of* $\mathbb{R}^i$ *defined by* $\cup_{j=1}^i L_1^{[j]} |_{k=p_l} \bigcup \cup_{j=1}^i L_2^{[j]} |_{k=p_l}$. *If there exists* l $(1 \leq l \leq m+1)$ *such that*

(1) $\forall X_n \in C_{ln}, f(X_n, p_l) \geq 0$; *and*

(2) $\forall i(0 \leq i \leq n-1) \forall g \in L_1^{[i]} \forall X_i^1, X_i^2 \in C_{li}, g(X_i^1, p_l) g(X_i^2, p_l) \geq 0$,

then for any $0 \leq i \leq n$ *and* $g_i(\boldsymbol{x}^{[i]}, k)$ *in* $L_1^{[i]}$, $g_i(\boldsymbol{x}^{[i]}, k)$ *is semi-definite on* $\mathbb{R}^i \times (k_{l-1}, k_l)$.

Proof. We prove it by induction on i. When $i = 0$, the conclusion is obvious. When $i = 1$, by Theorem 7.6, it is also true. Assume the conclusion is true when $i = j - 1 (j \geq 2)$. For any polynomial $g_j(\boldsymbol{x}^{[j]}, k)$ in $L_1^{[j]}$, notice that $\mathrm{Np}_1(g_j) \subseteq L_1^{[j-1]}$, $\mathrm{Np}_2(g_j) \subseteq L_2^{[j-1]}$. By the assumption of induction, we know that every polynomial in $\mathrm{Np}_1(g_j)$ is semi-definite on $\mathbb{R}^{j-1} \times (k_{l-1}, k_l)$. By Theorem 7.6, $g_j(\boldsymbol{x}^{[j]}, k)$ is semi-definite on $\mathbb{R}^j \times (k_{l-1}, k_l)$. That finishes the induction. $\qquad\square$

Theorem 7.10. *Algorithm 7.8 is correct, i.e. the output, K_f, is indeed* $\{\alpha \in \mathbb{R} | \forall X_n \in \mathbb{R}^n, f(X_n, \alpha) \geq 0\}$.

Proof.   Denote $\{\alpha \in \mathbb{R} | \forall X_n \in \mathbb{R}^n, f(X_n, \alpha) \geq 0\}$ by $\widetilde{K_f}$.

Algorithm 7.8 Findk

Input: A polynomial $f(\boldsymbol{x}, k) \in \mathbb{Z}[\boldsymbol{x}, k]$
Output: A set K_f such that $K_f = \{\alpha \in \mathbb{R} | \forall X_n \in \mathbb{R}^n, f(X_n, \alpha) \geq 0\}$
1: $K_f \leftarrow \{\}$;
2: $(L_1, L_2, B) \leftarrow \text{NKproj}(f)$; (with an ordering $k \prec x_1 \prec \cdots \prec x_n$)
3: Suppose $\text{V}_{\mathbb{R}}(L_1^{[0]} \cup L_2^{[0]}) = \cup_{i=1}^m \{k_i\}$ with $k_1 < \cdots < k_m$. Let $k_0 = -\infty$,
  $k_{m+1} = +\infty$.
4: **for** l from 1 to $m + 1$ **do**
5:   Choose a sample point $p_l \in (k_{l-1}, k_l) \setminus \bigcup_{h \in B} \text{V}_{\mathbb{R}}(h)$;
6:   $v \leftarrow 1$;
7:   **for** i from 1 to n **do**
8:     $C_{li} \leftarrow$ an   open   CAD   of $\mathbb{R}^i$ defined   by $\cup_{j=1}^i L_1^{[j]}$ $|_{k=p_l}$
      $\bigcup \cup_{j=1}^i L_2^{[j]} |_{k=p_l}$;
9:     **if** $i = n$ and there exists $X_n \in C_{ln}$ such that $f(X_n, p_l) < 0$ **then**
10:       $v \leftarrow 0$;
11:     **else**
12:       **if** there exist $X_i^1, X_i^2 \in C_{li}$ and $g \in L_1^{[i]}$ such that
        $g(X_i^1, p_l)g(X_i^2, p_l) < 0$ **then**
13:         $v \leftarrow 0$;
14:         **break**
15:       **end if**
16:     **end if**
17:   **end for**
18:   **if** $v = 1$ **then**
19:     $K_f \leftarrow K_f \bigcup (k_{l-1}, k_l)$;
20:   **end if**
21: **end for**
22: **for** α in $\{k_1, \ldots, k_m\} \cup \bigcup_{h \in B} \text{V}_{\mathbb{R}}(h) \setminus K_f$ **do**
23:   **if** $\text{DPS}(f(\boldsymbol{x}, \alpha))$ **then**
24:     $K_f \leftarrow K_f \bigcup \{\alpha\}$;
25:   **end if**
26: **end for**
27: **return** K_f

We first prove that $K_f \subseteq \widetilde{K_f}$. Suppose $(k_{l-1}, k_l) \subseteq K_f$. Since $\mathrm{sqrfree}_1(f) \subseteq L_1^{[n]}$, the semi-definiteness of f on $\mathbb{R}^n \times (k_{l-1}, k_l)$ follows from Lemma 7.8. Because we check the positive definiteness of f on sample points, $(k_{l-1}, k_l) \subseteq \widetilde{K_f}$.

We then prove that $\widetilde{K_f} \subseteq K_f$. It is sufficient to prove that if there exists $k' \in (k_{l-1}, k_l) \setminus \bigcup_{h \in B} V_{\mathbb{R}}(h)$ such that $\forall X_n \in \mathbb{R}^n, f(X_n, k') \geq 0$, then $(k_{l-1}, k_l) \in K_f$.

It is obviously true when $n = 1$. When $n \geq 2$, for any $g_n \in \mathrm{sqrfree}_1(f)$, $g_n(\boldsymbol{x}, k')$ is semi-definite by Lemma 7.7 since $f(\boldsymbol{x}, k')$ is semi-definite and $k' \notin \bigcup_{h \in B} V_{\mathbb{R}}(h)$. For any

$$g_{n-1}(\boldsymbol{x}^{[n-1]}, k) \in \mathrm{Np}_1(g_n(\boldsymbol{x}, k), x_n) = \mathrm{Oc}(g_n, x_n) \cup \mathrm{Od}(g_n, x_n),$$

we have

$$\mathrm{sqrfree}_1(g_{n-1}(\boldsymbol{x}^{[n-1]}, k')) \subseteq \mathrm{Np}_1(g_n(\boldsymbol{x}, k'), x_n)$$

because $k' \notin \bigcup_{h \in B} V_{\mathbb{R}}(h)$. By Theorem 7.7, $g_{n-1}(\boldsymbol{x}^{[n-1]}, k')$ is semi-definite on $\mathbb{R}^{n-1}$. Hence, for any polynomial $g_{n-1}(\boldsymbol{x}^{[n-1]}, k)$ in $L_1^{[n-1]}$, $g_{n-1}(\boldsymbol{x}^{[n-1]}, k')$ is semi-definite. In a similar way, we know that for any $1 \leq j \leq n-1$ and any polynomial $g_j(\boldsymbol{x}^{[j]}, k)$ in $L_1^{[j]}$, $g_j(\boldsymbol{x}^{[j]}, k')$ is semi-definite on $\mathbb{R}^j$. Therefore, for any $0 \leq i \leq n$ and any polynomial $g_i(\boldsymbol{x}^{[i]}, k)$ in $L_1^{[i]}$, $g_i(\boldsymbol{x}^{[i]}, k)$ is semi-definite on $\mathbb{R}^i \times (k_{l-1}, k_l)$ by Lemma 7.8. Hence, no matter what point $p_l \in (k_{l-1}, k_l)$ is chosen as the sample point of this open interval, (k_{l-1}, k_l) will be in the output of Algorithm 7.8, *i.e.* $(k_{l-1}, k_l) \in K_f$. The proof is complete. $\qquad\square$

For solving the global optimum problem: find the global infimum $\inf f(\mathbb{R}^n)$ for $f \in \mathbb{R}[\boldsymbol{x}]$, we only need to modify Algorithm **Findk** slightly and get the algorithm **Findinf**.

Theorem 7.11. *The output of Algorithm 7.9 is the global infimum* $\inf f(\mathbb{R}^n)$.

Proof. We only need to prove that if there exists $k' \in (k_{l-1}, k_l)$ such that $f(\boldsymbol{x}) \geq k'$ on $\mathbb{R}^n$, then $f(\boldsymbol{x}) \geq k_l$ on $\mathbb{R}^n$.

The result is obviously true when $n = 1$. When $n \geq 2$, we can find a "good" value $k'' \in (k_{l-1}, k') \setminus \mathrm{Bad}(f - k, k)$ because the bad values are finite according to Lemma 7.7. Since $f(\boldsymbol{x}) \geq \bar{k}$ for $\bar{k} \in (k_{l-1}, k')$, $f(\boldsymbol{x}) - k'' \geq 0$. Then, by Lemma 7.7 (3), $h(\boldsymbol{x}, k'')$ is semi-definite on $\mathbb{R}^n$ for any $h \in \mathrm{sqrfree}_1(f(\boldsymbol{x}) - k)$. In a similar way, we know that for any $1 \leq j \leq n-1$ and any polynomial $g_j(\boldsymbol{x}^{[j]}, k)$ in $LI_1^{[j]}$, $g_j(\boldsymbol{x}^{[j]}, k'')$ is semi-definite on $\mathbb{R}^j$.

Algorithm 7.9 Findinf

Input: A square-free polynomial $f \in \mathbb{Z}[\boldsymbol{x}]$

Output: $k \in \mathbb{R}$ such that $k = \inf_{\boldsymbol{x} \in \mathbb{R}^n} f(\boldsymbol{x})$

1: $(LI_1, LI_2) \leftarrow \mathrm{Np}(f(\boldsymbol{x}) - k)$ (with an ordering $k \prec x_1 \prec \cdots \prec x_n$)

2: Suppose $V_{\mathbb{R}}(LI_1^{[0]} \cup LI_2^{[0]}) = \cup_{i=1}^m \{k_i\}$ with $k_1 < \cdots < k_m$. Let $k_0 = -\infty$, $k_{m+1} = +\infty$.

3: **for** l from 1 to $m+1$ **do**

4:     Choose a sample point p_l of (k_{l-1}, k_l)

5:     $v \leftarrow 1$;

6:     **for** i from 1 to n **do**

7:         $C_{li} \leftarrow$ an open CAD of $\mathbb{R}^i$ defined by $\cup_{j=1}^i LI_1^{[j]} \big|_{k=p_l} \bigcup \cup_{j=1}^i LI_2^{[j]} \big|_{k=p_l}$;

8:         **if** $i = n$ and there exists $X_n \in C_{ln}$ such that $f(X_n) - p_l < 0$ **then**

9:             $v \leftarrow 0$;

10:         **else**

11:             **if** there exist $X_i^1, X_i^2 \in C_{li}, g \in LI_1^{[i]}$ such that $g(X_i^1, p_l)g(X_i^2, p_l) < 0$ **then**

12:                 $v \leftarrow 0$;

13:                 **break**

14:             **end if**

15:         **end if**

16:     **end for**

17:     **if** $v = 0$ **then**

18:         **return** k_{l-1}

19:     **end if**

20: **end for**

Therefore, for any $0 \le i \le n$ and any polynomial $g_i(\boldsymbol{x}^{[i]}, k)$ in $L_1^{[i]}$, $g_i(\boldsymbol{x}^{[i]}, k)$ is semi-definite on $\mathbb{R}^i \times (k_{l-1}, k_l)$ by Lemma 7.8. Hence, $f(\boldsymbol{x}) - k$ is positive semi-definite on $\mathbb{R}^n \times (k_{l-1}, k_l)$ by Theorem 7.6. $\qquad\square$

Remark 7.7. For $f, g \in \mathbb{R}[\boldsymbol{x}]$, if $g(\boldsymbol{x}) \ge 0$ on $\mathbb{R}^n$, Algorithm **Findinf** can also be applied to compute $\inf\{\frac{f(\boldsymbol{x})}{g(\boldsymbol{x})} \mid \boldsymbol{x} \in \mathbb{R}^n\}$. We just need to replace $(LI_1, LI_2) \leftarrow \mathrm{Np}(f(\boldsymbol{x}) - k)$ of Line 1 by $(LI_1, LI_2) \leftarrow \mathrm{Np}(f(\boldsymbol{x}) - kg(\boldsymbol{x}))$.

We show the different results of projection of Algorithm **Findinf** and Algorithm **SRes** by Example 7.6.

Example 7.6. Compute $\inf_{x,y,z\in\mathbb{R}} h(x,y,z)$, where

$$h = \frac{(x^2 - x + 1)(y^2 - y + 1)(z^2 - z + 1)}{(xyz)^2 - xyz + 1}.$$

Let $f = (x^2 - x + 1)(y^2 - y + 1)(z^2 - z + 1), g = (xyz)^2 - xyz + 1$. Since $g \geq 0$ for any $x, y, z \in \mathbb{R}$, this problem can be solved either by Algorithm `Findinf` or by Algorithm `SRes`.

If we apply Algorithm `Findinf`, after $\mathrm{Np}(f - kg)$ with an ordering $k \prec z \prec y \prec x$, we will get a polynomial in k,

$$p = (k - \frac{2}{3})(k^2 + 6k - 3)(k - \frac{279}{256})k(k - 1)(k - \frac{3}{4})(k - \frac{9}{16})(k - 9),$$

which has 9 distinct real roots. After sampling and checking signs, we finally know that $\inf_{x,y,z\in\mathbb{R}} h(x,y,z)$ is the real root of $k^2 + 6k - 3$ in $(\frac{1}{4}, \frac{1}{2})$.

If we apply Algorithm `SRes`, after `BMprojection`$(f - kg)$ with an ordering $k \prec z \prec y \prec x$, we will get a polynomial in k,

$$q = \frac{1}{614656}(614656k^4 - 4409856k^3 + 11013408k^2 - 11477376k + 4021893)\cdot$$

$$(k^4 - 294k^3 + 1425k^2 - 2277k + 1089)(k - \frac{9}{4}) \cdot p,$$

which has 14 distinct real roots. After sampling and checking signs, we finally know that $\inf_{x,y,z\in\mathbb{R}} h(x,y,z)$ is the real root of $k^2 + 6k - 3$ in $(\frac{1}{4}, \frac{1}{2})$.

Obviously, the scale of projection with the new projection is smaller. The polynomial in k calculated through the successive resultant method has three extraneous factors.

7.4 Open Weak CAD

In this section, we introduce the concept of *open weak CAD*. Open weak CADs might not have cylindrical structure in the sense of classical CAD. However, open weak CADs preserve some geometric information, *i.e.* they are *open weak delineable* (see Definition 7.12). As a result, every open CAD is an open weak CAD. On the contrary, an open weak CAD is not necessarily an open CAD.

An algorithm for computing open weak CADs is proposed. The key idea is to compute the intersection of projection factor sets produced by different projection orders. The resulting open weak CAD often has smaller number of sample points than open CADs.

The algorithm can be used for computing sample points for all open connected components of $f \neq 0$ for a given polynomial f. It can also be used for many other applications, such as testing semi-definiteness of polynomials and polynomial optimization. In fact, we solved several difficult semi-definiteness problems efficiently by using the algorithm.

7.4.1 *Concepts*

First, let us introduce the concept of *open weak delineable*.

Definition 7.12 (Open weak delineable). *Let S be an open set of $\mathbb{R}^j (1 \leq j < n)$. The polynomial $f \in \mathbb{R}[x_1, \ldots, x_n]$ is said to be* open weak delineable *on S if, for any maximal open connected set $U \subseteq \mathbb{R}^n$ defined by $f \neq 0$, we have*

$$(S \times \mathbb{R}^{n-j}) \bigcap U \neq \emptyset \iff \forall \alpha \in S, \ (\alpha \times \mathbb{R}^{n-j}) \bigcap U \neq \emptyset.$$

Let $h \in \mathbb{R}[x_1, \ldots, x_j]$ where $1 \leq j < n$. We say the polynomial f is open weak delineable *over h in $\mathbb{R}^j$, if f is open weak delineable on any open connected component of $h \neq 0$ in $\mathbb{R}^j$.*

Note that if f is *analytic delineable* ([Collins (1975)]) on S then f is also *open weak delineable* on S. However the converse is not necessarily true.

Example 7.7. Let

$$f = x_1 - x_2^4 + 10x_2^3 - 35x_2^2 + 50x_2 - 24 \in \mathbb{R}[x_1, x_2].$$

The plot of $f = 0$ is given by Fig. 7.1.

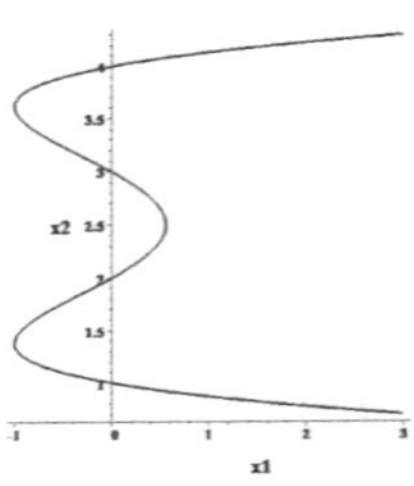

Fig. 7.1

Note that f is analytically delineable and also open weak delineable on the set $(-\infty, -1)$. Note that f is *not* analytically delineable but *is* open weak delineable on the set $(-1, \infty)$. Note also that f is open weak delineable over $h = x_1 + 1$ in $\mathbb{R}$.

Definition 7.13 (Open weak CAD). *Let $f \in \mathbb{R}[x_1,\ldots,x_n]$. A decomposition of $\mathbb{R}^j$, $1 \leq j < n$, is called an* open weak CAD *of f in $\mathbb{R}^j$ if and only if f is open weak delineable on every j-dimensional open set in the decomposition.*

Example 7.8. Let f be the polynomial from Example 7.7. Then

$$\{(-\infty,-1),[-1,-1],(-1,\infty)\}$$

is an open weak CAD of f.

We are now ready to state the problem precisely.

Problem. (Projection polynomials of Open Weak CAD) Devise an algorithm with the following specification.

In: $f \in \mathbb{Z}[x_1,\ldots,x_n]$
Out: $h_1, h_2, \ldots, h_{n-1}$ where $h_j \in \mathbb{Z}[x_1,\ldots,x_j]$ such that f is open weak delineable over h_j in $\mathbb{R}^j$.

We call the number of open components in $\mathbb{R}^j$ defined by $h_j \neq 0$, the *scale* of the open weak CAD of f defined by h_j in $\mathbb{R}^j$.

Remark 7.8. The output of the above problem is a list of "projection" polynomials, not an open weak CAD. However there are standard methods to compute sample points of an open weak CAD of f from the projection polynomials. Thus, sometimes we will call the above problem "Open Weak CAD".

Example 7.9. Consider the following polynomial.

In: $f = (x_3^2 + x_2^2 + x_1^2 - 1)(4x_3 + 3x_2 + 2x_1 - 1) \in \mathbb{Z}[x_1,x_2,x_3]$
Out: $h_1 = (x_1 - 1)(x_1 + 1)(29x_1^2 - 4x_1 - 24)((20x_1^2 - 4x_1 - 15)^2 + (13x_1^2 - 4x_1 - 8)^2) \in \mathbb{Z}[x_1]$
$h_2 = (x_2^2 + x_1^2 - 1)(25x_2^2 + 12x_2x_1 + 20x_1^2 - 6x_2 - 4x_1 - 15) \in \mathbb{Z}[x_1,x_2]$.

The left plot in Fig. 7.2 shows the open weak CAD of f produced by h_1 and h_2. The factor $(20x_1^2 - 4x_1 - 15)^2 + (13x_1^2 - 4x_1 - 8)^2$ in h_1 does not have real root and thus it does not contribute to the open weak CAD.

Remark 7.9. For comparison, if we apply an open CAD algorithm on the above f, one would obtain the following output

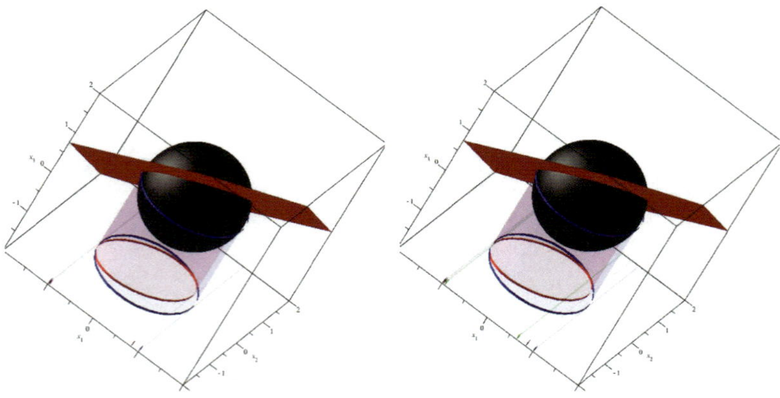

Fig. 7.2   open weak CAD vs. open CAD

Out: $h_1 = (x_1 - 1)(x_1 + 1)(29x_1^2 - 4x_1 - 24)(13x_1^2 - 4x_1 - 8) \in \mathbb{Z}[x_1]$
$h_2 = (x_2^2 + x_1^2 - 1)(25x_2^2 + 12x_2x_1 + 20x_1^2 - 6x_2 - 4x_1 - 15) \in \mathbb{Z}[x_1, x_2]$

The right plot in Figure 7.2 shows the open CAD of f produced by h_1 and h_2. Note that it has more cells than the open weak CAD (on the left).

Remark 7.10. It is natural to wonder whether the multivariate discriminants of f always produce open weak CADs. Unfortunately, this is not true since the discriminant $\mathrm{discrim}(f, [x_n, \ldots, x_{j+1}])$ may vanish identically and thus does not always produce an open weak CAD of $\mathbb{R}^j$. One may also wonder whether the multivariate discriminants of f would be the smallest open weak CADs if they do produce open weak CADs. Unfortunately this is not true, either. In Example 7.7, it has been shown that $x_1 + 1$ produces an open weak CAD of $\mathbb{R}$ with 2 open intervals. But the discriminant $\mathrm{discrim}(f, x_2) = -16(16x_1 - 9)(x_1 + 1)^2$ produces an open weak CAD of $\mathbb{R}$ with 3 open intervals.

7.4.2   *Projection Operator* Hp

In this section, we describe an algorithm (Algorithm 7.10) for computing open weak CAD and prove its correctness (Theorem 7.12). Before we go into the formal definitions, let us show the basic idea behind the new projection operator first by a simple example.

Example 7.10. Let $f = x^4 - 2x^2y^2 + 2x^2z^2 + y^4 - 2y^2z^2 + z^4 + 2x^2 + 2y^2 - 4z^2 - 4 \in \mathbb{Z}[x, y, z]$.

We first compute the projection polynomials by Brown's operator. Take the order $z \succ y \succ x$. Step 1, compute the projection polynomial (up to a nonzero constant)

$$f_z = \mathrm{res}(\mathrm{sqrfree}(f), \tfrac{\partial}{\partial z}\mathrm{sqrfree}(f), z)$$
$$= (x^4 - 2x^2y^2 + y^4 + 2x^2 + 2y^2 - 4)(3x^2 - y^2 - 4)^2.$$

Step 2, compute the projection polynomial (up to a nonzero constant)

$$f_{zy} = \mathrm{res}(\mathrm{sqrfree}(f_z), \tfrac{\partial}{\partial y}\mathrm{sqrfree}(f_z), y)$$
$$= (3x^2 - 4)(x^4 + 2x^2 - 4)(4x^2 - 5)^2(x - 1)^8(x + 1)^8.$$

In other words, $\mathtt{BMprojection}(f)$ under the order $z \succ y \succ x$ will give $\{f, f_z, f_{zy}\}$. Now, if we make use of the projection polynomials to compute an open CAD defined by $f \neq 0$ in $\mathbb{R}^3$, we will finally get 113 sample points of $f \neq 0$ in $\mathbb{R}^3$.

We then compute projection polynomials of *open weak CAD* defined by f in $\mathbb{R}^3$ by the new projection operator. Step 1, take the order $z \succ y \succ x$ and compute the projection polynomial f_{zy} as above. Step 2, take another order $y \succ z \succ x$ and we can similarly obtain a projection polynomial (up to a constant)

$$f_{yz} = (3x^2 - 4)^2(x^4 + 2x^2 - 4)(4x^2 - 5)(6x^2 - 7)^8.$$

Step 3, compute

$$g(x) = \gcd(f_{yz}, f_{zy}) = (3x^2 - 4)(x^4 + 2x^2 - 4)(4x^2 - 5).$$

In other words, the new projection polynomials are $\{f, f_z, g\}$ (or $\{f, f_y, g\}$).

Now, if we make use of the new projection polynomials to compute an open weak CAD of f in $\mathbb{R}^3$, we will finally get 87 sample points of $f \neq 0$ in $\mathbb{R}^3$.

Now, we give a formal definition of the new projection operator.

Definition 7.14 (Open weak CAD projection operator). Let $f \in \mathbb{Z}[x_1, \ldots, x_n]$. For given $m(1 \leq m \leq n)$, denote $[\boldsymbol{y}] = [y_1, \ldots, y_m]$ where $y_i \in \{x_1, \ldots, x_n\}$ for $1 \leq i \leq m$ and $y_i \neq y_j$ for $i \neq j$. For $1 \leq i \leq m$, $\mathtt{Hp}(f, [\boldsymbol{y}], y_i)$ and $\mathtt{Hp}(f, [\boldsymbol{y}])$ are defined recursively as follows.

$$\mathtt{Hp}(f, [\boldsymbol{y}], y_i) = \mathrm{BMproj}(\mathtt{Hp}(f, [\hat{\boldsymbol{y}}]_i), y_i),$$
$$\mathtt{Hp}(f, [\boldsymbol{y}]) = \gcd(\mathtt{Hp}(f, [\boldsymbol{y}], y_1), \ldots, \mathtt{Hp}(f, [\boldsymbol{y}], y_m)),$$
$$\mathtt{Hp}(f, [\,]) = f,$$

where $[\hat{\boldsymbol{y}}]_i = [y_1, \ldots, y_{i-1}, y_{i+1}, \ldots, y_m]$.

Example 7.11. We have

$$\mathrm{Hp}(f,[x_1,x_2]) = \gcd\left(\mathrm{Hp}\left(f,[x_1,x_2],x_1\right), \mathrm{Hp}\left(f,[x_1,x_2],x_2\right)\right)$$
$$\mathrm{Hp}\left(f,[x_1,x_2],x_1\right) = \mathrm{BMproj}(\mathrm{Hp}(f,[x_2]),x_1)$$
$$\mathrm{Hp}\left(f,[x_1,x_2],x_2\right) = \mathrm{BMproj}(\mathrm{Hp}(f,[x_1]),x_2)$$
$$\mathrm{Hp}(f,[x_2]) = \mathrm{Hp}(f,[x_2],x_2)$$
$$\mathrm{Hp}(f,[x_1]) = \mathrm{Hp}(f,[x_1],x_1)$$
$$\mathrm{Hp}(f,[x_2],x_2) = \gcd(\mathrm{BMproj}(\mathrm{Hp}(f,[\,]),x_2))$$
$$\mathrm{Hp}(f,[x_1],x_1) = \gcd(\mathrm{BMproj}(\mathrm{Hp}(f,[\,]),x_1))$$
$$\mathrm{Hp}(f,[\,]) = f$$

Condensing the above expressions, we have

$$\mathrm{Hp}(f,[x_1,x_2])$$
$$= \gcd\left(\mathrm{BMproj}(\mathrm{BMproj}(f,x_2),x_1), \mathrm{BMproj}(\mathrm{BMproj}(f,x_1),x_2)\right)$$

The Algorithm 7.10 (`OWCProj`) based on the new operator `Hp` solves the problem proposed in Section 7.4.1.

Algorithm 7.10 OWCProj (Open Weak CAD Projection)

Input: $f \in \mathbb{Z}[x_1,\ldots,x_n]$

Output: $h_1,h_2,\ldots,h_{n-1}$ where $h_j \in \mathbb{Z}[x_1,\ldots,x_j]$ such that each h_j produces an open weak CAD of $\mathbb{R}^j$ with respect to f

1: For all $1 \le j < n$ and $j < t \le n$, compute $\mathrm{Hp}(f,[x_n,\ldots,x_{j+1}],x_t)$ by Definition 7.14.

2: Compute $h_1(f), h_2(f), \ldots, h_{n-1}(f)$ where

$$h_j(f) = \mathrm{sqrfree}\left(\sum_{t=j+1}^{n} \mathrm{Hp}(f,[x_n,\ldots,x_{j+1}],x_t)^2\right).$$

Example 7.12. We illustrate Algorithm 7.10 using the polynomial f from Example 7.9.

In: $f = (x_3^2 + x_2^2 + x_1^2 - 1)(4x_3 + 3x_2 + 2x_1 - 1) \in \mathbb{Z}[x_1,x_2,x_3]$

1: For all $1 \le j < 3$ and $j < t \le 3$, compute $\mathrm{Hp}(f,[x_3,\ldots,x_{j+1}],x_t)$ by Definition 7.14.

$$\begin{aligned}
\mathtt{Hp}(f,[x_3],x_3) \quad &= \mathrm{BMproj}(\mathtt{Hp}(f,[\,]),[x_3]) = \mathrm{BMproj}(f,[x_3])\\
&= (x_2^2 + x_1^2 - 1)(25x_2^2 + 12x_2x_1 + 20x_1^2 - 6x_2 - 4x_1 - 15),\\
\mathtt{Hp}(f,[x_3,x_2],x_2) &= \mathrm{BMproj}(\mathtt{Hp}(f,[x_3]),[x_2]) = \mathrm{BMproj}(\mathtt{Hp}(f,[x_3],x_3),[x_2])\\
&= (x_1 - 1)(x_1 + 1)(29x_1^2 - 4x_1 - 24)(13x_1^2 - 4x_1 - 8),\\
\mathtt{Hp}(f,[x_3,x_2],x_3) &= \mathrm{BMproj}(\mathtt{Hp}(f,[x_2]),[x_3]) = \mathrm{BMproj}(\mathtt{Hp}(f,[x_2],x_2),[x_3])\\
&= (x_1 - 1)(x_1 + 1)(29x_1^2 - 4x_1 - 24)(20x_1^2 - 4x_1 - 15).
\end{aligned}$$

2: Compute $h_1(f), h_2(f)$ where

$$h_j(f) = \mathrm{sqrfree}\left(\sum_{t=j+1}^{n} \mathtt{Hp}(f,[x_3,\ldots,x_{j+1}],x_t)^2 \right).$$

$$\begin{aligned}
h_1 &= \mathrm{sqrfree}((\mathtt{Hp}(f,[x_3,x_2],x_2))^2 + (\mathtt{Hp}(f,[x_3,x_2],x_3))^2)\\
&= (x_1 - 1)(x_1 + 1)(29x_1^2 - 4x_1 - 24)\\
&\quad ((20x_1^2 - 4x_1 - 15)^2 + (13x_1^2 - 4x_1 - 8)^2),\\
h_2 &= \mathrm{sqrfree}((\mathtt{Hp}(f,[x_3],x_3))^2)\\
&= (x_2^2 + x_1^2 - 1)(25x_2^2 + 12x_2x_1 + 20x_1^2 - 6x_2 - 4x_1 - 15).
\end{aligned}$$

Out: $h_1 = (x_1 - 1)(x_1 + 1)(29x_1^2 - 4x_1 - 24)((20x_1^2 - 4x_1 - 15)^2 + (13x_1^2 - 4x_1 - 8)^2)$,
$h_2 = (x_2^2 + x_1^2 - 1)(25x_2^2 + 12x_2x_1 + 20x_1^2 - 6x_2 - 4x_1 - 15)$.

Remark 7.11. Although H_j is more complicated than any $\mathtt{Hp}(f,[x_n,\ldots,x_{j+1}],x_t)$,

$$V_{\mathbb{R}}(H_j) \subseteq V_{\mathbb{R}}(\mathtt{Hp}(f,[x_n,\ldots,x_{j+1}],x_{j+1})) \subseteq V_{\mathbb{R}}(\mathrm{BMproj}(f,[x_n,\ldots,x_{j+1}])).$$

That means, for every open cell C' of open CAD produced by Brown's operator, there exists an open cell C of open weak CAD produced by H_j such that $C' \subseteq C$. Thus, the scale of open weak CAD is no bigger than that of open CAD.

Remark 7.12. In Algorithm 7.10, the scale of the open weak CAD of f defined by h_j in $\mathbb{R}^j$ is not always the smallest. For example, let f be the polynomial in Example 7.7, then $h_1 = (16x_1 - 9)(x_1 + 1)$, and f is open weak delineable over $x_1 + 1$, as mentioned earlier.

Definition 7.15. (Open sample) A set of sample points $S_f \subseteq \mathbb{R}^k \setminus V_{\mathbb{R}}(f)$ is said to be an *open sample* defined by $f(\boldsymbol{x}^{[k]}) \in \mathbb{Z}[\boldsymbol{x}^{[k]}]$ in $\mathbb{R}^k$ if it has the following property: for every open connected set $U \subseteq \mathbb{R}^k$ defined by $f \neq 0$, $S_f \cap U \neq \emptyset$.

Suppose $g(\boldsymbol{x}^{[k]})$ is another polynomial. If S_f is an open sample defined by $f(\boldsymbol{x}^{[k]})$ in $\mathbb{R}^k$ such that $g(\alpha) \neq 0$ for any $\alpha \in S_f$, then we denote the open sample by $S_{f,g}$.

As a corollary of Theorems 7.1 and 7.2, a property of open CAD is that at least one sample point can be taken from every highest dimensional cell via the open CAD lifting phase. So, an open CAD is indeed an open sample.

Obviously, there are various efficient ways to compute $S_{f,g}$ for two given *univariate* polynomials $f, g \in \mathbb{Z}[x]$. For example, we may choose one rational point from every open interval defined by the real roots of f such that g does not vanish at this point. Therefore, we only describe the specification of such algorithms SPOne here and omit the details of the algorithms.

Algorithm 7.11 SPOne

Input: Two univariate polynomials $f, g \in \mathbb{Z}[x]$

Output: $S_{f,g}$, an open sample defined by $f(x)$ in $\mathbb{R}$ such that $g(\alpha) \neq 0$ for any $\alpha \in S_{f,g}$

Algorithm 7.12 OpenSP

Input: Two lists of polynomials $L_1 = [f_n(\boldsymbol{x}), \dots, f_j(\boldsymbol{x}^{[j]})]$, $L_2 = [g_n(\boldsymbol{x}), \dots, g_j(\boldsymbol{x}^{[j]})]$, and a set of points S in $\mathbb{R}^j$

Output: A set of sample points in $\mathbb{R}^n$

1:   $O \leftarrow S$;
2:   **for** i from $j + 1$ to n **do**
3:      $P \leftarrow \emptyset$
4:      **for** α in O **do**
5:         $P \leftarrow P \bigcup (\alpha \boxplus \text{SPOne}(f_i(\alpha, x_i), g_i(\alpha, x_i)))$;   // Definition 7.16
6:      **end for**
7:      $O \leftarrow P$;
8:   **end for**
9:   **return** O

Definition 7.16. Let $\boldsymbol{a}_j = (a_1, \dots, a_j) \in \mathbb{R}^j$ and $S \subseteq \mathbb{R}$ be a finite set, define

$$\boldsymbol{a}_j \boxplus S = \{(a_1, \dots, a_j, b) \mid b \in S\}.$$

Remark 7.13. For a polynomial $f(\boldsymbol{x}) \in \mathbb{Z}[\boldsymbol{x}]$, let
$$B_1 = [f, \mathrm{BMproj}(f, [x_n]), \ldots, \mathrm{BMproj}(f, [x_n, \ldots, x_2])],$$
$$B_2 = [1, \ldots, 1],$$
$$S = \mathtt{SPOne}(\mathrm{BMproj}(f, [x_n, \ldots, x_2]), 1),$$
then $\mathtt{OpenSP}(B_1, B_2, S)$ is an open CAD (an open sample) defined by $f(\boldsymbol{x})$.

We will provide in this section a method which computes two lists C_1 and C_2 where the polynomials in C_1 are factors of corresponding polynomials in B_1 and will prove that any $\mathtt{OpenSP}(C_1, C_2, S_{f_j, g_j})$ is an open sample of $\mathbb{R}^n$ defined by $f(\boldsymbol{x})$ for any open sample S_{f_j, g_j} in $\mathbb{R}^j$ where $f_j \in C_1$ and $g_j \in C_2$.

Remark 7.14. The output of $\mathtt{OpenSP}(L_1, L_2, S)$ is dependent on the method of choosing sample points in Algorithm $\mathtt{SPOne}$. In the following, when we use the terminology "any $\mathtt{OpenSP}(L_1, L_2, S)$", we mean "no matter which method is used in Algorithm $\mathtt{SPOne}$ for choosing sample points".

Definition 7.17. (Open delineable) Let
$$L_1 = [f_n(\boldsymbol{x}), f_{n-1}(\boldsymbol{x}^{[n-1]}), \ldots, f_j(\boldsymbol{x}^{[j]})], \tag{7.1}$$
$$L_2 = [g_n(\boldsymbol{x}), g_{n-1}(\boldsymbol{x}^{[n-1]}), \ldots, g_j(\boldsymbol{x}^{[j]})] \tag{7.2}$$
be two polynomial lists, S an open set of $\mathbb{R}^s$ ($s \leq j$) and $S' = S \times \mathbb{R}^{j-s}$. The polynomial $f_n(\boldsymbol{x})$ is said to be *open delineable* on S with respect to L_1 and L_2, if $\mathcal{A} \bigcap U \neq \emptyset$ for any maximal open connected set $U \subseteq \mathbb{R}^n$ defined by $f_n \neq 0$ with $U \bigcap (S' \times \mathbb{R}^{n-j}) \neq \emptyset$ and any $\mathcal{A} = \mathtt{OpenSP}(L_1, L_2, \{\alpha\})$ where $\alpha \in S'$ is any point such that $f_j(\alpha)g_j(\alpha) \neq 0$.

Remark 7.15. Let $s = j$ in Definition 7.17, it could be shown that if $f_n(\boldsymbol{x})$ is open delineable on S' with respect to L_1 and L_2, then $f_n(\boldsymbol{x})$ is open weak delineable on $S' \backslash \mathrm{V}_{\mathbb{R}}\{f_j g_j\}$.

Suppose $f_n(\boldsymbol{x})$ is a square-free polynomial in $\mathbb{Z}[\boldsymbol{x}]$ of positive degree and $S \subseteq \mathbb{R}^{n-1}$ is an open connected set in which $\mathrm{BMproj}(f_n, [x_n])$ is sign-invariant. According to Theorem 7.1 and Theorem 7.2, f_n is analytic delineable on S. It is easy to see that f_n is open delineable on S with respect to $[f_n, \mathrm{BMproj}(f_n, [x_n])]$ and $[f_n, \mathrm{BMproj}(f_n, [x_n])]$.

Open delineability has the following four properties.

Proposition 7.4. (*open sample property*) *Let L_1, L_2 be as in Definition 7.17. If $f_n(\boldsymbol{x})$ is open delineable on every open connected set of $f_j(\boldsymbol{x}^{[j]}) \neq 0$ with respect to L_1 and L_2, then for any open sample S_{f_j, g_j} in $\mathbb{R}^j$, any $\mathcal{A} = \mathtt{OpenSP}(L_1, L_2, S_{f_j, g_j})$ is an open sample defined by $f_n(\boldsymbol{x})$ in $\mathbb{R}^n$.*

Proof. For any open connected set $U \subseteq \mathbb{R}^n$ defined by $f_n \neq 0$, there exists at least one open connected set $S \subseteq \mathbb{R}^j$ defined by $f_j \neq 0$ such that $U \cap (S \times \mathbb{R}^{n-j}) \neq \emptyset$. Since f_n is open delineable on S with respect to L_1 and L_2, we have $\mathcal{A} \cap U \neq \emptyset$ for any $\mathcal{A} = \mathtt{OpenSP}(L_1, L_2, S_{f_j, g_j})$. $\square$

Proposition 7.5. (*transitive property*) *Let L_1, L_2, S, S' be as in Definition 7.17. Suppose that there exists $k(j \leq k \leq n)$ such that $f_k(\boldsymbol{x}^{[k]})$ is open delineable on S with respect to $L_1'' = [f_k(\boldsymbol{x}^{[k]}), \ldots, f_j(\boldsymbol{x}^{[j]})]$ and $L_2'' = [g_k(\boldsymbol{x}^{[k]}), \ldots, g_j(\boldsymbol{x}^{[j]})]$, and $f_n(\boldsymbol{x})$ is open delineable on every open connected set of $f_k(\boldsymbol{x}^{[k]}) \neq 0$ with respect to $L_1' = [f_n(\boldsymbol{x}), \ldots, f_k(\boldsymbol{x}^{[k]})]$ and $L_2' = [g_n(\boldsymbol{x}), \ldots, g_k(\boldsymbol{x}^{[k]})]$. Then $f_n(\boldsymbol{x})$ is open delineable on S with respect to L_1 and L_2.*

Proof. Let $\alpha \in S'$ be any point such that $f_j(\alpha)g_j(\alpha) \neq 0$, for any $\mathcal{A} = \mathtt{OpenSP}(L_1, L_2, \{\alpha\})$, we have $\mathcal{A} = \mathtt{OpenSP}(L_1', L_2', \mathcal{A}')$ where $\mathcal{A}' = \mathtt{OpenSP}(L_1'', L_2'', \{\alpha\})$. For any open connected set $U \subseteq \mathbb{R}^n$ defined by $f_n \neq 0$ with $U \cap (S' \times \mathbb{R}^{n-j}) \neq \emptyset$, there exists an open connected set $V \subseteq \mathbb{R}^k$ defined by $f_k \neq 0$ with $U \cap (V \times \mathbb{R}^{n-k}) \neq \emptyset$ and $V \cap (S \times \mathbb{R}^{k-s}) \neq \emptyset$. Now we have $\mathcal{A}' \cap V \neq \emptyset$ since $f_k(\boldsymbol{x}^{[k]})$ is open delineable on S with respect to L_1'' and L_2''. And then, $\mathcal{A} \cap U \neq \emptyset$ is implied by $U \cap (V \times \mathbb{R}^{n-k}) \neq \emptyset$ since $f_n(\boldsymbol{x})$ is open delineable on V with respect to L_1' and L_2'. $\square$

Proposition 7.6. (*nonempty intersection property*) *Let L_1, L_2 be as in Definition 7.17. For two open sets S_1 and S_2 of $\mathbb{R}^s$ ($s \leq j$) with $S_1 \cap S_2 \neq \emptyset$, if $f_n(\boldsymbol{x})$ is open delineable on both S_1 and S_2 with respect to L_1 and L_2, then $f_n(\boldsymbol{x})$ is open delineable on $S_1 \cup S_2$ with respect to L_1 and L_2.*

Proof. For any $\alpha_1 \in S_1$, $\alpha_2 \in S_2$, $\alpha_3 \in S_1 \cap S_2$ with $f_j(\alpha_i)g_j(\alpha_i) \neq 0$, any $\mathcal{A}_i = \mathtt{OpenSP}(L_1, L_2, \{\alpha_i\})$, and open connected set $U \subseteq \mathbb{R}^n$ defined by $f_n \neq 0$, we have $U \cap (S_1 \times \mathbb{R}^{n-s}) \neq \emptyset \iff \mathcal{A}_1 \cap U \neq \emptyset \iff \mathcal{A}_3 \cap U \neq \emptyset \iff \mathcal{A}_2 \cap U \neq \emptyset \iff U \cap (S_2 \times \mathbb{R}^{n-s}) \neq \emptyset$. $\square$

Notation 7.2. Let SP_n be the symmetric permutation group of $x_1, \ldots, x_n$. Define $SP_{n,i}$ to be the subgroup of SP_n, where any element σ of $SP_{n,i}$ fixes $x_1, \ldots, x_{i-1}$, i.e. $\sigma(x_j) = x_j$ for $j = 1, \ldots, i-1$.

Proposition 7.7. (*union property*) *Let L_1, L_2 be as in Definition 7.17. For $\sigma \in SP_{n,j+1}$, denote $\boldsymbol{y}_n = (y_1, \ldots, y_n) = \sigma(\boldsymbol{x})$ and $\boldsymbol{y}_i = (y_1, \ldots, y_i)$. Let $L_1' = [f_n(\boldsymbol{x}), p_{n-1}(\boldsymbol{y}_{n-1}), \ldots, p_j(\boldsymbol{y}_j)]$ and $L_2' = [q_n(\boldsymbol{x}), q_{n-1}(\boldsymbol{y}_{n-1}), \ldots, q_j(\boldsymbol{y}_j)]$ where $p_i(\boldsymbol{y}_i)$ and $q_i(\boldsymbol{y}_i)$ are polynomials in i variables.*

For two open sets S_1 and S_2 of $\mathbb{R}^j$, if
(a) $f_n(\boldsymbol{x})$ is open delineable on both S_1 and S_2 with respect to L_1 and L_2;
(b) $f_n(\boldsymbol{x})$ is open delineable on $S_1 \bigcup S_2$ with respect to L_1' and L_2', and
(c) $p_j(\boldsymbol{y}_j)q_j(\boldsymbol{y}_j)$ vanishes at no points in $S_1 \bigcup S_2$,
then $f_n(\boldsymbol{x})$ is open delineable on $S_1 \bigcup S_2$ with respect to L_1 and L_2.

Proof. Let $\alpha_1 \in S_1, \alpha_2 \in S_2$ be two points such that $g_j p_j q_j(\alpha_t) \neq 0$ for $t = 1, 2$. Let $\mathcal{A}_t = \mathtt{OpenSP}(L_1, L_2, \{\alpha_t\})$ and $\mathcal{A'}_t = \mathtt{OpenSP}(L_1', L_2', \{\alpha_t\})$. For any open connected set U defined by $f_n \neq 0$ with $U \bigcap (\alpha_1 \times \mathbb{R}^{n-j}) \neq \emptyset$, then $\mathcal{A}_1 \bigcap U \neq \emptyset$ and $\mathcal{A'}_1 \bigcap U \neq \emptyset$. Since $f_n(\boldsymbol{x})$ is open delineable on $S_1 \bigcup S_2$ with respect to L_1' and L_2', we have $\mathcal{A'}_2 \bigcap U \neq \emptyset$ which implies that $U \bigcap (S_2 \times \mathbb{R}^{n-j}) \neq \emptyset$ and $\mathcal{A}_2 \bigcap U \neq \emptyset$. Therefore, $f_n(\boldsymbol{x})$ is open delineable on $(S_1 \bigcup S_2) \backslash V_{\mathbb{R}}(p_j q_j)$ with respect to L_1 and L_2. Since $p_j q_j$ does not vanish at any point of $S_1 \bigcup S_2$, $f_n(\boldsymbol{x})$ is open delineable on $S_1 \bigcup S_2$ with respect to L_1 and L_2. $\qquad\square$

Example 7.13. We illustrate Proposition 7.7 using the polynomial f from Example 7.9.

$$f = (x_3^2 + x_2^2 + x_1^2 - 1)(4x_3 + 3x_2 + 2x_1 - 1) \in \mathbb{Z}[x_1, x_2, x_3]$$

Let

$$\begin{aligned}
L_1 &= [f, f_{x_3}, f_{x_3 x_2}] \\
L_1' &= [f, f_{x_2}, f_{x_2 x_3}] \\
L_2 &= L_1 \\
L_2' &= L_1'
\end{aligned}$$

where

$$\begin{aligned}
f_{x_3} &= (x_2^2 + x_1^2 - 1)(25x_2^2 + 12x_2 x_1 + 20x_1^2 - 6x_2 - 4x_1 - 15), \\
f_{x_3 x_2} &= (x_1 - 1)(x_1 + 1)(29x_1^2 - 4x_1 - 24)(13x_1^2 - 4x_1 - 8), \\
f_{x_2} &= (x_3^2 + x_1^2 - 1)(25x_3^2 + 16x_3 x_1 + 13x_1^2 - 8x_3 - 4x_1 - 8), \\
f_{x_2 x_3} &= (x_1 - 1)(x_1 + 1)(29x_1^2 - 4x_1 - 24)(20x_1^2 - 4x_1 - 15).
\end{aligned}$$

Let

$$S_1 = \left(\frac{2}{13} - \frac{6\sqrt{3}}{13}, \frac{2}{13} + \frac{6\sqrt{3}}{13} \right),$$

$$S_2 = \left(\frac{2}{13} + \frac{6\sqrt{3}}{13}, \frac{1}{10} + \frac{\sqrt{19}}{5} \right)$$

be two open intervals in x_1-axis, where $x_1 = \frac{2}{13} + \frac{6\sqrt{3}}{13}$ is one of the real roots of the equation $f_{x_3x_2} = 0$. By typical CAD methods, S_1 and S_2 are two different cells in x_1-axis.

It could be deduced easily by Theorem 7.1, Theorem 7.2, and Proposition 7.5 that the conditions (a) and (b) of Proposition 7.7 are satisfied. Since $f_{x_2x_3}$ vanishes at no points in $S_1 \bigcup S_2$, condition (c) is also satisfied.

By Proposition 7.7, f is open delineable on $S_1 \bigcup S_2$ with respect to L_1 and L_2. Roughly speaking, the real root of $x_1 - \frac{2}{13} - \frac{6\sqrt{3}}{13}$ would not affect the open delineability, thus we could combine the two cells S_1 and S_2.

For the same reason, the real roots of

$$(13x_1^2 - 4x_1 - 8)(20x_1^2 - 4x_1 - 15)$$

would not affect the open delineability either.

Lemma 7.9. *Let $f = \gcd(f_1, \ldots, f_m)$ where $f_i \in \mathbb{Z}[\boldsymbol{x}]$, $i = 1, \ldots, m$. Suppose f has no real zeros in a connected open set $U \subseteq \mathbb{R}^n$, then the open set $V = U \backslash V_{\mathbb{R}}(f_1, \ldots, f_m)$ is also connected.*

Proof. Without loss of generality, we can assume that $f = 1$. If $m = 1$, the result is obvious. The result of case $m = 2$ is just the claim of Lemma 7.4. For $m \geq 3$, let $g = \gcd(f_1, \ldots, f_{m-1})$ and $g_i = f_i/g$ $(i = 1, \ldots, m-1)$, then $\gcd(f_m, g) = 1$ and $\gcd(g_1, \ldots, g_{m-1}) = 1$. Let $A = V_{\mathbb{R}}(f_1, \ldots, f_m)$, $B = V_{\mathbb{R}}(g_1, \ldots, g_{m-1}) \bigcup V_{\mathbb{R}}(g, f_m)$. Since $A \subseteq B$, we have $U \backslash B \subseteq U \backslash A$. Notice that the closure of $U \backslash B$ equals the closure of $U \backslash A$, it suffices to prove that $U \backslash B$ is connected, which follows directly from Lemma 7.4 and induction. $\square$

Definition 7.18. Define

$$\overline{\mathrm{Hp}}(f, i) = \{f, \mathrm{Hp}(f, [x_n]), \ldots, \mathrm{Hp}(f, [x_n, \ldots, x_i])\},$$
$$\widetilde{\mathrm{Hp}}(f, i) = \{f, \mathrm{Hp}(f, [x_n], x_n), \ldots, \mathrm{Hp}(f, [x_n, \ldots, x_i], x_i)\}.$$

As a corollary of Theorem 7.1 and Theorem 7.2, we have

Proposition 7.8. *Let $f \in \mathbb{Z}[\boldsymbol{x}]$ be a square-free polynomial of class n. Then f is open delineable on every open connected set defined by $\mathrm{BMproj}(f, x_n) \neq 0$ in $\mathbb{R}^{n-1}$ with respect to $\overline{\mathrm{Hp}}(f, n)$ and $\widetilde{\mathrm{Hp}}(f, n)$.*

We now prove the correctness of Algorithm 7.10.

Theorem 7.12. *Let j be an integer and $2 \leq j \leq n$. For any given polynomial $f(\boldsymbol{x}) \in \mathbb{Z}[\boldsymbol{x}]$ and any open connected set $U \subseteq \mathbb{R}^{j-1}$ of*

$\mathtt{Hp}(f, [x_n, \ldots, x_j]) \neq 0$, *let*

$$S = U \backslash V_\mathbb{R}(\{\mathtt{Hp}(f, [x_n, \ldots, x_j], x_t) \mid t = j, \ldots, n\}).$$

Then $f(\boldsymbol{x})$ is open delineable on the open connected set S with respect to $\overline{\mathtt{Hp}}(f, j)$ and $\widetilde{\mathtt{Hp}}(f, j)$. As a result, f is open weak delineable over $H_{j-1}(f)$ in $\mathbb{R}^{j-1}$, and Algorithm 7.10 is correct.

Proof. First, by Lemma 7.9, S is open connected. We prove the theorem by induction on $k = n - j$. When $k = 0$, it is obviously true from Proposition 7.8. Suppose the theorem is true for all polynomials $g(\boldsymbol{x}^{[k]}) \in \mathbb{Z}[\boldsymbol{x}^{[k]}]$ with $k = 0, 1, \ldots, n - i - 1$. We now consider the case $k = n - i$. Let $[\mathbb{Z}] = [x_n, \ldots, x_i]$. For any given polynomial $f(\boldsymbol{x}) \in \mathbb{Z}[\boldsymbol{x}]$, let $U \subseteq \mathbb{R}^{i-1}$ be an open connected set of $\mathtt{Hp}(f, [\mathbb{Z}]) \neq 0$ and $S = U \backslash V_\mathbb{R}(\{\mathtt{Hp}(f, [\mathbb{Z}], x_t) \mid t = i, \ldots, n\})$.

For any point $\alpha \in S$ with $\mathtt{Hp}(f, [\mathbb{Z}], x_i)(\alpha) \neq 0$, there exists an open connected set $S_\alpha \subseteq \mathbb{R}^{i-1}$ such that $\alpha \in S_\alpha$ and $0 \notin \mathtt{Hp}(f, [\mathbb{Z}], x_i)(S_\alpha)$. By induction, $\mathtt{Hp}(f, [x_n, \ldots, x_{i+1}])$ is open delineable on S_α with respect to $\{\mathtt{Hp}(f, [\mathbb{Z}])\}$ and $\{\mathtt{Hp}(f, [\mathbb{Z}], x_i)\}$. By induction again and the transitive property of open delineable (Proposition 7.5), f is open delineable on S_α with respect to $\overline{\mathtt{Hp}}(f, i)$ and $\widetilde{\mathtt{Hp}}(f, i)$.

For any point $\alpha \in S$ with $\mathtt{Hp}(f, [\mathbb{Z}], x_i)(\alpha) = 0$, there exists an i' such that $n \geq i' \geq i + 1$ and $\mathtt{Hp}(f, [\mathbb{Z}], x_{i'})(\alpha) \neq 0$. Thus there exists an open connected set S'_α of $\mathbb{R}^{i-1}$ such that $\alpha \in S'_\alpha$ and $0 \notin \mathtt{Hp}(f, [\mathbb{Z}], x_{i'})(S'_\alpha)$. Let $\sigma \in SP_{n,i}$ with $\sigma(x_i) = x_{i'}$, in such case, $f(\sigma(\boldsymbol{x}))$ is open delineable on S'_α with respect to $\overline{\mathtt{Hp}}(f(\sigma(\boldsymbol{x})), i)$ and $\widetilde{\mathtt{Hp}}(f(\sigma(\boldsymbol{x})), i)$. For any $\beta \in S'_\alpha$ with $\mathtt{Hp}(f, [\mathbb{Z}], x_i)(\beta) \neq 0$, there exists an open connected set $S''_\alpha \subseteq S'_\alpha$ and f is open delineable on S''_α with respect to $\overline{\mathtt{Hp}}(f, i)$ and $\widetilde{\mathtt{Hp}}(f, i)$. From union property of open delineable (Proposition 7.7), f is open delineable on S'_α with respect to $\overline{\mathtt{Hp}}(f, i)$ and $\widetilde{\mathtt{Hp}}(f, i)$.

To summarize, the above discussion shows that for any point $\alpha \in S$, there exists an open connected set $S_\alpha \subseteq S$ such that $\alpha \in S_\alpha$ and f is open delineable on S_α with respect to $\overline{\mathtt{Hp}}(f, i)$ and $\widetilde{\mathtt{Hp}}(f, i)$. By the nonempty intersection property of open delineable (Proposition 7.6) and the fact that S is connected, $f(\boldsymbol{x})$ is open delineable on S with respect to $\overline{\mathtt{Hp}}(f, i)$ and $\widetilde{\mathtt{Hp}}(f, i)$ as desired.

Therefore, the theorem is proved by induction. The last statement of the theorem follows from the fact

$$V_\mathbb{R}(\{\mathtt{Hp}(f, [x_n, \ldots, x_j], x_t) \mid t = j, \ldots, n\}) = V_\mathbb{R}(H_{j-1}(f)).$$

$\square$

7.4.3   *Computing Open Sample*

As a direct application of Theorem 7.12, we show how to compute open sample based on Algorithm 7.10.

Definition 7.19. A set of sample points in $\mathbb{R}^n$ obtained through Algorithm 7.13 is called a *reduced open CAD* of $f(\boldsymbol{x})$ with respect to $[x_n, \ldots, x_{j+1}]$.

Algorithm 7.13 ReducedOpenCAD

Input: $f(\boldsymbol{x}) \in \mathbb{Z}[\boldsymbol{x}]$ and $S_{\mathrm{Hp}(f,[x_n,\ldots,x_{j+1}]),\mathrm{Hp}(f,[x_n,\ldots,x_{j+1}],x_{j+1})}$ (an open sample) in $\mathbb{R}^j$

Output: A set of sample points in $\mathbb{R}^n$

  1: $O \leftarrow S_{\mathrm{Hp}(f,[x_n,\ldots,x_{j+1}]),\mathrm{Hp}(f,[x_n,\ldots,x_{j+1}],x_{j+1})}$;

  2: **for** i from $j+2$ to $n+1$ **do**

  3:    $P \leftarrow \emptyset$;

  4:    **for** α in O **do**

  5:       **if** $i \leq n$ **then**

  6:         $P \leftarrow P \bigcup (\alpha \boxplus$

              $\mathtt{SPOne}(\mathrm{Hp}(f, [x_n, \ldots, x_i])(\alpha, x_{i-1}), \mathrm{Hp}(f, [x_n, \ldots, x_i], x_i)(\alpha, x_{i-1})))$;

  7:       **else**

  8:         $P \leftarrow P \bigcup (\alpha \boxplus \mathtt{SPOne}(f(\alpha, x_n), f(\alpha, x_n)))$;

  9:       **end if**

10:    **end for**

11:    $O \leftarrow P$;

12: **end for**

13: **return** O

The following corollary of Theorem 7.12 shows that the reduced open CAD owns the property of open delineability.

Corollary 7.1. *A reduced open CAD of $f(\boldsymbol{x})$ with respect to $[x_n, \ldots, x_{j+1}]$ is an open sample defined by $f(\boldsymbol{x})$.*

Example 7.14. We illustrate the main steps of Algorithm 7.13. using the polynomial f from Examples 7.9 and 7.12.

In: $f = (x_3^2 + x_2^2 + x_1^2 - 1)(4x_3 + 3x_2 + 2x_1 - 1) \in \mathbb{Z}[x_1, x_2, x_3]$,

    $S_{\mathrm{Hp}(f,[x_3,x_2]),\mathrm{Hp}(f,[x_3,x_2],x_2)} = \{-2, -\frac{27}{32}, 0, \frac{63}{64}, 2\}$ in $\mathbb{R}$.

  1: $O \leftarrow \{-2, -\frac{27}{32}, 0, \frac{63}{64}, 2\}$ // (O has 5 elements, $\alpha_1, \ldots, \alpha_5$)

  3: $P \leftarrow \emptyset$

6: $P \leftarrow P \bigcup (\alpha_1 \boxplus \mathtt{SPOne}(\mathtt{Hp}(f,[x_3])(\alpha_1, x_2), \mathtt{Hp}(f,[x_3], x_3)(\alpha_1, x_2)))$
$\qquad P \leftarrow P \bigcup (\alpha_2 \boxplus \mathtt{SPOne}(\mathtt{Hp}(f,[x_3])(\alpha_2, x_2), \mathtt{Hp}(f,[x_3], x_3)(\alpha_2, x_2)))$
$\qquad \vdots$
$\qquad P \leftarrow P \bigcup (\alpha_5 \boxplus \mathtt{SPOne}(\mathtt{Hp}(f,[x_3])(\alpha_5, x_2), \mathtt{Hp}(f,[x_3], x_3)(\alpha_5, x_2)))$
11: $O \leftarrow P$ $\qquad$ // (O now has 13 elements, $\alpha_1, \ldots, \alpha_{13}$)
 3: $P \leftarrow \emptyset$
 8: $P \leftarrow P \bigcup (\alpha_1 \boxplus \mathtt{SPOne}(\mathtt{Hp}(f)(\alpha_1, x_3), \mathtt{Hp}(f)(\alpha_1, x_3)))$
$\qquad P \leftarrow P \bigcup (\alpha_2 \boxplus \mathtt{SPOne}(\mathtt{Hp}(f)(\alpha_2, x_3), \mathtt{Hp}(f)(\alpha_2, x_3)))$
$\qquad \vdots$
$\qquad P \leftarrow P \bigcup (\alpha_{13} \boxplus \mathtt{SPOne}(\mathtt{Hp}(f)(\alpha_{13}, x_3), \mathtt{Hp}(f)(\alpha_{13}, x_3)))$
11: $O \leftarrow P$ $\qquad$ // (O now has 36 elements, $\alpha_1, \ldots, \alpha_{36}$)
Out: O

Algorithm 7.14 $\mathtt{HpTwo}$

Input: A polynomial $f \in \mathbb{Z}[\boldsymbol{x}]$ of class n.

Output: An open sample defined by f, *i.e.* a set of sample points which contains at least one point from each connected component of $f \neq 0$ in $\mathbb{R}^n$

1: $g \leftarrow f;\ L_1 \leftarrow \{\};\ L_2 \leftarrow \{\};\ i \leftarrow n;$
2: **while** $i \geq 3$ **do**
3: $\quad L_1 \leftarrow L_1 \bigcup \overline{\mathtt{Hp}}(g, i-1);$
4: $\quad L_2 \leftarrow L_2 \bigcup \widetilde{\mathtt{Hp}}(g, i-1);$
5: $\quad g \leftarrow \mathtt{Hp}(g, [x_i, x_{i-1}]);$
6: $\quad i \leftarrow i - 2;$
7: **end while**
8: **if** $i = 2$ **then**
9: $\quad L_1 \leftarrow L_1 \bigcup \overline{\mathtt{Hp}}(g, i);$
10: $\quad L_2 \leftarrow L_2 \bigcup \widetilde{\mathtt{Hp}}(g, i);$
11: $\quad g \leftarrow \mathtt{Hp}(g, [x_i]);$
12: **end if**
13: $S \leftarrow \mathtt{SPOne}(L_1^{[1]}, L_2^{[1]});$
14: $C \leftarrow \mathtt{OpenSP}(L_1, L_2, S);$
15: **return** C

Remark 7.16. As an application of Theorem 7.12, we could design a CAD-like method to get an open sample defined by $f(\boldsymbol{x})$ for a given polynomial $f(\boldsymbol{x})$. Roughly speaking, if we have already got an open sample defined by

$\mathrm{Hp}(f, [x_n, \ldots, x_j])$ in $\mathbb{R}^{j-1}$, according to Theorem 7.12, we could obtain an open sample defined by f in $\mathbb{R}^n$. That process could be done recursively.

In the definition of Hp, we first choose m variables from $\{x_1, \ldots, x_n\}$, compute all projection polynomials under all possible orders of those m variables, and then compute the gcd of all those projection polynomials. Therefore, Theorem 7.12 provides us many ways for designing various algorithms for computing open samples. For example, we may set $m = 2$ and choose $[x_n, x_{n-1}]$, $[x_{n-2}, x_{n-3}]$, etc. successively in each step. Because there are only two different orders for two variables, we compute the gcd of two projection polynomials under the two orders in each step. Algorithm 7.14 (HpTwo) is based on this choice.

Remark 7.17. If $\mathrm{Hp}(f, [x_n, x_{n-1})] \neq \mathrm{BMproj}(f, [x_n, x_{n-1}])$ and $n > 3$, it is obvious that the scale of projection in Algorithm 7.14 is smaller than that of open CAD in Definition 7.8.

Remark 7.18. It should be mentioned that there are some non-CAD methods for computing sample points in semi-algebraic sets, such as critical point method. For related work, see for example, [Basu *et al.* (1998); Safey El Din and Schost (2003); Hong and Safey El Din (2012)].

7.4.4 *Combining* Hp *and* Np

In this section, we combine the idea of Hp and the simplified CAD projection operator Np we introduced previously in Section 7.3, to get a new algorithm for testing semi-definiteness of polynomials.

Definition 7.20. Let $f \in \mathbb{Z}[x_1, \ldots, x_n]$ of class n. Denote $[\boldsymbol{y}] = [y_1, \ldots, y_m]$ where $y_i \in \{x_1, \ldots, x_n\}$ for $1 \leq i \leq m$ and $y_i \neq y_j$ for $i \neq j$. Define

$$\mathrm{Np}(f, [x_i]) = \mathrm{Np}_2(f, [x_i]), \quad \mathrm{Np}(f, [x_i], x_i) = \prod_{g \in \mathrm{Np}_1(f, [x_i])} g.$$

For $m(m \geq 2)$ and $i(1 \leq i \leq m)$, $\mathrm{Np}(f, [\boldsymbol{y}], y_i)$ and $\mathrm{Np}(f, [\boldsymbol{y}])$ are defined recursively as follows.

$$\mathrm{Np}(f, [\boldsymbol{y}], y_i) = \mathrm{BMproj}(\mathrm{Np}(f, [\hat{\boldsymbol{y}}]_i), y_i),$$
$$\mathrm{Np}(f, [\boldsymbol{y}]) = \gcd(\mathrm{Np}(f, [\boldsymbol{y}], y_1), \ldots, \mathrm{Np}(f, [\boldsymbol{y}], y_m)),$$

where $[\hat{\boldsymbol{y}}]_i = [y_1, \ldots, y_{i-1}, y_{i+1}, \ldots, y_m]$. Define

$$\overline{\mathrm{Np}}(f, i) = \{f, \mathrm{Np}(f, [x_n]), \ldots, \mathrm{Np}(f, [x_n, \ldots, x_i])\},$$

and

$$\widetilde{\mathrm{Np}}(f, i) = \{f, \mathrm{Np}(f, [x_n], x_n), \ldots, \mathrm{Np}(f, [x_n, \ldots, x_i], x_i)\}.$$

Now, we can rewrite Theorem 7.5 in another way.

Proposition 7.9. *Let $f \in \mathbb{Z}[\boldsymbol{x}]$ be a square-free polynomial of class n and U a connected component of $\mathrm{Np}(f, [x_n]) \neq 0$ in $\mathbb{R}^{n-1}$. If the polynomials in $\mathrm{Np}_1(f, [x_n])$ are semi-definite on U, then f is open delineable on U with respect to $\overline{\mathrm{Np}}(f, n)$ and $\widetilde{\mathrm{Np}}(f, n)$.*

Theorem 7.13. *Let j be an integer and $2 \leq j \leq n$. For any given polynomial $f(\boldsymbol{x}) \in \mathbb{Z}[\boldsymbol{x}]$, and any open connected set U of $\mathrm{Np}(f, [x_n, \ldots, x_j]) \neq 0$ in $\mathbb{R}^{j-1}$, let*

$$S = U \backslash \mathrm{V}_{\mathbb{R}}(\{\mathrm{Np}(f, [x_n, \ldots, x_j], x_t) \mid t = j, \ldots, n\}).$$

If the polynomials in $\bigcup_{i=0}^{n-j} \mathrm{Np}_1(f, [x_{n-i}])$ are all semi-definite on $U \times \mathbb{R}^{n-j}$, $f(\boldsymbol{x})$ is open delineable on S with respect to $\overline{\mathrm{Np}}(f, j)$ and $\widetilde{\mathrm{Np}}(f, j)$.

Proof. Notice that the proof of Theorem 7.12 only uses the properties of open delineable (Propositions 7.4-7.7) and Proposition 7.8 (f is open delineable on every open connected set defined by $\mathrm{BMproj}(f, [x_n]) \neq 0$ in $\mathbb{R}^{n-1}$ with respect to $\overline{\mathrm{Hp}}(f, n)$ and $\widetilde{\mathrm{Hp}}(f, n)$). According to Proposition 7.9, f is open delineable on every open connected set defined by $\mathrm{Np}(f, [x_n]) \neq 0$ in $\mathbb{R}^{n-1}$ with respect to $\overline{\mathrm{Np}}(f, n)$ and $\widetilde{\mathrm{Np}}(f, n)$. The same proof of Theorem 7.12 will yield the conclusion of the theorem. $\square$

Theorem 7.13 and Theorem 7.7 provide us a new way to decide the non-negativity of a polynomial as stated in the next theorem.

Theorem 7.14. *Let $f \in \mathbb{Z}[\boldsymbol{x}]$ be a square-free polynomial of class n and U a connected open set of $\mathrm{Np}(f, [x_n, \ldots, x_j]) \neq 0$ in $\mathbb{R}^{j-1}$. Denote*

$$S = U \backslash \mathrm{V}_{\mathbb{R}}(\{\mathrm{Np}(f, [x_n, \ldots, x_j], x_t) \mid t = j, \ldots, n\}).$$

The necessary and sufficient condition for $f(\boldsymbol{x})$ to be positive semi-definite on $U \times \mathbb{R}^{n-j+1}$ is the following two conditions hold.
(1) The polynomials in $\bigcup_{i=0}^{n-j} \mathrm{Np}_1(f, [x_{n-i}])$ are all semi-definite on $U \times \mathbb{R}^{n-j}$.
(2) There exists a point $\alpha \in S$ such that $f(\alpha, x_j, \ldots, x_n)$ is positive semi-definite on $\mathbb{R}^{n-j+1}$.

Proposition 7.10. *Given a positive integer $n \geq 3$. Let $f \in \mathbb{Z}[\boldsymbol{x}]$ be a square-free polynomial of class n and U a connected open set of $\mathrm{Np}(f, [x_n, x_{n-1}]) \neq 0$ in $\mathbb{R}^{n-2}$. Denote*

$$S = U \backslash \mathrm{V}_{\mathbb{R}}(\mathrm{Np}(f, [x_n, x_{n-1}], x_n), \mathrm{Np}(f, [x_n, x_{n-1}], x_{n-1})).$$

The necessary and sufficient condition for $f(\boldsymbol{x})$ to be positive semi-definite on $U \times \mathbb{R}^2$ is that the following two conditions hold.

(1) The polynomials in either $\mathtt{Np}_1(f, [x_n])$ or $\mathtt{Np}_1(f, [x_{n-1}])$ are semi-definite on $U \times \mathbb{R}$.

(2) There exists a point $\alpha \in S$ such that $f(\alpha, x_{n-1}, x_n)$ is positive semi-definite on $\mathbb{R}^2$.

Based on the above theorems, it is easy to design some different algorithms (depending on the choice of j) to prove polynomial inequalities. For example, Algorithm $\mathtt{PSD\text{-}HpTwo}$ for deciding the nonnegativity of polynomials is based on Theorem 7.14 when $j = n - 1$ (Proposition 7.10).

Algorithm 7.15 $\mathtt{PSD\text{-}HpTwo}$

Input: An irreducible polynomial $f \in \mathbb{Z}[\boldsymbol{x}]$.

Output: Whether or not $\forall \alpha_n \in \mathbb{R}^n (f(\alpha_n) \geq 0)$.

 1: **if** $n \leq 2$ **then**

 2:    **if** $\mathrm{DPS}(f(x_n)) = \mathtt{false}$ **then**

 3:       **return** $\mathtt{false}$

 4:    **end if**

 5: **else**

 6:    $L_1 \leftarrow \mathtt{Np}_1(f, [x_n]) \bigcup \mathtt{Np}_1(f, [x_{n-1}])$;

 7:    $L_2 \leftarrow \mathtt{Np}(f, [x_n, x_{n-1}])$;

 8:    **for** g in L_1 **do**

 9:       **if** $\mathtt{PSD\text{-}HpTwo}(g) = \mathtt{false}$ **then**

10:         **return** $\mathtt{false}$

11:       **end if**

12:    **end for**

13:    $C_{n-2} \leftarrow$ A reduced open CAD of L_2 w.r.t. $[x_{n-2}, \ldots, x_2]$, which satisfies that $\mathrm{V}_{\mathbb{R}}(\mathtt{Np}(f, [x_n, x_{n-1}], x_n), \mathtt{Np}(f, [x_n, x_{n-1}], x_{n-1})) \cap C_{n-2} = \emptyset$;

14:    **if** $\exists \alpha_{n-2} \in C_{n-2}$ such that $\mathrm{DPS}(f(\alpha_{n-2}, x_{n-1}, x_n)) = \mathtt{false}$ **then**

15:       **return** $\mathtt{false}$

16:    **end if**

17: **end if**

18: **return** $\mathtt{true}$

7.4.5 *Examples*

Example 7.15. [Strzeboński (2000)]

$$f = ax^3 + (a + b + c)x^2 + (a^2 + b^2 + c^2)x + a^3 + b^3 + c^3 - 1.$$

Under the order $a \prec b \prec c \prec x$, an open CAD defined by f has 132 sample points, while an open sample obtained by Algorithm `HpTwo` has 15 sample points.

Example 7.16. For 100 random polynomials $f(x, y, z)$ with degree 8, generated by `randpoly([x,y,z],degree=8)` in Maple 15, Fig. 7.3 shows the numbers of real roots of $\mathrm{BMproj}(f, [z, y])$, $\mathrm{BMproj}(f, [y, z])$ and $\mathrm{Hp}(f, [y, z])$, respectively. It is clear that the number of real roots of $\mathrm{Hp}(f, [y, z])$ is always less than those of $\mathrm{BMproj}(f, [z, y])$ and $\mathrm{BMproj}(f, [y, z])$.

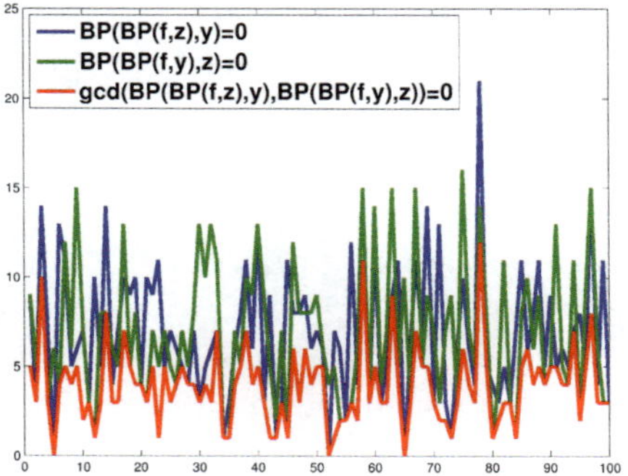

Fig. 7.3   The number of real roots.

Example 7.17. In this example, we compare the performance of Algorithm `HpTwo` with open CAD on randomly generated polynomials. All the data in this example were obtained on a PC with Intel(R) Core(TM) i5 3.20GHz CPU, 8GB RAM, Windows 7 and Maple 17.

In the following table, we list the average time of projection phase and lifting phase, and the average number of sample points on 30 random polynomials with 4 variables and degree 4 generated by `randpoly([x,y,z,w],degree=4)-1`.

	Projection	Lifting	Sample points
HpTwo	0.13	0.29	262
open CAD	0.19	3.11	486

If we get random polynomials with 5 variables and degree 3 by the command

$$\texttt{randpoly}([\texttt{seq}(\texttt{x[i]}, \texttt{i} = 1..5)], \texttt{degree} = 3),$$

then the degrees of some variables are usually one. That makes the computation very easy for both HpTwo and open CAD. Therefore, we run the command

$$\texttt{randpoly}([\texttt{seq}(\texttt{x[i]}, \texttt{i} = 1..5)], \texttt{degree} = 3) + \texttt{add}(\texttt{x[i]}^2, \texttt{i} = 1..5) - 1$$

ten times to generate 10 random polynomials with 5 variables and degree 3. The data on the 10 polynomials are listed in the following table.

	Projection	Lifting	Sample points
HpTwo	2.87	3.51	2894
open CAD	0.76	12.01	7802

For many random polynomials with 4 variables and degree greater than 4 (or 5 variables and degree greater than 3), neither HpTwo nor open CAD can finish computation in reasonable time.

A main application of the new projection operator Hp is testing semi-definiteness of polynomials. Now, we illustrate the performance of our implementation of Algorithm PSD-HpTwo with several non-trivial examples. For more examples, visit the homepage of Jingjun Han[1].

We report the timings of the programs PSD-HpTwo and DPS, the function PartialCylindricalAlgebraicDecomposition (PCAD) in Maple 15, function FindInstance (FI) in Mathematica 9 and QEPCAD B (QEPCAD) on these examples.

QEPCAD was performed on a PC with Intel(R) Core(TM) i5 3.20GHz CPU, 4GB RAM and ubuntu. The other computations were performed on a laptop with Inter Core(TM) i5-3317U 1.70GHz CPU, 4GB RAM, Windows 8 and Maple 15.

Example 7.18. [Han (2011)] Prove that

$$f(\boldsymbol{x}) = (\sum_{i=1}^{n} x_i^2)^2 - 4\sum_{i=1}^{n} x_i^2 x_{i+1}^2 \geq 0,$$

---

[1]`https://sites.google.com/site/jingjunhan/home/software`

where $x_{n+1} = x_1$.

Hereafter "∞" means either the running time is over 4000 seconds or the software fails to get an answer.

n	5	8	11	17	23
PSD $-$ HpTwo	0.28	0.95	6.26	29.53	140.01
DPS	0.29	∞	∞	∞	∞
FI	0.10	∞	∞	∞	∞
PCAD	0.26	∞	∞	∞	∞
QEPCAD	0.10	∞	∞	∞	∞

We then test the semi-definiteness of the polynomials (in fact, all $f(\boldsymbol{x})$ are indefinite)

$$g(\boldsymbol{x}) = f(\boldsymbol{x}) - \frac{1}{10^{10}} x_n^4.$$

The timings are reported in the following table.

n	PSD–HpTwo	DPS	FI	PCAD	QEPCAD
20	3.828	∞	∞	∞	∞
30	13.594	∞	∞	∞	∞

Example 7.19. Prove that

$$B(\boldsymbol{x}_{3m+2}) = \left(\sum_{i=1}^{3m+2} x_i^2 \right)^2 - 2 \sum_{i=1}^{3m+2} x_i^2 \sum_{j=1}^{m} x_{i+3j+1}^2 \geq 0,$$

where $x_{3m+2+r} = x_r$. If $m = 1$, it is equivalent to the case $n = 5$ of Example 7.18.

m	PSD–HpTwo	DPS	FI	PCAD	QEPCAD
1	0.296	0.297	0.1	0.26	0.104
2	1.390	23.094	∞	∞	∞
3	9.672	∞	∞	∞	∞
4	∞	∞	∞	∞	∞

Remark 7.19. For some special examples like Example 7.18, PSD-HpTwo could solve problems with more than 30 variables efficiently. Of course, there also exist some other examples on which PSD-HpTwo performs badly. For example, PSD-HpTwo could not solve the problems in [Kaltofen *et al.* (2009)] within 4000 seconds while they can be solved by RAGlib efficiently.

As showed by Example 7.17, according to our experiments, the application of HpTwo and PSD-HpTwo is limited at 3-4 variables and low degrees generally. It is not difficult to see that, if the input polynomial $f(x)$ is symmetric, the new projection operator Hp cannot reduce the projection scale and the number of sample points. Thus, it is reasonable to conclude that the complexity of PSD-HpTwo is still doubly exponential.

Chapter 8

Dimension-Decreasing Algorithm

To prove an algebraic inequality with radicals, a common method is to introduce new variables for the radicals and transform the inequality to a polynomial inequality with some equational constraints. However, the new variables increase the dimension (number of variables) of the problem and thus often cause the problem intractable in practice. In this chapter, we introduce an algorithm, called dimension-decreasing algorithm, which can deal with radical inequalities efficiently and keep the dimension unchange.

8.1   Inequalities with Radicals

Consider an SAS S of the form (3.1). All the equations and inequalities in S can be seen as atom formulas and the system is the conjunction of the atom formulas. Let Φ denote the conjunction. If Φ_0 is a polynomial equation (inequality), then

$$\Phi \Rightarrow \Phi_0$$

is a proposition of real algebra or real geometry. Obviously, the proposition is true if and only if the following formula (an SAS)

$$\Phi \wedge \neg\Phi_0$$

is inconsistent. Herein, $\neg\Phi_0$ stands for the negation of Φ_0.

If the proposition contains some polynomial equations, it is natural to think about eliminating some variables by the equations. That is just the basic idea of real root classification which is elaborated in Chapter 6. To eliminate (some) variables, we triangularize the system into TSAs and compute border polynomial to partition the parametric space because the main variables of the triangular set cannot be solved explicitly in general.

233

However, if every equation in the TSAs is of degree no more than 2 with respect to its main variable, those main variables can be solved explicitly as algebraic functions in parameters with square roots.

Fortunately, the equations in premise of propositions in the so-called *constructible geometric theorems* are all of degree no more than 2 with respect to their main variables. So, the main variables can be expressed by parameters explicitly and the problem is reduced to proving inequalities with radicals (square roots).

Example 8.1. Suppose real numbers $x, y, z, u_1, u_2, u_3, u_4, u_5, u_6$ satisfy the following 15 conditions:

$$\begin{cases} (xy + yz + xz)^2 u_1^2 - x^3(y + z)(xy + xz + 4yz) = 0, \\ (xy + yz + xz)^2 u_2^2 - y^3(x + z)(xy + yz + 4xz) = 0, \\ (xy + yz + xz)^2 u_3^2 - z^3(x + y)(yz + xz + 4xy) = 0, \\ (x + y + z)(u_4^2 - x^2) - xyz = 0, \\ (x + y + z)(u_5^2 - y^2) - xyz = 0, \\ (x + y + z)(u_6^2 - z^2) - xyz = 0, \\ x > 0, \ y > 0, \ z > 0, \\ u_1 > 0, \ u_2 > 0, \ u_3 > 0, \ u_4 > 0, \ u_5 > 0, \ u_6 > 0. \end{cases} \tag{8.1}$$

Prove that $u_1 + u_2 + u_3 \leq u_4 + u_5 + u_6$.

If we view x, y, z as parameters, then the 6 equations of system (8.1) are all of degree 2 with respect to their main variables $u_1, \ldots, u_6$. After expressing u_i with x, y, z, we transform the above proposition to the following one with radicals.

Example 8.2. Suppose $x > 0, \ y > 0, \ z > 0$. Prove that

$$\frac{\sqrt{x^3(y + z)(xy + xz + 4yz)}}{xy + yz + xz} + \frac{\sqrt{y^3(x + z)(xy + yz + 4xz)}}{xy + yz + xz}$$
$$+ \frac{\sqrt{z^3(x + y)(yz + xz + 4xy)}}{xy + yz + xz}$$
$$\leq \sqrt{x^2 + \frac{xyz}{x + y + z}} + \sqrt{y^2 + \frac{xyz}{x + y + z}} + \sqrt{z^2 + \frac{xyz}{x + y + z}}. \tag{8.2}$$

The inequality contains 3 variables and 6 radicals while the original problem involves 9 variables.

Yang (1999); Yang and Xia (2000); Yang and Zhang (2001) proposed a so-called dimension-decreasing algorithm which could deal with radical inequalities effectively and reduce the dimension (number of variables) as many as possible. The algorithm has been implemented and updated by Lu Yang and Shihong Xia as a Maple program BOTTEMA that has verified more than 1000 algebraic and geometric inequalities including 100 more open problems on personal computers. The total machine time for proving 100 basic inequalities (including some classical inequalities such as Euler's inequality, Finsler-Hadwiger's inequality, Gerretsen's inequality and so on) using BOTTEMA on a PC (Intel Core i5-3470 CPU @ 3.20GHz, 8G memory, Windows 7 OS) with Maple 17 is no more than 1.5 seconds.

8.2 Dimension-Decreasing Algorithm

8.2.1 *Concepts*

We introduce some concepts and illustrate them by examples.

Definition 8.1. Suppose $l(x, y, z, \ldots)$ and $r(x, y, z, \ldots)$ are continuous algebraic function in $x, y, z, \ldots$. We call

$$l(x, y, z, \ldots) \leq r(x, y, z, \ldots) \quad \text{or} \quad l(x, y, z, \ldots) < r(x, y, z, \ldots)$$

an *algebraic inequality* and $l(x, y, z, \ldots) = r(x, y, z, \ldots)$ *algebraic equality* with respect to $x, y, z, \ldots$.

Definition 8.2. Suppose Φ is an algebraic equality or inequality with respect to $x, y, z, \ldots$. A polynomial $L(T)$ is said to be a *left polynomial* of Φ if

- $L(T)$ is a polynomial in T and its coefficients are polynomials in $\mathbb{Q}[x, y, z, \ldots]$; and
- the left-hand side of Φ is a root of $L(T) = 0$.

The following requirement on $L(T)$ is not necessary but can help reduce the computation.

- $L(T)$ has the least degree among all those polynomials satisfying the above two conditions.

By the definition, if the left-hand side of Φ is the zero polynomial, then $L(T) = T$. Analogously, we may define the *right polynomial $R(T)$* of Φ.

Definition 8.3. Suppose Φ is an algebraic inequality or equality with respect to $x, y, z, \ldots$ and $L(T)$ and $R(T)$ are its left and right polynomial, respectively. Let $P(x, y, \ldots)$ be the Sylvester resultant of $L(T)$ and $R(T)$ with respect to T and call it the *critical polynomial* of Φ. The surface defined by $P(x, y, \ldots) = 0$ is called the *critical surface* of Φ.

The definitions of left and right polynomials are necessary for computing critical surface effectively. For Example 8.2, let

$$
\begin{aligned}
f_1 &= (xy + yz + xz)^2 u_1^2 - x^3(y + z)(xy + xz + 4\,yz), \\
f_2 &= (xy + yz + xz)^2 u_2^2 - y^3(x + z)(xy + yz + 4\,xz), \\
f_3 &= (xy + yz + xz)^2 u_3^2 - z^3(x + y)(yz + xz + 4\,xy), \\
f_4 &= (x + y + z)(u_4^2 - x^2) - xyz, \\
f_5 &= (x + y + z)(u_5^2 - y^2) - xyz, \\
f_6 &= (x + y + z)(u_6^2 - z^2) - xyz,
\end{aligned}
$$

then the left and right polynomials of inequality (8.2) can be computed by successive resultants:

$$
\begin{aligned}
&\mathrm{res}(\mathrm{res}(\mathrm{res}(u_1 + u_2 + u_3 - T, f_1, u_1), f_2, u_2), f_3, u_3), \\
&\mathrm{res}(\mathrm{res}(\mathrm{res}(u_4 + u_5 + u_6 - T, f_4, u_4), f_5, u_5), f_6, u_6).
\end{aligned}
$$

Deleting the factors without T, we have

$$
\begin{aligned}
L(T) = {}& (x\,y + x\,z + y\,z)^8\,T^8 - 4(x^4\,y^2 + 2\,x^4\,y\,z + x^4\,z^2 + 4\,x^3\,y^2\,z + 4\,x^3\,y\,z^2 \\
& + x^2\,y^4 + 4\,x^2\,y^3\,z + 4\,x^2\,y\,z^3 + x^2\,z^4 + 2\,x\,y^4\,z + 4\,x\,y^3\,z^2 + 4\,x\,y^2\,z^3 \\
& + 2\,x\,y\,z^4 + y^4\,z^2 + y^2\,z^4)(x\,y + x\,z + y\,z)^6\,T^6 + \cdots,
\end{aligned}
$$

$$
\begin{aligned}
R(T) = {}& (x + y + z)^4\,T^8 - 4(x^3 + x^2\,y + x^2\,z + x\,y^2 + 3\,x\,y\,z + x\,z^2 + y^3 + y^2\,z \\
& + y\,z^2 + z^3)\,(x + y + z)^3\,T^6 + 2(16\,x\,y\,z^4 + 14\,x\,y^2\,z^3 + 14\,x\,y^3\,z^2 + 16\,x\,y^4\,z \\
& + 14\,x^2\,y\,z^3 + 14\,x^2\,y^3\,z + 14\,x^3\,y\,z^2 + 14\,x^3\,y^2\,z + 16\,x^4\,y\,z + 3\,x^6 + 5\,x^4\,y^2 \\
& + 5\,x^4\,z^2 + 5\,x^2\,y^4 + 5\,x^2\,z^4 + 5\,y^4\,z^2 + 5\,y^2\,z^4 + 21\,x^2\,y^2\,z^2 + 3\,y^6 + 3\,z^6 \\
& + 6\,x^5\,y + 6\,x^5\,z + 4\,x^3\,y^3 + 4\,x^3\,z^3 + 6\,x\,y^5 + 6\,x\,z^5 + 6\,y^5\,z + 4\,y^3\,z^3 + 6\,y\,z^5) \\
& (x + y + z)^2\,T^4 \\
& - 4(x + y + z)(x^6 - x^4\,y^2 - x^4\,z^2 + 2\,x^3\,y^2\,z + 2\,x^3\,y\,z^2 - x^2\,y^4 + 2\,x^2\,y^3\,z \\
& + 7\,x^2\,y^2\,z^2 + 2\,x^2\,y\,z^3 - x^2\,z^4 + 2\,x\,y^3\,z^2 + 2\,x\,y^2\,z^3 + y^6 - y^4\,z^2 - y^2\,z^4 + z^6) \\
& (x^3 + 3\,x^2\,y + 3\,x^2\,z + 3\,x\,y^2 + 7\,x\,y\,z + 3\,x\,z^2 + y^3 + 3\,y^2\,z + 3\,y\,z^2 + z^3)\,T^2 \\
& + (-6\,x\,y^2\,z^3 - 6\,x\,y^3\,z^2 - 6\,x^2\,y\,z^3 - 6\,x^2\,y^3\,z - 6\,x^3\,y\,z^2 - 6\,x^3\,y^2\,z + x^6 \\
& - x^4\,y^2 - x^4\,z^2 - x^2\,y^4 - x^2\,z^4 - y^4\,z^2 - y^2\,z^4 - 9\,x^2\,y^2\,z^2 + y^6 + z^6 + 2\,x^5\,y \\
& + 2\,x^5\,z - 4\,x^3\,y^3 - 4\,x^3\,z^3 + 2\,x\,y^5 + 2\,x\,z^5 + 2\,y^5\,z - 4\,y^3\,z^3 + 2\,y\,z^5)^2.
\end{aligned}
$$

The time consumed by computing $L(T)$ and $R(T)$ above on a PC (Intel Core i5-3470 CPU @ 3.20GHz, 8G memory, Windows 7 OS) with Maple 17 is 0.03 and 0.001 seconds, respectively. Computing the critical polynomial $(\mathrm{res}(L(T), R(T), T))$ consumes 0.22 seconds. After deleting positive factors, the square-free part of the resulting polynomial has 2691 terms and degree 100.

Otherwise, if we transform (8.2) equivalently into

$$\frac{\sqrt{x^3(y+z)(xy+xz+4\,yz)}}{xy+yz+xz} + \frac{\sqrt{y^3(x+z)(xy+yz+4\,xz)}}{xy+yz+xz}$$
$$+\frac{\sqrt{z^3(x+y)(yz+xz+4\,xy)}}{xy+yz+xz} - \sqrt{x^2+\frac{xyz}{x+y+z}} - \sqrt{y^2+\frac{xyz}{x+y+z}}$$
$$\leq \sqrt{z^2+\frac{xyz}{x+y+z}},$$

and then compute similarly as follows:

```
f:=u1+u2+u3-u4-u5-T;
for i to 5 do f:=resultant(f,f||i,u||i) od;
```

the computation cannot be completed within 3 hours.

We can also try to compute critical polynomial like this:

```
f:=u1+u2+u3-u4-u5-u6;
for i to 6 do f:=resultant(f,f||i,u||i) od;
```

The computation cannot be completed within 3 hours, either.

Example 8.3. Given an algebraic inequality with respect to x, y, z:

$$m_a + m_b + m_c \leq 2\,s \tag{8.3}$$

where

$$m_a = \frac{1}{2}\sqrt{2\,(x+y)^2 + 2\,(x+z)^2 - (y+z)^2},$$
$$m_b = \frac{1}{2}\sqrt{2\,(y+z)^2 + 2\,(x+y)^2 - (x+z)^2},$$
$$m_c = \frac{1}{2}\sqrt{2\,(x+z)^2 + 2\,(y+z)^2 - (x+y)^2},$$
$$s = x+y+z$$

and $x > 0$, $y > 0$, $z > 0$. Compute its left and right polynomials and critical polynomial.

Let

$$f_1 = 4\,m_a^2 + (y+z)^2 - 2\,(x+y)^2 - 2\,(x+z)^2,$$
$$f_2 = 4\,m_b^2 + (x+z)^2 - 2\,(y+z)^2 - 2\,(x+y)^2,$$
$$f_3 = 4\,m_c^2 + (x+y)^2 - 2\,(x+z)^2 - 2\,(y+z)^2.$$

Compute the successive resultants

$$\mathrm{res}(\mathrm{res}(\mathrm{res}(m_a + m_b + m_c - T, f_1, m_a), f_2, m_b), f_3, m_c),$$

which gives the left polynomial of (8.3)

$$
\begin{aligned}
T^8 &- 6\,(x^2 + y^2 + z^2 + x\,y + y\,z + z\,x)\,T^6 + 9(x^4 + 2\,x\,y\,z^2 + y^4 + 2\,x\,z^3 \\
&+ 2\,x^3\,y + z^4 + 3\,y^2\,z^2 + 2\,y^2\,z\,x + 2\,y^3\,z + 2\,y\,z^3 + 3\,x^2\,z^2 + 2\,x^3\,z + 2\,x^2\,y\,z \\
&+ 2\,x\,y^3 + 3\,x^2\,y^2)T^4 - (72\,x^4\,y\,z + 78\,x^3\,y\,z^2 + 4\,x^6 + 4\,y^6 + 4\,z^6 + 12\,x\,y^5 \\
&- 3\,x^4\,y^2 - 3\,x^2\,z^4 - 3\,x^2\,y^4 - 3\,y^4\,z^2 - 3\,y^2\,z^4 - 3\,x^4\,z^2 - 26\,x^3\,y^3 - 26\,x^3\,z^3 \\
&- 26\,y^3\,z^3 + 12\,x\,z^5 + 12\,y^5\,z + 12\,y\,z^5 + 12\,x^5\,z + 12\,x^5\,y + 84\,x^2\,y^2\,z^2 \\
&+ 72\,x\,y\,z^4 + 72\,x\,y^4\,z + 78\,x\,y^3\,z^2 + 78\,x\,y^2\,z^3 + 78\,x^2\,y\,z^3 + 78\,x^3\,y^2\,z \\
&+ 78\,x^2\,y^3\,z)T^2 + 81\,x^2\,y^2\,z^2\,(x + y + z)^2.
\end{aligned}
\tag{8.4}
$$

Because the right-hand side of the inequality has no radicals, its right polynomial can be obtained easily. That is

$$
T - 2\,(x + y + z).
\tag{8.5}
$$

Compute the resultant of (8.4) and (8.5) with respect to T:

$$
\begin{aligned}
(144\,x^5\,y &+ 144\,x^5\,z + 780\,x^4\,y^2 + 1056\,x^4\,y\,z + 780\,x^4\,z^2 + 1288\,x^3\,y^3 \\
&+ 3048\,x^3\,y^2\,z + 3048\,x^3\,y\,z^2 + 1288\,x^3\,z^3 + 780\,x^2\,y^4 + 3048\,x^2\,y^3\,z \\
&+ 5073\,x^2\,y^2\,z^2 + 3048\,x^2\,y\,z^3 + 780\,x^2\,z^4 + 144\,x\,y^5 + 1056\,x\,y^4\,z \\
&+ 3048\,x\,y^3\,z^2 + 3048\,x\,y^2\,z^3 + 1056\,x\,y\,z^4 + 144\,x\,z^5 + 144\,y^5\,z + 780\,y^4\,z^2 \\
&+ 1288\,y^3\,z^3 + 780\,y^2\,z^4 + 144\,y\,z^5)(x + y + z)^2.
\end{aligned}
$$

Deleting a positive factor $(x + y + z)^2$, we obtain the critical surface as

$$
\begin{aligned}
144\,x^5\,y &+ 144\,x^5\,z + 780\,x^4\,y^2 + 1056\,x^4\,y\,z + 780\,x^4\,z^2 + 1288\,x^3\,y^3 \\
&+ 3048\,x^3\,y^2\,z + 3048\,x^3\,y\,z^2 + 1288\,x^3\,z^3 + 780\,x^2\,y^4 + 3048\,x^2\,y^3\,z \\
&+ 5073\,x^2\,y^2\,z^2 + 3048\,x^2\,y\,z^3 + 780\,x^2\,z^4 + 144\,x\,y^5 + 1056\,x\,y^4\,z \\
&+ 3048\,x\,y^3\,z^2 + 3048\,x\,y^2\,z^3 + 1056\,x\,y\,z^4 + 144\,x\,z^5 + 144\,y^5\,z + 780\,y^4\,z^2 \\
&+ 1288\,y^3\,z^3 + 780\,y^2\,z^4 + 144\,y\,z^5 = 0.
\end{aligned}
$$

Remark 8.1. The critical polynomial for a concrete problem may be identically 0. A simple example illustrating this case is:

$$
\sqrt{(x + y)^2} \le \sqrt{x^2} + \sqrt{y^2}.
$$

In this case, the so-called dimension-decreasing method is not applicable without making any adjustments. Note that the definition of critical polynomial depends on the left and right polynomials. The case may be handled via changing the left and right polynomials of a given inequality. Han proposed a modified method to deal with this situation in [Han (2016)]. In this chapter, the critical polynomials of all the examples are not identically 0.

8.2.2 *Algorithm*

We first state the problem that we will discuss in this subsection.

Problem. Devise an algorithm with the following specification.

In: A sentence of the form

$$\forall(x_1,\ldots,x_n)(\Phi_1 \wedge \Phi_2 \wedge \cdots \wedge \Phi_s \Rightarrow \Phi_0), \tag{8.6}$$

where Φ_0, Φ_1, ..., Φ_s are algebraic inequalities with respect to $x_1,\ldots,x_n$, Φ_0 is a *non-strict* inequality and the premise, *i.e.* $\Phi_1 \wedge \Phi_2 \wedge \cdots \wedge \Phi_s$ defines either an open set (not necessarily connected) or an open set together with part or whole boundary.

Out: true or **false.**

Example 8.2 can be restated as

$$\forall(x, y, z)(x > 0 \wedge y > 0 \wedge z > 0 \Rightarrow (8.2)),$$

where $x > 0 \wedge y > 0 \wedge z > 0$ defines an open set in $\mathbb{R}^3$. Thus Example 8.2 belongs to the class of the problem. So does Example 8.3. The class of problems stated above covers a major part of inequalities listed in *"Geometric Inequalities"* [Bottema *et al.* (1969)] and *"Recent Advances in Geometric Inequalities"* [Mitrinović *et al.* (1989)]. In fact, Example 8.3 is a geometric inequality expressed by variables x, y, z [Bottema *et al.* (1969)].

Algorithm 8.1 DimDec

Input: A sentence of the form (8.6)
Output: true or **false**
  1: Compute the critical polynomials of $\Phi_0, \Phi_1, \ldots, \Phi_s$ and let $p(x_1, \ldots, x_n)$ be the product of the critical polynomials;
  2: Compute an open CAD $\mathcal{D}$ defined by $p(x_1, \ldots, x_n)$;
  3: Verify the sentence at each of the sample points in $\mathcal{D}$;
  4: **if** the sentence is true at all the sample points **then**
  5:    **return true**
  6: **else**
  7:    **return false**
  8: **end if**

Theorem 8.1. *Algorithm* `DimDec` *is correct.*

Proof.   For each $\mu = 0, \ldots, s$, let $l_\mu(\boldsymbol{x}), r_\mu(\boldsymbol{x})$ and $p_\mu(\boldsymbol{x}) = 0$ denote the left and right hand side and critical surface of inequality Φ_μ, respectively. Set

$$\delta_\mu(\boldsymbol{x}) = l_\mu(\boldsymbol{x}) - r_\mu(\boldsymbol{x}), \ \ \delta(\boldsymbol{x}) = \delta_0 \cdots \delta_s.$$

Let Δ denote the complement of $V_{\mathbb{R}}(\delta(\boldsymbol{x}))$. We first prove that each connected component Δ_i of Δ contains at least one sample point of $\mathcal{D}$. Because $V_{\mathbb{R}}(\delta_\mu) \subseteq V_{\mathbb{R}}(p_\mu)$, every connected component of the complement of $V_{\mathbb{R}}(p_\mu)$ is contained in some Δ_i. Therefore each Δ_i contains at least one sample point of $\mathcal{D}$.

We then prove that, for each δ_μ and each Δ_i, δ_μ has nonzero constant sign on Δ_i. Suppose α is a sample point in Δ_i and there exists a point β in Δ_i such that $\delta_\mu(\alpha)\delta_\mu(\beta) < 0$. Then any path $\Gamma \subseteq \Delta_i$ joining α and β will have a point γ such that $\delta_\mu(\gamma) = 0$. A contradiction!

So, by a bit abuse of notation, $\mathcal{D}$ is also a *sign-invariant open CAD* of $\delta(\boldsymbol{x})$. The conclusion follows. $\square$

8.3   Inequalities on Triangles

There are several hundreds of inequalities in [Bottema *et al.* (1969)], most of which are inequalities on triangles. In the literature, one can find thousands of inequalities on triangles.

The variables of inequalities on triangles are usually geometric invariants instead of Cartesian coordinates. For a triangle ABC, we list below some frequently used notations in this chapter and BOTTEMA.

A,B,C;	interior angles
a, b, c;	side-lengths
s;	half perimeter, $s = (a + b + c)/2$,
x, y, z;	$x = s - a, \ y = s - b, \ z = s - c$
S;	area
R;	circumradius
r;	inradius
r_a, r_b, r_c;	radii of escribed circles
h_a, h_b, h_c;	altitudes
m_a, m_b, m_c;	lengths of medians
w_a, w_b, w_c;	lengths of internal bisectors

Usually, x, y, z are chosen as parameters (free variables) and the others as (constrained) variables. From the viewpoint of reducing the degrees of polynomials involved, we might have better choice.

An algebraic inequality $\Phi(x, y, z)$ can be viewed as a geometric inequality on a triangle if

- $x > 0$, $y > 0$, $z > 0$;
- $l(x, y, z)$ and $r(x, y, z)$, the left and right hand side of Φ, are homogeneous polynomials; and
- $l(x, y, z)$ and $r(x, y, z)$ have the same degree.

The first condition means that the sum of any two sides of a triangle is greater than the other one. The second and third conditions together mean that the truth of a proposition does not change under similar transformation. For example, the left and right hand side of inequality (8.3), $m_a + m_b + m_c$ and $2s$, are both homogeneous polynomials in x, y, z of degree 1.

In addition, suppose $l(x, y, z)$ and $r(x, y, z)$ are symmetric with respect to x, y, z. Then replacing x, y, z with $x' = \rho x$, $y' = \rho y$, $z' = \rho z$ where $\rho > 0$ in $l(x, y, z)$ and $r(x, y, z)$ will not change the truth of the proposition.

Obviously, the left polynomial $L(T, x', y', z')$ and right polynomial $R(T, x', y', z')$ of $\Phi(x', y', z')$ are both symmetric with respect to x', y', z' and thus can be expressed by elementary symmetric polynomials of x', y', z', *i.e.*

$$L(T, x', y', z') = H_l(T, \sigma_1, \sigma_2, \sigma_3), \quad R(T, x', y', z') = H_r(T, \sigma_1, \sigma_2, \sigma_3),$$

where $\sigma_1 = x' + y' + z'$, $\sigma_2 = x'y' + y'z' + z'x'$, $\sigma_3 = x'y'z'$.

Let $\rho = \sqrt{\frac{x+y+z}{xyz}}$, we have $x'y'z' = x' + y' + z'$, *i.e.* $\sigma_3 = \sigma_1$. Moreover, letting

$$s = \sigma_1(= \sigma_3), \qquad p = \sigma_2 - 9,$$

we can transform $L(T, x', y', z')$ and $R(T, x', y', z')$ into polynomials in T, p, s, denoted by $F(T, p, s)$ and $G(T, p, s)$, respectively. Especially, if the terms containing s in F and G are all of even degrees in s, F and G can be further transformed into polynomials in T, p and q, where $q = s^2 - 4p - 27$. Usually, these polynomials in T, p, q have less terms and lower degrees than $L(T, x, y, z)$ and $R(T, x, y, z)$. Therefore, this kind of transformation leads to smaller critical polynomials in only two variables (p, s or p, q) and decreases the computational complexity greatly.

Example 8.4. [Bottema *et al.* (1969)] Prove that

$$w_b w_c + w_c w_a + w_a w_b \leq s^2.$$

It is easy to know that

$$w_a = 2\,\frac{\sqrt{x\,(x+y)(x+z)(x+y+z)}}{2\,x+y+z},$$

$$w_b = 2\,\frac{\sqrt{y\,(x+y)(y+z)(x+y+z)}}{2\,y+x+z},$$

$$w_c = 2\,\frac{\sqrt{z\,(x+z)(y+z)(x+y+z)}}{2\,z+x+y},$$

and $s = x + y + z$. By successive resultant computation as we did in the last section, we will get a left polynomial of degree 20 with 557 terms. The right polynomial is $T - (x+y+z)^2$. The critical polynomial $p(x,\,y,\,z)$ has 136 terms and degree 15.

However, the left and right polynomials in $p,\,q$ are

$$(9\,p + 2\,q + 64)^4\,T^4$$
$$-32(4\,p + q + 27)\,(p+8)\,(4\,p^2 + pq + 69\,p + 10\,q + 288)\,(9\,p + 2\,q + 64)^2\,T^2$$
$$-\,512\,(4\,p + q + 27)^2\,(p+8)^2\,(9\,p + 2\,q + 64)^2\,T + 256(4\,p + q + 27)^3$$
$$(p+8)^2\,(-1024 - 64\,p + 39\,p^2 - 128\,q - 12\,pq - 4\,q^2 + 4\,p^3 + p^2\,q)$$

and $T - 4\,p - q - 27$, respectively. And the critical polynomial is

$$\begin{aligned}
Q(p,\,q) = {} & 5600256\,p^2\,q + 50331648\,p + 33554432\,q + 5532160\,p^3 \\
& + 27246592\,p^2 + 3604480\,q^2 + 22872064\,pq + 499291\,p^4 + 16900\,p^5 \\
& + 2480\,q^4 + 16\,q^5 + 143360\,q^3 + 1628160\,pq^2 + 22945\,p^4\,q \\
& + 591704\,p^3\,q + 11944\,p^3\,q^2 + 2968\,p^2\,q^3 + 242568\,p^2\,q^2 + 41312\,pq^3 \\
& + 352\,pq^4
\end{aligned}$$

which has 20 terms and degree 5.

8.4   BOTTEMA

BOTTEMA is a program which implements Algorithm 8.1 and other algorithms for inequality proving and polynomial optimization using Maple language. In this section, we introduce three main functions of BOTTEMA, *i.e.* **prove**, **xprove** and **yprove**.

prove

Aim: Prove geometric inequalities on triangles or equivalent algebraic inequalities

Calling Sequence:

> prove(ineq);   or   prove(ineq, [ineqs]);

Parameter:

- ineq - An inequality to be proven which is expressed by geometric invariants listed in the last section
- ineqs - Some inequalities as constraints expressed by geometric invariants listed in the last section

Description:

- The inequality to be proven has to be a non-strict one and the constraints define an open set or an open set plus part or whole boundary.
- **prove** is applicable to such algebraic inequality proving where ineqs and ineq are expressed by homogeneous rational functions or radicals in x, y, z $(x > 0, y > 0, z > 0)$.

Remark 8.2. Since the constraints "ineqs" should define an open set or an open set plus part or whole boundary, it is not correct to input $P \geq Q$ and $P \leq Q$ in "ineqs" to stand for an equational constraint $P = Q$.

xprove

Aim: Prove algebraic inequalities whose variables are all nonnegative

Calling Sequence:

> xprove(ineq);   or   xprove(ineq, [ineqs]);

Parameter:

- ineq - An algebraic inequality whose variables are all nonnegative
- ineqs - Some algebraic inequalities as constraints whose variables are all nonnegative

Description:

- The inequality to be proven has to be a non-strict one and the constraints define an open set or an open set plus part or whole boundary.
- ineqs and ineq contain only rational functions and radicals.
- All variables are viewed nonnegative by default.

yprove

Aim: Prove algebraic inequalities
Calling Sequence:

> yprove(ineq);   or   yprove(ineq, [ineqs]);

Parameter:

- ineq - An algebraic inequality to be proven.
- ineqs - Some algebraic inequalities as constraints.

Description:

- The inequality to be proven has to be a non-strict one and the constraints define an open set or an open set plus part or whole boundary.
- ineqs and ineq contain only rational functions and radicals.

8.4.1 *Inequality Proving with BOTTEMA*

We list a few representative examples proved by BOTTEMA in this section. There are a tremendous amount of this kind of inequalities in the literature. See, for example, [Kuang (2010); Wang (2011)], which are extremely rich in content. All the timings in this section are taken from a PC (Intel Core i5-3470 CPU @ 3.20GHz, 8G memory, Windows 7 OS) with Maple 17.

The famous Janous Inequality [Janous (1986)] was proposed as an open problem in 1986 and was closed in 1988.

Example 8.5. Prove that

$$\frac{1}{m_a} + \frac{1}{m_b} + \frac{1}{m_c} \geq \frac{5}{s}.$$

The inequality is difficult since its left-hand side contains implicitly three radicals.

After loading BOTTEMA, we type in: `prove(1/ma+1/mb+1/mc>=5/s);` The inequality is verified in 0.39 seconds.

The next example was an open problem in *Amer. Math. Monthly,* **93**:(1986), 299, labeled as E. 3146*.

Example 8.6. Determine whether

$$2\,s(\sqrt{s-a}+\sqrt{s-b}+\sqrt{s-c}) \leq 3\,(\sqrt{bc(s-a)}+\sqrt{ca(s-b)}+\sqrt{ab(s-c)}).$$

BOTTEMA proves the inequality in 1.295 seconds.

Example 8.7. Determine whether

$$\sqrt[3]{r_a r_b r_c} \leq \frac{1}{3}(w_a + w_b + w_c),$$

i.e. whether the geometric mean of r_a, r_b, r_c is less than or equal to the arithmetic mean of w_a, w_b, w_c. The right-hand side of the inequality contains implicitly three radicals.

We type in

```
prove((ra*ra*rb)^(1/3)<=(wa+wb+wc)/3);
```

BOTTEMA proves the inequality in 0.858 seconds while the following inequality takes 3.557 seconds.

Example 8.8. [Shan (1996)]

$$a\, m_a + b\, m_b + c\, m_c \le \frac{2}{\sqrt{3}}\,(w_a^2 + w_b^2 + w_c^2).$$

In 1985, Garfunkel first proposed the following conjecture in *Crux Math.* Afterwards, it appeared again as an open problem in [Mitrinović *et al.* (1989)].

Example 8.9.

$$\cos\frac{B-C}{2} + \cos\frac{C-A}{2} + \cos\frac{A-B}{2}$$
$$\le \frac{1}{\sqrt{3}}\,(\cos\frac{A}{2} + \cos\frac{B}{2} + \cos\frac{C}{2} + \sin A + \sin B + \sin C\,).$$

It takes 14.757 seconds for BOTTEMA to prove the inequality.

To answer the question raised by P. Erdös, A. Oppenheim studied the following inequality [Mitrinović *et al.* (1989)].

Example 8.10. If $c = \min\{a,\, b,\, c\}$, then

$$2\, m_a + 2\, m_b + 2\, m_c \le 2\, a + 2\, b + (3\sqrt{3} - 4)\, c.$$

We type in

```
prove(2*ma+2*mb+2*mc<=2*a+2*b+(3*sqrt(3)-4)*c, [c<=a,c<=b]);
```

It takes 11.1 seconds to prove the inequality. If we only type in

```
prove(2*ma+2*mb+2*mc<=2*a+2*b+(3*sqrt(3)-4)*c);
```

the output is "*The inequality does not hold*" together with a counterexample
$$[a = 203,\, b = 706,\, c = 505].$$

Example 8.11. [Liu (2003)] Assume $x > 0,\, y > 0,\, z > 0$. Prove
$$2187(y^4 z^4 (y+z)^4 (2x+y+z)^8 + x^4 z^4 (x+z)^4 (x+2y+z)^8$$
$$+ x^4 y^4 (x+y)^4 (x+y+2z)^8) - 256(x+y+z)^8 (x+y)^4 (x+z)^4 (y+z)^4 \ge 0.$$

Because the left-hand side is a homogeneous symmetric polynomial in three positive variables, according to the discussion before, we may call **prove** or **xprove**. The polynomial has 201 terms and the greatest absolute value of its coefficients is 181394432. The command **prove** may make use of the property and prove the inequality much faster (0.11 seconds) than **xprove** (7.223 seconds).

The famous Euler inequality $R \geq 2\,r$ and the inequality $m_a \geq w_a$ have been used to illustrate various provers based on different principles [Chou *et al.* (1992); Wu (1994b, 1998)]. In the following example, we compare $R - 2\,r$ to $m_a - w_a$.

Example 8.12. It takes BOTTEMA 0.437 seconds to prove that

$$m_a - w_a \leq R - 2\,r.$$

Of course, BOTTEMA is applicable to not only inequalities on triangles but also algebraic inequalities such as the so-called *"Ptolemy Inequality"*, where the variables are Cartesian coordinates instead of geometric invariants.

Example 8.13. Given four points A, B, C, D in a plane. Denote by AB, AC, AD, BC, BD, CD the distances between six pairs of points. Prove that

$$AB \cdot CD + BC \cdot AD \geq AC \cdot BD.$$

Set $A = (-\frac{1}{2}, 0)$, $B = (x, y)$, $C = (\frac{1}{2}, 0)$, $D = (u, v)$. The inequality is transformed to

$$\sqrt{(-\tfrac{1}{2} - x)^2 + y^2}\sqrt{(\tfrac{1}{2} - u)^2 + v^2} + \sqrt{(x - \tfrac{1}{2})^2 + y^2}\sqrt{(-\tfrac{1}{2} - u)^2 + v^2}$$
$$\geq \sqrt{(x - u)^2 + (y - v)^2}.$$

We type in **yprove(%);** where **%** stands for the above formula. The inequality is proven in 0.359 seconds.

8.4.2 *Non-linear Optimization with BOTTEMA*

BOTTEMA implements the successive resultant method (Algorithm 7.2) in Chapter 7. So, it can be applied to the problem of finding the optimal value of a parameter such that an algebraic/geometric inequality holds. There are several functions in BOTTEMA for solving optimization problems. In this section, we introduce two main functions: **cmin (cmax)** and **xmin (xmax)**.

For other functions and more details, refer to [Yang and Xia (2000, 2008)].

cmin

Calling Sequence:
>   cmin(ineq, [ineqs], var);

Parameter:

- ineq - A *geometric* inequality containing one parameter "var". If var is specified, the resulting inequality can be handled by function `prove`.
- ineqs - Some *geometric* inequalities as constraints which can be handled by function `prove`.
- var - The parameter.

Description: Find the infimum of var such that the inequality holds under the constraints. The output is an algebraic number defined by a polynomial and an interval if the infimum exists.

Example:

```
> cmin( wa^2+wb^2+wc^2 <= 4*R^2+11*r^2+k*r*(R-2*r),[ ],k );
```

xmin

Calling Sequence:
>   xmin(ineq, [ineqs], var);

Parameter:

- ineq - An *algebraic* inequality containing one parameter "var". If var is specified, the resulting inequality can be handled by function `xprove`.
- ineqs - Some *algebraic* inequalities as constraints which can be handled by function `xprove`.
- var - The parameter.

Description: Find the infimum of var such that the inequality holds under the constraints. The output is an algebraic number defined by a polynomial and an interval if the infimum exists.

Example:

```
> ineq:=1/5*(x2^3+1)^(1/3)*x2^2*x3^2<=k;
> ineqs:=x2^3+1 <= 12167/1000, x3 <= 32/10,
```

```
   10-(x2^3+1)^(2/3)-x2^2-x3^2-2/5*x2*x3 >= 0,
   sqrt(250-25*(x2^3+1)^(2/3)-25*x2^2-25*x3^2+10*x2*x3)
   +sqrt(250-25*(x2^3+1)^(2/3)-25*x2^2-25*x3^2-10*x2*x3)
   <= 32;
> xmin(ineq,[ineqs],k);
```

Analogously, the functions `cmax` and `xmax` are used to find supremum of parameter in geometric and algebraic inequalities, respectively.

Example 8.14. Find the supremum of the following function

$$g_{13} = \cos^{13} A + \cos^{13} B + \cos^{13} C,$$

where $A + B + C = \pi,\ \ A > 0, B > 0, C > 0$.

The degree of the function is rather high. With BOTTEMA, we type in

```
> cmin( cos(A)^13+cos(B)^13+cos(C)^13-k<=0, [ ], k);
```

Within 247.2 seconds, BOTTEMA outputs the supremum is the only real root in $[\frac{13}{11}, \frac{3}{2}]$ ($\approx 1.1973580278 \cdots$) of the following equation:

$$913438523331814323877303020447676887284957839\,36\,T^{23}$$
$$- 11150372599265311570767859136324180752990208\,0\,T^{22}$$
$$- 53515255055028613521187227061893377664693043\,2\,T^{21}$$
$$- 15114835584147807272875379923008485177530777\,6000\,T^{20}$$
$$- 78665379462953640048233727031905222865715200\,00\,T^{19}$$
$$- 23428988895592433605929901825217426316580618\,240\,T^{18}$$
$$+ 34488383979821012965716421660743079075799629\,824\,T^{17}$$
$$- 14509476362229950348920184177554458355475415\,04\,T^{16}$$
$$- 30760921513972673918718015649890682178245754\,88\,T^{15}$$
$$- 24372078245608100090019481186441261588099891\,2\,T^{14}$$
$$+ 24881955820083904961119343538287967666700288\,T^{13}$$
$$- 20933223389001054063703055683181010442485\,76\,T^{12}$$
$$- 300352733449075426757245834187011606992\,21\,T^{11}$$
$$- 12255981065722736074095622868778095957\,37\,T^{10}$$
$$+ 632051201590034874768819023028211986\,9\,T^{9}$$
$$- 45556666989990221887268934958342199\,T^{8}$$

$$- 256756835432746423776218266451858 \, T^7$$
$$- 18730045022295170528778810178 6 \, T^6 + 7395190620535790603758962 18 \, T^5$$
$$+ 84113490857050133159826 \, T^4 - 36635332206811455433 \, T^3$$
$$- 339106861240027837 \, T^2 - 124596035635879 \, T - 45812984491$$
$$= 0.$$

It is interesting that the supremum is not attained by equilateral triangles, *i.e.* $A = B = C = \frac{\pi}{3}$, as one may conjecture. It is attained by an isosceles triangle with sides $a = b = 1$ and $c = 1.963765212\cdots$ which is the only real root of the following equation

$$c^{23} + c^{22} - 23\,c^{21} - 23\,c^{20} + 241\,c^{19} + 241\,c^{18} - 1519\,c^{17} - 1519\,c^{16} + 6401\,c^{15}$$
$$+6401\,c^{14} - 18943\,c^{13} - 18943\,c^{12} + 40193\,c^{11} + 40192\,c^{10} - 61184\,c^9$$
$$-61184\,c^8 + 65536\,c^7 + 65536\,c^6 - 47104\,c^5 - 47104\,c^4 + 20480\,c^3 + 20480\,c^2$$
$$-4096\,c - 4096 = 0.$$

Example 8.15. Find the supremum of λ such that

$$4\sqrt{3\,x\,y\,z\,(x+y+z)} \le (y+z)^2 + (z+x)^2 + (x+y)^2$$
$$- \lambda\left(\frac{(2\,z+x+y)\,(y-x)^2}{x+y} + \frac{(2\,x+y+z)\,(z-y)^2}{y+z}\right.$$
$$\left.+ \frac{(2\,y+z+x)\,(x-z)^2}{z+x}\right)$$

under the constraints that $x > 0, y > 0, z > 0$.

Because the functions are homogeneous and symmetric with respect to x, y, z, the degree and dimension of the problem can be decreased under suitable transformation. BOTTEMA can do this kind of transformation automatically. The supremum is obtained within 52.0 seconds, which is the only real root of $32\lambda^3 - 9\lambda^2 - 6\lambda - 1 = 0$, *i.e.*

$$\lambda_{\max} = \frac{1}{32}\frac{(827 + 384\sqrt{2})^{2/3} + 73 + 3\,(827 + 384\sqrt{2})^{1/3}}{(827 + 384\sqrt{2})^{1/3}} = 0.6462266581\cdots.$$

In Example 8.12 we compare $m_a - w_a$ to $R - 2\,r$. In the following example, we consider a more general problem.

Example 8.16. Find the infimum of λ such that

$$m_a - w_a \le \lambda\,(R - 2\,r).$$

This is equivalent to finding the infimum of λ such that

$$2\sqrt{x^4\,y^2 + 2\,x^3\,y + x^2 - 2\,x^3\,y^3 + 12\,x^2\,y^2 - 2\,x\,y + x^2\,y^4 + 2\,x\,y^3 + y^2}$$
$$-8\,\frac{\sqrt{y^2+1}\,\sqrt{x^2+1}\,x\,y}{x\,y+1} \le \lambda\,(x^2\,y^2 + x^2 - 8\,x\,y + 9 + y^2)$$

under constraints $x > 0$, $x\,y > 1$.

We obtain $\lambda_{min} = 1$ by BOTTEMA within 138.6 seconds. We also find a beautiful inequality $m_a - w_a \le R - 2\,r$ as a byproduct.

Chapter 9

SOS Decomposition

Let us begin with an example. Prove that

$$f(x,y) = x^6 + y^6 + 2x^5y + 5y^4x^2 + 4xy^5 \geq 0$$

for all $x, y \in \mathbb{R}$. We may use the algorithms and tools presented in previous chapters to prove the inequality. However, can we give a proof which can be easily checked by readers? For some inequalities, SOS (Sum Of Squares of polynomials) decomposition can give such proofs. For example, if we express $f(x, y)$ as an SOS like

$$f(x,y) = \left(y^3 + 2xy^2 - \frac{1}{2}x^3 \right)^2 + \left(xy^2 + \frac{1}{2}x^2y - \frac{1}{6}x^3 \right)^2$$
$$+ 3\left(\frac{5}{6}x^2y + \frac{13}{30}x^3 \right)^2 + \frac{143}{900}x^6,$$

the correctness of the inequality can be easily checked by expanding the right-hand side and comparing to $f(x, y)$. Such a proof is sometimes called "readable proof" and the expression is a *certificate*.

The study on sum of squares was initiated by Hilbert (1888). In that paper, Hilbert proved that a positive semi-definite homogeneous polynomial of degree m in n variables can be expressed as a sum of squares of homogeneous polynomials if (and only if)

  (i) $n \leq 2$, or
 (ii) $m = 2$, or
(iii) $n = 3$, $m = 4$.

In 1893, Hilbert proved that any given positive semi-definite homogeneous polynomial in 3 variables can be represented as a quotient of two sums of squares of polynomials and conjectured that the result is valid for general positive semi-definite homogeneous polynomials.

In 1900, Hilbert gave the famous lecture at ICM in Paris [Hilbert (1901)], raising 23 famous problems. The 17th problem is:

For any $f \in \mathbb{R}[\boldsymbol{x}]$, is it true that f being positive semi-definite on $\mathbb{R}^n$ implies f is a sum of squares of rational functions?

It is well-known that Artin proved Hilbert's conjecture [Artin (1927)], *i.e.* gave a positive answer to Hilbert's 17th problem. Artin's theory and method, which is now called the Artin-Schreier theory, lead to the development of modern real algebra and real algebraic geometry. Two major milestones of the subject are quantifier elimination theory over real closed fields established by Tarski [Tarski (1951)] and the Positivstellensatz discovered by Krivine [Krivine (1964)] and Stengle [Stengle (1974)].

In 1940, Habicht made use of Pólya's theorem [Pólya (1928)] to construct effective representations of sums of squares of rational functions for *positive* polynomials [Habicht (1940)]. Habicht's work started the constructive research on Hilbert's 17th problem.

Although the problem of determining whether a polynomial can be written as sum of squares of rational functions is still unsolved, there are now effective algorithms for determining whether a polynomial can be represented as sum of squares of polynomials (SOS), see for example [Choi *et al.* (1995); Powers and Wörmann (1998); Parrilo (2000); Lasserre (2001); Parrilo and Sturmfels (2003)].

If a polynomial can be written as an SOS, it is obviously a certificate for the polynomial to be positive semi-definite. On the other hand, we know that there are positive semi-definite polynomials which cannot be represented as SOS. Actually, the number of positive semi-definite polynomials is much larger than that of SOS polynomials [Blekherman (2006)]. A first such example is the famous Motzkin's polynomial [Motzkin (1967)]:

$$z^6 + x^4 y^2 + y^4 x^2 - 3x^2 y^2 z^2.$$

It is not hard to prove that the polynomial cannot be SOS while a sum of squares of rational functions representation for the polynomial is (p.6 of [Marshall (2008)]):

$$\frac{(x^2 y^2 (x^2 + y^2 + z^2)(x^2 + y^2 - 2z^2)^2 + z^6 (x^2 - y^2)^2}{(x^2 + y^2)^2}$$

$$= \left(\frac{x^2 y (x^2 + y^2 - 2z^2)}{x^2 + y^2}\right)^2 + \left(\frac{xy^2 (x^2 + y^2 - 2z^2)}{x^2 + y^2}\right)^2$$

$$+ \left(\frac{xyz(x^2 + y^2 - 2z^2)}{x^2 + y^2}\right)^2 + \left(\frac{z^3 (x^2 - y^2)}{x^2 + y^2}\right)^2.$$

From an algorithmic point of view, writing a multivariate polynomial as an SOS is a crucial part of many applications, see for example [Vandenberghe and Boyd (1996); Lasserre (2001); Parrilo (2003); Kim *et al.* (2005); Schweighofer (2005)]. By Gram matrix representation [Choi *et al.* (1995)], the problem of SOS decomposition can be transformed as a problem of semi-definite programming (SDP), which can be solved symbolically or numerically. Numerical algorithms for SOS decompositions can handle big scale problems and may be used to get exact results [Kaltofen *et al.* (2008)]. Actually, there exist some well-known free available SOS solvers which are based on numerical SDP solvers [Papachristodoulou *et al.* (2013); Lofberg (2004); Seiler (2013)].

In this chapter, SOS always means sum of squares of *polynomials*. If not specified, "polynomials" in this chapter are polynomials with real coefficients. The main content of this chapter is from [Dai and Xia (2015)].[a]

9.1 Preliminary

The symbol $\mathbb{Z}_+$ denotes the set of natural numbers. We use $\boldsymbol{x}, \boldsymbol{y}$ to denote the variable vectors $(x_1, \ldots, x_n), (y_1, \ldots, y_n)$, respectively. A hyperplane in $\mathbb{R}^n$ is denoted by $\pi(\boldsymbol{x}) = 0$.

Consider a polynomial

$$p(\boldsymbol{x}) = \sum_{\alpha \in \mathrm{P}} c_\alpha \boldsymbol{x}^\alpha \tag{9.1}$$

in the variable vector $\boldsymbol{x} \in \mathbb{R}^n$ with a *support* $\mathrm{P} \subseteq \mathbb{Z}_+^n$ and real coefficients $c_\alpha \neq 0$ $(\alpha \in \mathrm{P})$. Denote by $\mathsf{S}(p)$ the support of a polynomial p. For example, if $p = 1 + x_1^2 + x_2^3$, then $n = 2, \mathsf{S}(p) = \{(0,0), (2,0), (0,3)\}$. When $p = 0$, define $\mathsf{S}(p) = \emptyset$.

For any $T \subseteq \mathbb{R}^n$ and $k \in \mathbb{R}$, denote by kT the set $\{k\alpha \mid \alpha \in T\}$, where $k(a_1, \ldots, a_n) = (ka_1, \ldots, ka_n)$, and by $\mathrm{conv}(T)$ the convex hull of T. Let P^e be the set of $\alpha \in \mathrm{P}$ whose coordinates α_k $(k = 1, 2, \ldots, n)$ are all even non-negative integers, *i.e.* $\mathrm{P}^e = \mathrm{P} \cap (2\mathbb{Z}_+^n)$.

Obviously, $p(\boldsymbol{x})$ can be represented in terms of a sum of squares of polynomials or in short, p is SOS, if and only if there exist polynomials $q_1(\boldsymbol{x}), \ldots, q_s(\boldsymbol{x}) \in \mathbb{R}[\boldsymbol{x}]$ such that

$$p(\boldsymbol{x}) = \sum_{i=1}^{s} q_i(\boldsymbol{x})^2. \tag{9.2}$$

To find both s and polynomials $q_1(\boldsymbol{x}), \ldots, q_s(\boldsymbol{x})$, it is necessary to estimate and decide the supports of unknown polynomials $q_i(\boldsymbol{x})(i = 1, \ldots, s)$. Let Q_i

---

[a]Dai, L. and Xia, B., Smaller SDP for SOS decomposition, *Journal of Global Optimization* **63**, 343-361, © 2015 Springer, with permission of Springer.

be an unknown support of the polynomial $q_i(\boldsymbol{x})$ $(i = 1, \ldots, s)$. Then each polynomial $q_i(\boldsymbol{x})$ is represented as $q_i(\boldsymbol{x}) = \sum_{\alpha \in Q_i} c_{(i,\alpha)} \boldsymbol{x}^\alpha$ with nonzero coefficients $c_{(i,\alpha)}$ $(\alpha \in Q_i,\ i = 1, \ldots, s)$.

Lemma 9.1. *[Reznick (1978)] Suppose $p(\boldsymbol{x})$ is of the form (9.2), then* $P \subseteq \mathrm{conv}(P^e)$ *and*

$$\bigcup_{i=1}^{s} \mathsf{S}(q_i) \subseteq \frac{1}{2}\mathrm{conv}(P^e). \tag{9.3}$$

Definition 9.1. For a polynomial p, a set $Q \subseteq \mathbb{Z}_+^n$ is said to satisfy the relation $\mathrm{SOSS}(p, Q)$ (SOSS stands for SOS support) if

$$p \text{ is SOS} \implies \exists q_i(i = 1, \ldots, s) \text{ such that } p = \sum_{i=1}^{s} q_i^2 \text{ and } \mathsf{S}(q_i) \subseteq Q.$$

By Lemma 9.1,

$$Q^0 = \left(\frac{1}{2}\mathrm{conv}(P^e)\right) \cap \mathbb{Z}_+^n \tag{9.4}$$

satisfies $\mathrm{SOSS}(p, Q^0)$. Therefore we can confine effective supports of unknown polynomials $q_1(\boldsymbol{x}), \ldots, q_s(\boldsymbol{x})$ to Q^0.

Theorem 9.1. *A real symmetric matrix M is positive semi-definite if and only if one of the following conditions holds.*
    (1) All the roots of the characteristic polynomial of M are non-negative.
    (2) There exists a real matrix V such that $M = VV^{\mathrm{T}}$.
    (3) All the principal minors of M are non-negative.

Proof.   A well-known result in linear algebra. $\square$

Theorem 9.2. *[Powers and Wörmann (1998)] Suppose $p \in \mathbb{R}[\boldsymbol{x}]$ is of even degree. Then p is SOS if and only if there exist a positive semi-definite matrix M and a support Q such that*

$$p(\boldsymbol{x}) = Q(\boldsymbol{x})^{\mathrm{T}} M Q(\boldsymbol{x}), \tag{9.5}$$

where $Q(\boldsymbol{x})$ is a vector of monomials corresponding to the support Q. Furthermore, if the rank of M is t, then we can construct polynomials $h_1, \ldots, h_t$ such that $p = \sum_{i=1}^{t} h_i^2$.

Proof.   Necessity. Suppose $p = \sum_{i=1}^{t} h_i^2$ and

$$Q(\boldsymbol{x}) = \bigcup_{i=1}^{t} \mathsf{S}(h_i) = (\boldsymbol{x}^{\beta_1}, \ldots, \boldsymbol{x}^{\beta_k})^{\mathrm{T}}.$$

Obviously, there exists a real matrix V such that $p = Q^T(VV^T)Q$, where V is a $k \times t$ matrix whose ith column is the coefficients of h_i corresponding to $Q(\boldsymbol{x})$. By Theorem 9.1, $M = VV^T$ is positive semi-definite.

Sufficiency. Suppose there exists a positive semi-definite matrix M of rank t and a support $Q(\boldsymbol{x}) = (\boldsymbol{x}^{\beta_1}, \ldots, \boldsymbol{x}^{\beta_k})^T$ such that $p = Q(\boldsymbol{x})^T M Q(\boldsymbol{x})$. By Theorem 9.1, there exist real matrix $V = (v_{ij})$ and real diagonal matrix

$$D = \mathrm{diag}(d_1, \ldots, d_t, 0, \ldots, 0) \quad (d_i > 0)$$

such that $M = V \cdot D \cdot V^T$. Then $p = Q(\boldsymbol{x})^T \cdot V \cdot D \cdot V^T \cdot Q(\boldsymbol{x})$. Set

$$h_i = \sqrt{d_i} \sum_{j=1}^{k} v_{ji} \boldsymbol{x}^{\beta_i},$$

we have $p = h_1^2 + \cdots + h_t^2$.
$\square$

The matrix $M = VV^T$ in the proof is called the *Gram matrix* of p with respect to $h_1, \ldots, h_t$.

By the above theorem, Eq. (9.2) is equivalent to the existence of a positive semi-definite matrix M such that (9.5) holds. So finding the SOS representation is equivalent to solving the feasibility problem of (9.5).

Notation 9.1. We denote by $\mathrm{SOS}(p, Q)$ an algorithm of finding positive semi-definite matrix M such that (9.5) holds.

$\mathrm{SOS}(p, Q)$ can be any algorithm which solves the SDP problem under the constraints defined by (9.5) either symbolically or numerically. Popular SOS tools all employ numerical SDP solvers so that they can solve large problems. To further illustrate Theorem 9.2, however, we first describe a symbolic procedure based on [Powers and Wörmann (1998)].

For a given $p \in \mathbb{R}[\boldsymbol{x}]$ and a support Q satisfying $\mathrm{SOSS}(p, Q)$,

Step 1. Set $p - Q(\boldsymbol{x})^T M Q(\boldsymbol{x}) = 0$ where M is a real symmetric matrix to be determined (with parametric entries). This is a set of linear equations. Solve the equations to get a representation of M in some parameters.

Step 2. Compute the characteristic polynomial $g(y)$ of M. By Proposition 4.1, obtain a condition on the parameters such that all the roots of $g(y)$ are non-negative. Denote the condition by S which is an SAS.

Step 3. Compute a sample point $D(\mathrm{S})$ of S.

Step 4. If $D(\mathrm{S})$ is empty, p is not SOS. Otherwise,

Step 5. Substitute the sample in M and decompose M into $M = VV^T$.

Step 6. Compute $\| Q(\boldsymbol{x})^T V \|^2$ and output the SOS representation.

Example 9.1. Let $p(x,y) = x^6 + y^6 + 2x^5 y + 5y^4 x^2 + 4xy^5$ and $Q(\boldsymbol{x})^{\mathrm{T}} = [y^3, y^2 x, x^2 y, x^3]$.

Assume

$$M = \begin{pmatrix} a_{11} & a_{12} & a_{13} & a_{14} \\ a_{12} & a_{22} & a_{23} & a_{24} \\ a_{13} & a_{23} & a_{33} & a_{34} \\ a_{14} & a_{24} & a_{34} & a_{44} \end{pmatrix}.$$

Solving $p(x,y) - Q(\boldsymbol{x})^{\mathrm{T}} M Q(\boldsymbol{x}) = 0$, we have

$$M = \begin{pmatrix} 1 & 2 & w & -s \\ 2 & -2w+5 & s & v \\ w & s & -2v & 1 \\ -s & v & 1 & 1 \end{pmatrix}$$

where v, s, w are parameters.

Find a sample for v, s, w such that M is positive semi-definite. For example, we may have $v = -\frac{7}{6}, s = \frac{1}{2}, w = 0$. Substitute the values in M and decompose it as VV^{T} where

$$V^{\mathrm{T}} = \begin{pmatrix} 1 & 2 & 0 & \frac{-1}{2} \\ 0 & 1 & \frac{1}{2} & -\frac{1}{6} \\ 0 & 0 & \frac{5\sqrt{3}}{6} & \frac{13\sqrt{3}}{30} \\ 0 & 0 & 0 & \frac{\sqrt{143}}{30} \end{pmatrix}.$$

Then $p(x,y) = Q(\boldsymbol{x})^{\mathrm{T}} V V^{\mathrm{T}} Q(\boldsymbol{x}) = \|Q(\boldsymbol{x})^{\mathrm{T}} V\|^2$. Finally, we have

$$p(x,y) = \left(y^3 + 2xy^2 - \frac{1}{2}x^3 \right)^2 + \left(xy^2 + \frac{1}{2}x^2 y - \frac{1}{6}x^3 \right)^2$$
$$+ 3 \left(\frac{5}{6}x^2 y + \frac{13}{30}x^3 \right)^2 + \frac{143}{900}x^6.$$

Remark 9.1. To improve the efficiency of SDP based SOS solvers, a key problem is how to reduce the scales of corresponding SDP problems. A commonly used method is to prune more unnecessary monomials from Q. In general, one can start from a coarse Q verifying SOSS(p, Q), keep eliminating elements of Q which does not satisfy some conditions, and finally obtain a smaller Q. Obviously, Q^0 of (9.4) satisfies SOSS(p, Q^0) for every given p. So one can start from Q^0.

In the next two sections, we focus on reducing the sizes of inputs to SOS(p, Q). Two types of polynomials, convex cover polynomials and split polynomials, are defined. A convex cover polynomial or a split polynomial can be decomposed into several smaller sub-polynomials such that the original polynomial is SOS if and only if the sub-polynomials are all SOS. Thus the original SOS problem can be decomposed equivalently into smaller sub-problems.

9.2 Convex Cover Polynomial

We introduce some necessary concepts about *Newton polytope* here. For formal definitions of the concepts and the theory and application of Newton polytope, please see for example [Sturmfels (1998)].

A *polytope* is a subset of $\mathbb{R}^n$ that is the convex hull of a finite set of points. A simple example is the convex hull of

$$\{(0,0,0),(0,1,0),(0,0,1),(0,1,1),(1,0,0),(1,1,0),(1,0,1)(1,1,1)\}$$

in $\mathbb{R}^3$; this is the regular 3-cube. A d-dimensional polytope has *faces*, which are again polytopes of various dimensions from 0 to $d-1$. The 0-dimensional faces are called *vertices*, the 1-dimensional faces are called *edges*, and the $(d-1)$-dimensional faces are called *facets*. For instance, the cube has 8 vertices, 12 edges, and 6 facets. If $d=2$ then the edges coincide with the facets. A 2-dimensional polytope is called a *polygon*.

For a given polynomial p, each term $\boldsymbol{x}^\alpha = x_1^{a_1} \cdots x_n^{a_n}$ appearing in p corresponds to an integer lattice point $(a_1, \ldots, a_n)$ in $\mathbb{R}^n$. The set of all these lattice points is called the *support* of p. The convex hull of the support, $\mathrm{conv}(\mathbf{S}(p))$, is defined as the *Newton polytope* of p and is denoted by $\mathbf{N}(p)$.

Definition 9.2. For a polynomial $p = \sum_\alpha c_\alpha \boldsymbol{x}^\alpha$ and a set $T \subseteq \mathbb{R}^n$, we denote by $\mathtt{Proj}(p, T)$ the polynomial obtained by deleting the terms $c_\alpha \boldsymbol{x}^\alpha$ of p where $\alpha \notin (T \cap \mathbb{Z}_+^n)$.

Example 9.2. $p = 2x_1^4 + 4x_2^4 - 3x_3^2 + 1$ and $T = \{(0,0,0),(1,0,0),(4,0,0)\}$, then $\mathtt{Proj}(p, T) = 2x_1^4 + 1$.

Since the results of the following Lemma 9.2 are either obvious or well known, we omit the proofs.

Lemma 9.2.

- *For any two polynomials f, g, two real numbers k_1, k_2 and any $T \subseteq \mathbb{Z}_+^n$,*

$$\mathtt{Proj}(k_1 f + k_2 g, T) = k_1 \mathtt{Proj}(f, T) + k_2 \mathtt{Proj}(g, T).$$

- *For any $T \subseteq \mathbb{Z}_+^n$ and any $k \in \mathbb{R} \setminus \{0\}$, we have*

$$k(\frac{1}{k}T \cap \mathbb{Z}_+^n) \subseteq T.$$

- *Suppose N is an n-dimensional polytope. For any face F of N, there is an $(n-1)$-dimensional hyperplane $\pi(\boldsymbol{y}) = 0$ such that $\pi(\alpha) = 0$ for any $\alpha \in F$ and $\pi(\beta) > 0$ for any $\beta \in N \setminus F$.*
- *Suppose $\pi(\boldsymbol{y}) = 0$ is a hyperplane and $F \subseteq \mathbb{Z}_+^n \cap (\pi(\boldsymbol{y}) = 0)$. For any polynomial $p = \sum_\alpha c_\alpha \boldsymbol{x}^\alpha$ in n variables, we have*

$$\mathtt{S}(\mathtt{Proj}(p, F)) \subseteq \mathtt{S}(\mathtt{Proj}(p, \mathtt{S}(p) \cap (\pi(\boldsymbol{y}) = 0))).$$

- *If f, g are two polynomials and $\mathtt{S}(f) \cap \mathtt{S}(g) = \emptyset$, then*

$$\mathtt{S}(f + g) = \mathtt{S}(f) \cup \mathtt{S}(g).$$

- *Let $T_1 = \mathtt{S}(f)$ and $T_2 = \mathtt{S}(g)$ for two polynomials f and g. Then*

$$\mathtt{S}(fg) \subseteq T_1 + T_2,$$

where $T_1 + T_2$ is the Minkowski sum *of T_1 and T_2, which is defined as*

$$T_1 + T_2 = \{\alpha + \beta \mid \alpha \in T_1, \beta \in T_2\}.$$

- $\mathtt{N}(fg) = \mathtt{N}(f) + \mathtt{N}(g)$.
- *If γ is a vertex of $\mathtt{N}(fg)$, there exist unique vertices $\gamma_1 \in \mathtt{N}(f), \gamma_2 \in \mathtt{N}(g)$ such that $\gamma = \gamma_1 + \gamma_2$. Every edge of $\mathtt{N}(fg)$ is a parallel translate of an edge of $\mathtt{N}(f)$ or of an edge of $\mathtt{N}(g)$.*

Lemma 9.3. *Suppose $\pi(\boldsymbol{y}) = 0$ is a hyperplane, $T \subseteq \mathbb{Z}_+^n$ and f, g are two n-variate polynomials. Let $T_1 = \mathtt{S}(f), T_2 = \mathtt{S}(g)$. If $T \subseteq \{\boldsymbol{y} \mid \pi(\boldsymbol{y}) = 0\}, 2T_1 \subseteq \{\boldsymbol{y} \mid \pi(\boldsymbol{y}) \geq 0\}$ and $2T_2 \subseteq \{\boldsymbol{y} \mid \pi(\boldsymbol{y}) > 0\}$, then $\mathtt{Proj}(fg, T) = 0$.*

Proof. By Lemma 9.2, $\mathtt{S}(fg) \subseteq T_1 + T_2$. By the definition of Minkowski sum, for any $\alpha \in T_1 + T_2$ there exist $\alpha_1 \in T_1, \alpha_2 \in T_2$ such that $\alpha = \alpha_1 + \alpha_2$. Because $\pi(2\alpha_1) \geq 0$ and $\pi(2\alpha_2) > 0$,

$$\pi(\alpha) = \pi(\alpha_1 + \alpha_2) = \frac{1}{2}(\pi(2\alpha_1) + \pi(2\alpha_2)) > 0.$$

So $T_1 + T_2 \subseteq \{\boldsymbol{y} \mid \pi(\boldsymbol{y}) > 0\}$. Thus, $\mathtt{S}(fg) \cap (\pi(\boldsymbol{y}) = 0) = \emptyset$ which implies $\mathtt{Proj}(fg, T)) = 0$ by Lemma 9.2 and $T \subseteq (\pi(\boldsymbol{y}) = 0)$. $\qquad\square$

Lemma 9.4. *Suppose $p = \sum_{i=1}^s q_i^2$ and F is a face of $\mathtt{N}(p)$. Let $F_z = F \cap \mathbb{Z}_+^n, F_{\frac{z}{2}} = \frac{1}{2}F \cap \mathbb{Z}_+^n, q_i' = \mathtt{Proj}(q_i, F_{\frac{z}{2}}), q_i'' = q_i - q_i', T_i' = \mathtt{S}(q_i')$ and $T_i'' = \mathtt{S}(q_i'')$, then there is a hyperplane $\pi(\boldsymbol{y}) = 0$ such that*

(1) $F \subseteq \{\boldsymbol{y} \mid \pi(\boldsymbol{y}) = 0\}$,

(2) $2T_i' \subseteq \{y \mid \pi(y) = 0\}$, *and*

(3) $2T_i'' \subseteq \{y \mid \pi(y) > 0\}$.

Proof. By Lemma 9.2, there is a hyperplane $\pi(y) = 0$ such that $\forall \alpha \in F, \pi(\alpha) = 0$ and $\forall \alpha \in N(p) \setminus F, \pi(\alpha) > 0$. We prove that π is a hyperplane which satisfies the requirement. First, because $T_i' \subseteq F_{\frac{z}{2}}$, by Lemma 9.2, $2T_i' \subseteq 2F_{\frac{z}{2}} \subseteq F$ and thus $2T_i' \subseteq \{y \mid \pi(y) = 0\}$.

Second, it is obvious that $T_i'' \cap F_{\frac{z}{2}} = \emptyset$, $T_i' \cap T_i'' = \emptyset$ and $T_i' \cup T_i'' = T_i$ where $T_i = S(q_i)$. By Eq. (9.3), we have $T_i \subseteq \frac{1}{2}N(p)$ and $2T_i \subseteq N(p)$. Thus $2T_i'' \subseteq 2T_i \subseteq N(p) \subseteq \{y \mid \pi(y) \geq 0\}$. If there is an $\alpha \in T_i''$ such that $\pi(2\alpha) = 0$, then $\alpha \in F_{\frac{z}{2}}$, which contradicts with $T_i'' \cap F_{\frac{z}{2}} = \emptyset$. Therefore, $2T_i'' \subseteq \{y \mid \pi(y) > 0\}$. $\qquad\square$

Using the above lemmas, we prove Theorem 9.3 now.

Theorem 9.3. *If p is SOS, then* $\mathrm{Proj}(p, F)$ *is SOS for every face F of* $N(p)$.

Proof. Suppose $p = \sum_{i=1}^{s} q_i^2$ and F is a face of $N(p)$. Let $F_z = F \cap \mathbb{Z}_+^n$, $q_i' = \mathrm{Proj}(q_i, \frac{1}{2}F_z)$ and $q_i'' = q_i - q_i'$. Then

$$p = \sum_{i=1}^{s}(q_i' + q_i'')^2 = \sum_{i=1}^{s} q_i'^2 + 2\sum_{i=1}^{s} q_i' q_i'' + \sum_{i=1}^{s} q_i''^2.$$

By Lemma 9.2,

$$\mathrm{Proj}(p, F_z)$$
$$= \sum_{i=1}^{s}\mathrm{Proj}(q_i'^2, F_z) + 2\sum_{i=1}^{s}\mathrm{Proj}(q_i' q_i'', F_z) + \sum_{i=1}^{s}\mathrm{Proj}(q_i''^2, F_z).$$

By Lemma 9.4, there is a hyperplane $\pi(y) = 0$ such that

(1) $\forall \alpha \in F, \pi(\alpha) = 0$;

(2) $\forall \alpha \in N(p) \setminus F, \pi(\alpha) > 0$;

(3) for any q_i', $2S(q_i') \subseteq \{y \mid \pi(y) = 0\}$; and

(4) $2S(q_i'') \subseteq \{y \mid \pi(y) > 0\}$.

By Lemma 9.3, $\mathrm{Proj}(q_i' q_i'', F_z) = 0$ and $\mathrm{Proj}(q_i''^2, F_z) = 0$. Therefore

$$\mathrm{Proj}(p, F) = \mathrm{Proj}(p, F_z) = \sum_{i=1}^{s}\mathrm{Proj}(q_i'^2, F_z) = \sum_{i=1}^{s}(q_i')^2.$$

The last equality holds because $S(q_i'^2) \subseteq F_z$. $\qquad\square$

Remark 9.2. Theorem 9.3 is strongly related to Theorem 3.6 of [Reznick (1989)], which states that if p is positive semidefinite, then $\mathrm{Proj}(p, F)$ is positive semidefinite for every face F of $N(p)$.

Theorem 9.3 gives a necessary condition for a polynomial to be SOS.

Example 9.3. $p = x_1^4 + x_2^4 + x_3^4 - 1$.

Obviously, the polynomial in Example 9.3 is not SOS (*e.g.*, $p(0,0,0) = -1$). By Theorem 9.3, one necessary condition for p to be SOS is that $\mathtt{Proj}(p, \{(0,0,0)\}) = -1$ should be SOS which can be efficiently checked. On the other hand, if we use Newton polytope based method to construct Q in (9.5), the size of Q is $\binom{3+2}{2} = 10$. The number of constraints is $\binom{3+4}{4} = 35$.

Definition 9.3 (Convex cover polynomial). *A polynomial p is said to be a convex cover polynomial if there exist some pairwise disjoint faces $F_i (i = 1, \ldots, u)$ of $\mathtt{N}(p)$ such that $\mathtt{S}(p) \subseteq \cup_{i=1}^{u} F_i$.*

It is easy to get the following proposition by the definition of convex cover polynomial.

Proposition 9.1. *The support of a convex cover polynomial does not intersect the interior of its Newton polytope.*

The following theorem is a direct corollary of Theorem 9.3.

Theorem 9.4. *Suppose $F_i (i = 1, \ldots, u)$ are pairwise disjoint faces of $\mathtt{N}(p)$ such that $\mathtt{S}(p) \subseteq \cup_{i=1}^{u} F_i$, i.e. p is a convex cover polynomial. Let*

$$p_i = \mathtt{Proj}(p, F_i) \ for \ i = 1, \ldots, u.$$

Then p is SOS if and only if p_i is SOS for $i = 1, \ldots, u$.

We use the following example to demonstrate the benefit of Theorem 9.4.

Example 9.4. $p = x_1^6 + x_2^6 + x_1^4 - 2x_1^2 x_2^2 + x_2^4$.

$\mathtt{S}(p) = \{(6,0), (0,6), (4,0), (2,2), (0,4)\}$. Let $F_1 = \{(6,0)\}, F_2 = \{(0,6)\}$, $F_3 = \mathtt{conv}(\{(4,0),(2,2),(0,4)\})$ be three faces of $\mathtt{N}(p)$. Because F_1, F_2, F_3 satisfy the condition of Definition 9.3, p is a convex cover polynomial. Let $p_i = \mathtt{Proj}(p, F_i)$ for $i = 1, 2, 3$. Then, by Theorem 9.4, proving p is SOS is equivalent to proving p_i is SOS for $i = 1, 2, 3$. Therefore, the original problem is divided into three simpler sub-problems. When using Newton polytope based method to prove p is SOS, the size of Q is 7 and the number of constraints is 18, denoted by $(7, 18)$. However, for p_1, p_2, p_3, the corresponding data are $(1, 1)$, $(1, 1)$ and $(3, 5)$, respectively.

9.3 Split Polynomial

Definition 9.4. For a set Q of vectors and any $\alpha \in Q + Q$, define $\varphi_Q(\alpha) = \{\beta \in Q \mid \exists \gamma \in Q, \beta + \gamma = \alpha\}$.

Definition 9.5. Suppose Q satisfies $\mathrm{SOSS}(p, Q)$ (see Definition 9.1) for a polynomial p. Define $\mathcal{V}(p, Q)$ to be the set $\{\alpha \in Q \mid \varphi_Q(2\alpha) = \{\alpha\}\}$.

Definition 9.6. Suppose Q satisfies $\mathrm{SOSS}(p, Q)$ for a polynomial p. For any $\alpha \in Q + Q$ and $R \subseteq \mathbb{Z}_+^n$, define

$$H_Q(\alpha, R) = \begin{cases} \emptyset & \text{if } \alpha \in R; \\ \{\tfrac{1}{2}\alpha\} & \text{if } \alpha \notin R \text{ and } \varphi_Q(\alpha) = \{\tfrac{1}{2}\alpha\}; \\ \bigcup_{\beta,\gamma \in Q, \beta \neq \gamma, \beta+\gamma=\alpha} \left(H_Q(2\beta, R \cup \{\alpha\}) \cup H_Q(2\gamma, R \cup \{\alpha\}) \right) & \text{otherwise.} \end{cases}$$

If $\alpha \notin Q + Q$, define $H_Q(\alpha, R) = \emptyset$ for any R. Define

$$\psi_Q(\alpha) = H_Q(\alpha, \emptyset) \text{ for any } \alpha \in Q + Q.$$

By the definition of $H_Q(\alpha, R)$, the number of elements in R increases by one after a recursion. So the recursive depth of $H_Q(\alpha, R)$ is at most $|Q| + 1$ for any $\alpha \in Q + Q$ and any R. Therefore, $H_Q(\alpha, R)$ uniquely exists for any $\alpha \in \mathbb{Z}_+^n$ and $R \subseteq \mathbb{Z}_+^n$. And so does $\psi_Q(\alpha)$.

Lemma 9.5. *Suppose* Q *satisfies* $\mathrm{SOSS}(p, Q)$ *for a polynomial p and F is a face of* $\mathrm{conv}(Q+Q)$. *Let* $T = \{\alpha \mid \alpha \in \mathcal{V}(p, Q), 2\alpha \in F\}, Q_1 = (Q+Q) \cap F$. *Then* $\psi_Q(\alpha) \subseteq T$ *for any* $\alpha \in Q_1$.

Proof. For any $\beta \in \psi_Q(\alpha)$, by the definition of $\psi_Q(\alpha)$, there are $\beta_1, \ldots, \beta_k, \gamma_1, \ldots, \gamma_k \in Q$ such that $\beta_i \neq \gamma_i$ for $i = 1, \ldots, k-1$,

$$\alpha = \beta_1 + \gamma_1, 2\beta_1 = \beta_2 + \gamma_2, \ldots, 2\beta_{k-1} = \beta_k + \gamma_k, \beta_k = \gamma_k = \beta$$

and $\psi_Q(2\beta) = \{\beta\}$.

We prove $2\beta_i \in F$ by induction. Because $\alpha = \beta_1 + \gamma_1$ and $\alpha \in F$, we have $2\beta_1 \in F$. Assume that $2\beta_i \in F$ for $i < m$. If $i = m$, since $2\beta_{m-1} = \beta_m + \gamma_m$ and $2\beta_{m-1} \in F$, we have $2\beta_m \in F$. Then $2\beta = 2\beta_k \in F$ and hence, $\beta \in T$. $\qquad\square$

Definition 9.7. Suppose Q satisfies $\mathrm{SOSS}(p, Q)$ for a polynomial p and $T \subseteq \mathcal{V}(p, Q)$. Define $\sigma(T) = \{\gamma \mid \gamma \in Q, \psi_Q(2\gamma) \subseteq T\}$.

Lemma 9.6. *Suppose p is SOS, say $p = \sum_{i=1}^{s} h_i^2$, and $\mathsf{S}(h_i) \subseteq \mathsf{Q}$. For any $T \subseteq \mathcal{V}(p, \mathsf{Q})$ and any $\beta \in \mathsf{Q} + \mathsf{Q}$, if $\psi_{\mathsf{Q}}(\beta) \subseteq T$, then $\beta \notin \mathsf{S}((p - \sum_{i=1}^{s} \mathrm{Proj}(h_i, \sigma(T))^2))$.*

Proof. For any $\gamma_1, \gamma_2 \in \mathsf{Q}$ with $\gamma_1 + \gamma_2 = \beta$, we have $\psi_{\mathsf{Q}}(2\gamma_1) \subseteq \psi_{\mathsf{Q}}(\beta)$ and $\psi_{\mathsf{Q}}(2\gamma_2) \subseteq \psi_{\mathsf{Q}}(\beta)$ by the definition of ψ_{Q}. Since $\psi_{\mathsf{Q}}(\beta) \subseteq T$, we have $\gamma_1, \gamma_2 \in \sigma(T)$ by the definition of $\sigma(T)$. It is not difficult to see that the coefficient of the term $\boldsymbol{x}^\beta$ in $\sum_{i=1}^{s} \mathrm{Proj}(h_i, \sigma(T))^2$ equals that of the term $\boldsymbol{x}^\beta$ in $\sum_{i=1}^{s} h_i^2$. Thus, $\boldsymbol{x}^\beta$ does not appear in $p - \sum_{i=1}^{s} \mathrm{Proj}(h_i, \sigma(T))^2$ since $p - \sum_{i=1}^{s} h_i^2 = 0$. $\qquad\square$

Theorem 9.5. *Assume $p = \sum c_\alpha \boldsymbol{x}^\alpha$ is SOS, Q satisfies $\mathrm{SOSS}(p, \mathsf{Q})$ and $T \subseteq \mathcal{V}(p, \mathsf{Q})$. If $\psi_{\mathsf{Q}}(\alpha + \beta) \subseteq T$ for any $\alpha, \beta \in \sigma(T)$, then $p_1 = \sum_{\alpha \in \mathsf{S}(p), \psi_{\mathsf{Q}}(\alpha) \subseteq T} c_\alpha \boldsymbol{x}^\alpha$ is SOS.*

Proof. Suppose $p = \sum_{i=1}^{s} h_i^2$ and $p_1' = p - p_1$. Set $h_i' = \mathrm{Proj}(h_i, \sigma(T))$ and $h_i'' = h_i - h_i'$, then $p = \sum_{i=1}^{s} (h_i')^2 + 2 \sum_{i=1}^{s} h_i' h_i'' + \sum_{i=1}^{s} (h_i'')^2$. By Lemma 9.6, $\beta \notin \mathsf{S}(p - \sum_{i=1}^{s} (h_i')^2)$ for any $\beta \in \mathsf{S}(p_1)$, *i.e.* $\mathsf{S}(p_1) \cap \mathsf{S}(p - \sum_{i=1}^{s} (h_i')^2) = \emptyset$. Since $\psi_{\mathsf{Q}}(\alpha + \beta) \subseteq T$ for any $\alpha, \beta \in \sigma(T)$, by the definition of $\sigma(T)$, $\psi_{\mathsf{Q}}(\beta) \subseteq T$ for any $\beta \in \mathsf{S}(\sum_{i=1}^{s} (h_i')^2)$. Thus, $\mathsf{S}(p_1') \cap \mathsf{S}(\sum_{i=1}^{s} (h_i')^2) = \emptyset$. Summarizing the above, we have

(1) $p_1 + p_1' = \sum_{i=1}^{s} (h_i')^2 + (p - \sum_{i=1}^{s} (h_i')^2)$,
(2) $\mathsf{S}(p_1) \cap \mathsf{S}(p - \sum_{i=1}^{s} (h_i')^2) = \emptyset$, and
(3) $\mathsf{S}(p_1') \cap \mathsf{S}(\sum_{i=1}^{s} (h_i')^2) = \emptyset$.

Therefore, $p_1 = \sum_{i=1}^{s} (h_i')^2$. $\qquad\square$

Definition 9.8 (Split polynomial). *Let Q satisfies $\mathrm{SOSS}(p, \mathsf{Q})$ for a polynomial p. If there exist some pairwise disjoint nonempty subsets $T_i (i = 1, \ldots, u)$ of $\mathcal{V}(p, \mathsf{Q})$ such that*

(1) $\psi_{\mathsf{Q}}(\alpha + \beta) \subseteq T_i$ for any $\alpha, \beta \in \sigma(T_i)$ (see Definition 9.7) for any $i = 1, \ldots, u$, and

(2) for any $\alpha \in \mathsf{S}(p)$, there exist exact one T_i such that $\psi_{\mathsf{Q}}(\alpha) \subseteq T_i$,

then p is said to be a split polynomial with respect to $T_1, \ldots, T_u$.

If p is a split polynomial with respect to a non-empty set $T \subset \mathcal{V}(p, \mathsf{Q})$ and its complement in $\mathcal{V}(p, \mathsf{Q})$, for brevity, we simply say that p is a split polynomial with respect to T.

Theorem 9.6. *Suppose $p = \sum c_\alpha \boldsymbol{x}^\alpha$ is a split polynomial with respect to $T_1, \ldots, T_u$, then p is SOS if and only if each $p_i = \sum_{\alpha \in \mathsf{S}(p), \psi_\mathsf{Q}(\alpha) \subseteq T_i} c_\alpha \boldsymbol{x}^\alpha$ is SOS for $i = 1, \ldots, u$.*

Proof. Necessity is a direct corollary of Theorem 9.5. For sufficiency, note that the second condition of Definition 9.8 guarantees that $\mathsf{S}(p_i) \cap \mathsf{S}(p_j) = \emptyset$ for any $i \neq j$ and $p = \sum_{i=1}^{u} p_i$. $\qquad\square$

Now, we give the relation between convex cover polynomial and split polynomial, which indicates that split polynomial is a wider class of polynomials.

Theorem 9.7. *If p is a convex cover polynomial, then p is a split polynomial. The converse is not true.*

Proof. If p is a convex cover polynomial then there exist pairwise disjoint faces $F_i (i = 1, \ldots, u)$ of $\mathsf{N}(p)$ such that $\mathsf{S}(p) \subseteq \cup_{i=1}^{u} F_i$. Suppose $\mathrm{conv}(\mathsf{Q} + \mathsf{Q}) = \mathsf{N}(p)$ and Q satisfies $\mathrm{SOSS}(p, \mathsf{Q})$. Let $T_i = \{\alpha \in \mathcal{V}(p, \mathsf{Q}) \mid 2\alpha \in F_i\}, i = 1, \ldots, u$. We prove that p is a split polynomial with respect to $T_1, \ldots, T_u$.

We claim that $\sigma(T_j) = \{\gamma \in \mathsf{Q} \mid 2\gamma \in F_j\}$ for $j = 1, \ldots, u$. If there exist $\gamma_0 \in \sigma(T_j), 2\gamma_0 \notin F_j$, as F_j is a face of $\mathsf{N}(p)$, there exist a linear function π such that $\pi(2\gamma_0) > \pi(\alpha)$ for any $\alpha \in F_j$. By the Definition of $\psi_\mathsf{Q}(2\gamma_0)$, there exists $\beta_0 \in \psi_\mathsf{Q}(2\gamma_0) \subseteq T_j$ such that $\pi(2\beta_0) \geq \pi(2\gamma_0)$. This contradicts with $2\beta_0 \in F_j$. Thus, $\sigma(T_j) \subseteq \{\gamma \in \mathsf{Q} \mid 2\gamma \in F_j\}$.

We then prove that $\{\gamma \in \mathsf{Q} \mid 2\gamma \in F_j\} \subseteq \sigma(T_j)$. Assume that there exists $\gamma_0 \in \mathsf{Q}$ with $2\gamma_0 \in F_j$ such that $\gamma_0 \notin \sigma(T_j)$. Then there exists $\beta_0 \in \psi_\mathsf{Q}(2\gamma_0)$ such that $2\beta_0 \notin F_j$. Because F_j is a face of $\mathsf{N}(p)$, it is not difficult to see that if $\alpha_1 + \alpha_2 \in F_j$ where $\alpha_1, \alpha_2 \in \mathsf{Q}$, then $2\alpha_1 \in F_j, 2\alpha_2 \in F_j$. Therefore, $2\beta \in F_j$ for any $\beta \in \psi_\mathsf{Q}(2\gamma_0)$, which contradicts with $2\beta_0 \notin F_j$.

Now we have $\sigma(T_j) = \{\gamma \in \mathsf{Q} \mid 2\gamma \in F_j\}$. By Lemma 9.5, $\psi_\mathsf{Q}(\alpha + \beta) \subseteq T_j$ for any $\alpha, \beta \in \sigma(T_j)$. Since $\mathsf{S}(p) \subseteq \cup_{i=1}^{u} F_i$ and F_i are pairwise disjoint, there exists exactly one T_i such that $\psi_\mathsf{Q}(\alpha) \subseteq T_i$ for any $\alpha \in \mathsf{S}(p)$. As a result, p is a split polynomial with respect to $T_1, \ldots, T_u$.

Note that the Motzkin polynomial in Example 9.8 is a split polynomial but not a convex cover polynomial since $x_1^2 x_2^2$ lies in the interior of $\mathsf{N}(p)$ (see Proposition 9.1). $\qquad\square$

Remark 9.3. One may wonder under what condition a split polynomial is a convex cover polynomial. A reasonable conjecture may be as this:

Let Q be a finite set satisfying $\mathrm{SOSS}(p, Q)$ *with* $\mathrm{conv}(Q + Q) = \mathrm{N}(p)$ *for a polynomial p. If* $\mathcal{V}(p, Q)$ *contains only vertices of* $\mathrm{conv}(Q)$, *then p is a split polynomial if and only if p is a convex cover polynomial.*

Unfortunately, the conjecture is not true. For example, let

$$p = x_1^4 x_2^2 x_3^2 + x_1^2 x_2^4 x_3^2 - 2 x_1^2 x_2^2 x_3^2 + x_3^2 + x_1^2 x_2^2 + x_1^2 x_2^2 x_3^4,$$

then

$$Q = \{(2,1,1), (1,2,1), (1,1,1), (0,0,1), (1,1,0), (1,1,2)\},$$
$$\mathcal{V}(p, Q) = \{(2,1,1), (1,2,1), (0,0,1), (1,1,0), (1,1,2)\}.$$

Obviously, $\mathcal{V}(p, Q)$ contains only vertices of $\mathrm{conv}(Q)$. Set

$$T_1 = \{(2,1,1), (1,2,1), (0,0,1)\}, \ T_2 = \{(1,1,0), (1,1,2)\},$$

then it is easy to check that p is a split polynomial with respect to T_1, T_2. But p is not a convex cover polynomial by Proposition 9.1 because $x_1^2 x_2^2 x_3^2$ lies in the interior of $\mathrm{N}(p)$.

The example indicates that the relation between split polynomial and convex cover polynomial may be complicated. We do not find a good sufficient condition for a split polynomial to be a convex cover polynomial.

9.4   Algorithm

Existing SDP based SOS solvers consist of the following two main steps: computing a set Q satisfying $\mathrm{SOSS}(p, Q)$ for a given p; solving the feasibility problem of (9.5) related to Q by SDP solvers. In this section, we give a new algorithm (Algorithm 9.2) for SOS decomposition. The algorithm employs the following strategies. First, we give a different technique for computing an initial set Q which satisfies $\mathrm{SOSS}(p, Q)$ for a given p. Second, we check one necessary condition (Lemma 9.9) to refute quickly some non-SOS polynomials. Third, if the input polynomial is detected to be a split polynomial, we reduce the problem into several smaller sub-problems based on Theorem 9.6. This section is dedicated to describe the strategies in detail and the performance of the algorithm is reported in the next section.

We first describe the new technique for computing an initial set Q. The following lemma is a direct corollary of the result in [Reznick (1978)] (see also Eq. (9.3) in Section 9.1).

Lemma 9.7. *Suppose p is a polynomial and γ is a given vector. Let $c = \max_{\alpha \in \frac{1}{2} \mathrm{P}^e} \gamma^{\mathrm{T}} \alpha$. For any Q which satisfies $\mathrm{SOSS}(p, Q)$, after deleting every β in Q such that $\gamma^{\mathrm{T}} \beta > c$, $\mathrm{SOSS}(p, Q)$ still holds.*

By Lemma 9.7, it is easy to give a method for computing an initial set Q which satisfies SOSS(p, Q) for a given p. That is, first choose a coarse set Q which satisfies SOSS(p, Q), *e.g.*, the set defined by Eq. (9.4); then prune the superfluous elements in Q by choosing *randomly* γ. This is indeed a common method in existing work [Papachristodoulou *et al.* (2013); Lofberg (2004); Seiler (2013)].

We employ a different strategy to construct an initial Q satisfying SOSS(p, Q). The procedure is as follows. For a given polynomial p, firstly, we compute the set $\frac{1}{2}P^e$ (recall that $P^e = P \cap (2\mathbb{Z}_+^n)$ where P is the support of p) and an over approximation set Q of integer points in conv$(\frac{1}{2}P^e)$. Secondly, let B be the matrix whose columns are all the vectors of $\frac{1}{2}P^e$. We choose one by one the hyperplanes whose normal directions are the eigenvectors of BB^{T} to delete superfluous lattice points in Q by Lemma 9.7.

Notation 9.2. We denote by `PCAG`(p) the above procedure to compute an initial Q satisfying SOSS(p, Q) for a given polynomial p.

We cannot prove that the above strategy is better in general than the random one. However, inspired by principal component analysis (PCA), we believe in many cases the shape of conv$(\frac{1}{2}P^e)$ depends on eigenvectors of BB^{T}. On a group of randomly generated examples (see Example 9.5), we show that the size of Q obtained by using random hyperplanes to delete superfluous lattice points are 10% greater than that of the output of our algorithm `PCAG` (see Figure 9.1).

Example 9.5. $SQR(k, n, d, t) = g_1^2 + \cdots + g_k^2$ where $\deg(g_i) = d$, $\#(\mathsf{S}(g_i)) = t$, $\#(\mathrm{var}(g_i)) = n$.

Lemma 9.8. *[Kim* et al. *(2005); Seiler* et al. *(2013)] For a polynomial p and a set Q which satisfy SOSS(p, Q), after deleting every element α in Q which satisfies that $2\alpha \notin P^e$ and $\varphi_Q(2\alpha) = \{\alpha\}$, the relation SOSS$(p, Q)$ still holds.*

Notation 9.3. We denote by `EXACTG`(p) the procedure which deletes superfluous elements of the output of `PCAG`(p) based on Lemma 9.8.

The following lemma is a simple but very useful necessary condition which can detect non-SOS polynomials efficiently in many cases.

Lemma 9.9. *Suppose Q satisfies SOSS(p, Q) for a polynomial p. If p is SOS, then $\alpha \in Q + Q$ for any $\alpha \in \mathsf{S}(p)$.*

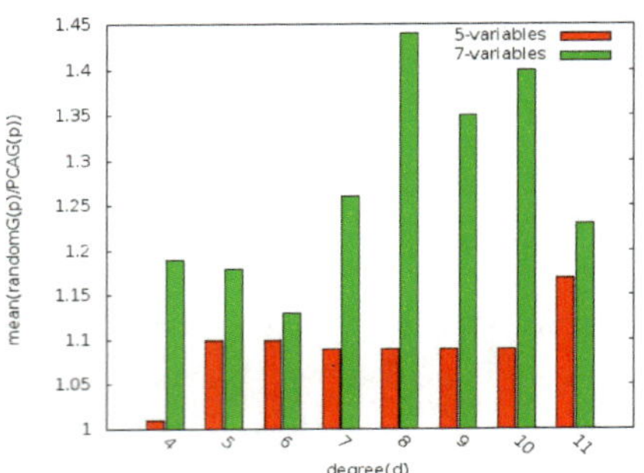

Fig. 9.1   Mean ratio of $\#(Q)$ between random algorithm and $\mathtt{PCAG}(p)$ on every random group $SQR(k, n, d, t)$. The red bars correspond to $k = 4, n = 5, t = 3$ and the green bars correspond to $k = 5, n = 7, t = 4$. For any given (k, n, d, t), we generate 10 polynomials randomly. (Adapted from [Dai and Xia (2015)].)

Proof.   If p is SOS, since p and Q satisfy relation $\mathrm{SOSS}(p, Q)$, there are $q_1, \ldots, q_s$ such that $p = \sum_{i=1}^{s} q_i^2$ and $\mathsf{S}(q_i) \subseteq Q$. Hence, for every monomial $\boldsymbol{x}^\alpha$ of p there are q_i, $\boldsymbol{x}^\beta, \boldsymbol{x}^\gamma$ such that $\boldsymbol{x}^\beta, \boldsymbol{x}^\gamma$ are monomials of q_i and $\boldsymbol{x}^\alpha = \boldsymbol{x}^\beta \boldsymbol{x}^\gamma$. Therefore, $\alpha \in Q + Q$ for any $\alpha \in \mathsf{S}(p)$. $\square$

Example 9.6. [Choi and Lam (1977)] Let $q(x, y, z) = 1 + x^2 y^2 + y^2 z^2 + z^2 x^2 - 4xyz$. It is easy to know that

$$\tfrac{1}{2}\mathrm{P}^e = \{(0, 0, 0), (1, 1, 0), (1, 0, 1), (0, 1, 1)\},$$
$$Q^0 = \{(0, 0, 0), (1, 0, 0), (0, 1, 0), (0, 0, 1), (1, 1, 0), (1, 0, 1), (0, 1, 1)\}.$$

By Lemma 9.8, after deleting $(1, 0, 0), (0, 1, 0), (0, 0, 1)$ from Q^0, we have

$$Q = \mathtt{EXACTG}(q) = \{(0, 0, 0), (1, 1, 0), (1, 0, 1), (0, 1, 1)\}$$

and $\mathrm{SOSS}(q, Q)$ holds. Since $(1, 1, 1) \notin Q + Q$, by Lemma 9.9, p is not SOS.

For an input polynomial p, by setting $Q = \mathtt{EXACTG}(p)$, we obtain a set Q satisfying $\mathrm{SOSS}(p, Q)$. Now, we check whether or not p is a split polynomial related to this Q. And if it is, the original problem can be reduced to several smaller sub-problems. The details are described formally as Algorithm 9.1 and Algorithm 9.2.

Example 9.7. We illustrate $\mathtt{QuickSOS}$ on the polynomial p in Example 9.4. First,

$$\mathsf{S}(p) = \{(0, 6), (6, 0), (0, 4), (4, 0), (2, 2)\},$$
$$Q = \{(0, 2), (0, 3), (1, 1), (1, 2), (2, 0), (2, 1), (3, 0)\},$$
$$\mathcal{V}(p, Q) = \{(0, 2), (0, 3), (2, 0), (3, 0)\}.$$

Algorithm 9.1 `MonomialRelation`

Input: $p \in \mathbb{Q}[\boldsymbol{x}]$

Output: The map ψ_{Q} defined by Definition 9.6

1: $\mathrm{Q} \leftarrow \mathrm{EXACTG}(p)$;
2: Let C be a map from Q to $\{\texttt{true}, \texttt{false}\}$;
3: **for** $\alpha \in \mathrm{Q}$ **do**
4:    $C(\alpha) \leftarrow \texttt{false}$;
5: **end for**
6: Let $\mathcal{V}(p, \mathrm{Q})$ be the set defined by Definition 9.5;
7: Initialize $\psi_{\mathrm{Q}}(\alpha) = \emptyset$ for any $\alpha \in \mathbb{Z}_+^n$;
8: **for** $\alpha \in \mathcal{V}(p, \mathrm{Q})$ **do**
9:    $\psi_{\mathrm{Q}}(2\alpha) \leftarrow \{\alpha\}$;
10:    $C(\alpha) \leftarrow \texttt{true}$;
11: **end for**
12: $\mathrm{run} \leftarrow \texttt{true}$;
13: **while** run **do**
14:    $\mathrm{run} \leftarrow \texttt{false}$;
15:    **for** $\alpha \in \mathrm{Q}$ **do**
16:      **if** $C(\alpha)$ **then**
17:        $C(\alpha) \leftarrow \texttt{false}$;
18:        **for** $\beta \in \mathrm{Q}$ **do**
19:          **if** $\psi_{\mathrm{Q}}(2\alpha) \nsubseteq \psi_{\mathrm{Q}}(\alpha + \beta)$ **then**
20:            $\psi_{\mathrm{Q}}(\alpha + \beta) \leftarrow \psi_{\mathrm{Q}}(\alpha + \beta) \cup \psi_{\mathrm{Q}}(2\alpha)$;
21:            **if** $\alpha + \beta \in 2\mathbb{Z}_+^n$ **then**
22:               $\mathrm{run} \leftarrow \texttt{true}$; $C((\alpha + \beta)/2) \leftarrow \texttt{true}$;
23:            **end if**
24:          **end if**
25:        **end for**
26:      **end if**
27:    **end for**
28: **end while**
29: **return** ψ_{Q}

Second, $\psi_{\mathrm{Q}}((0,4)) = \{(0,2)\}$, $\psi_{\mathrm{Q}}((0,6)) = \{(0,3)\}$, $\psi_{\mathrm{Q}}((4,0)) = \{(2,0)\}$, $\psi_{\mathrm{Q}}((6,0)) = \{(3,0)\}$, $\psi_{\mathrm{Q}}((2,2)) = \{(0,2),(2,0)\}$. Set $T = \psi_{\mathrm{Q}}((2,2)) = \{(2,0),(0,2)\}$, it is easy to see that p is a split polynomial with respect to T and $p_1 = x_1^4 - 2x_1^2 x_2^2 + x_2^4, p_2 = x_1^6 + x_2^6$.

Third, similarly, $\texttt{QuickSOS}(p_2)$ divides p_2 into $p_{21} = x_1^6, p_{22} = x_2^6$. Fi-

---

Algorithm 9.2 QuickSOS

Input: $p \in \mathbb{Q}[x]$

Output: false that means p is not SOS; or $\{q_1, \ldots, q_s\}$ where p, q_i satisfy
    Eq. (9.2) numerically

 1: Let ψ_Q be the output of MonomialRelation(p);
 2: **for** $\alpha \in S(p)$ **do**
 3:    **if** $\alpha \notin Q + Q$ **then**
 4:       **return** false;   // Lemma 9.9
 5:    **end if**
 6: **end for**
 7: **for** $\alpha \in S(p)$ **do**
 8:    **if** p is a split polynomial with respect to $\psi_Q(\alpha)$ **then**
 9:       Let p_1, p_2 be as in Theorem 9.6;
10:       Let R_1 be the output of QuickSOS(p_1);
11:       Let R_2 be the output of QuickSOS(p_2);
12:       **if** R_1 or R_2 is false **then**
13:          **return** false;
14:       **end if**
15:       **return** $R_1 \cup R_2$;
16:    **end if**
17: **end for**
18: **return** SOS(p, Q);   // Notation 9.1

---

nally, QuickSOS(p) outputs "$\{-1.00 * x_2^2 + 1.00 * x_1^2, 1.00 * x_2^3, 1.00 * x_1^3\}$".

Example 9.8 (Motzkin polynomial).
$$f = x_1^4 x_2^2 + x_1^2 x_2^4 - 3x_1^2 x_2^2 + 1.$$
   Because $S(f) = \{(4,2),(2,2),(2,4),(0,0)\}$ and $Q = \{(0,0),(1,1),$
$(2,1),(1,2)\}$, MonomialRelation(f) returns
$$\psi_Q((4,2)) = \{(2,1)\}, \psi_Q((2,4)) = \{(2,4)\},$$
$$\psi_Q((2,2)) = \{(1,1)\}, \psi_Q((0,0)) = \{(0,0)\}.$$
Then QuickSOS(f) will return false when it reaches line 10 for $\alpha = (2,2)$.

Remark 9.4. Let $Q = $ EXACTG(p). By Definition 9.8, to determine whether p is a split polynomial, one should check all the non-empty subsets of $\mathcal{V}(p, Q)$. However, this approach is obviously inefficient. Therefore, in Algorithm 9.2 we only check whether p is a split polynomial with respect to $\psi_Q(\alpha)$ for $\alpha \in S(p)$. Although this incomplete check may miss some split polynomials, it is effective in many cases, as is shown in the next section.

Tools	1	2	3	4	5	6	7	8	9	10
					$\#(Q)$					
				$k = 4, n = 5, d = 5, t = 3$						
QuickSOS	(2,15)	(1,44)	(2,11)	(4,4)	(4,4)	(1,25)	(4,5)	(3,9)	(2,8)	(2,20)
YALMIP	24	45	33	18	23	36	22	20	15	25
SOSTOOLS	24	45	33	18	23	36	22	20	15	25
				$k = 4, n = 5, d = 10, t = 3$						
QuickSOS	(4,3)	(4,3)	(4,10)	(3,6)	(2,7)	(4,4)	(4,3)	(4,3)	(4,3)	(2,26)
YALMIP	97	91	42	23	45	40	101	62	95	52
SOSTOOLS	104	94	36	23	48	41	109	70	104	52
				$k = 5, n = 7, d = 5, t = 4$						
QuickSOS	(4,7)	(5,5)	(2,13)	(5,4)	(4,11)	(5,4)	(4,7)	(3,12)	(5,5)	(4,10)
YALMIP	21	33	24	24	28	24	21	28	42	33
SOSTOOLS	wrong	33	24	24	28	24	21	28	42	33
				$k = 5, n = 7, d = 10, t = 4$						
QuickSOS	(5,4)	(5,4)	(5,5)	(5,5)	(5,4)	(5,4)	(5,4)	(5,4)	(3,11)	(5,4)
YALMIP	45	82	74	59	48	70	79	63	41	57
SOSTOOLS	wrong	wrong	wrong	63	57	76	wrong	67	wrong	wrong
				$k = 5, n = 7, d = 5, t = 6$						
QuickSOS	(1,26)	(1,29)	(1,28)	(1,72)	(1,37)	(1,30)	(1,27)	(4,7)	(2,14)	(1,61)
YALMIP	28	38	28	82	48	31	33	34	34	69
SOSTOOLS	wrong	38	28	82	wrong	31	33	wrong	34	wrong
				$k = 5, n = 7, d = 8, t = 6$						
QuickSOS	(4,7)	(4,6)	(4,7)	(4,7)	(4,6)	(4,6)	(4,6)	(2,24)	(4,6)	(4,6)
YALMIP	38	34	71	121	51	57	75	100	47	29
SOSTOOLS	39	wrong	wrong	128	67	67	78	111	52	31

9.5 Experiments

We use a class of examples, $SQR(k, n, d, t)$, which are sparse polynomials of the form SOS randomly generated by Maple's command **randpoly**, to test the effectiveness of our method.

The number of elements in a set T is denoted by $\#(T)$, $\deg(p)$ denotes the total degree of a polynomial p, $\mathrm{var}(p)$ denotes the set of variables occurring in a polynomial p.

Example 9.9. (see also Example 9.5)

$$SQR(k, n, d, t) = g_1^2 + \cdots + g_k^2$$

where $\deg(g_i) = d$, $\#(\mathrm{S}(g_i)) = t$, $\#(\mathrm{var}(g_i)) = n$.

Remark 9.5. As explained before, SQR is constructed in the form of SOS. But the polynomial is expanded before input to the tools.

We report the comparison of the size of Q computed by different tools. It is reasonable to believe that the total time of computing SOS decomposition becomes shorter as the size of Q getting smaller if we use the same SDP solver and the cost of computing smaller Q is not expensive.

We explain the notations in the table. Each (b, s) for **QuickSOS**'s $\#(Q)$ means **QuickSOS** divides the polynomial into b polynomials $p_1, \ldots, p_b$ and s is the largest number of $\#(Q_i)$ corresponding to p_i. The "——" denotes that there is no corresponding outputs.

Table 9.1 lists the results on examples SQR. We randomly generate 10 polynomials for every (k, n, d, t). All the outputs of **QuickSOS** and **YALMIP**

[Lofberg (2004)] are correct[1]. Some data corresponding to SOSTOOLS [Papachristodoulou *et al.* (2013)] are "wrong", which means that SOSTOOLS's output is wrong or there is an error occurred during its execution. For many examples of SQR, QuickSOS can divide the original polynomial into some simpler polynomials.

Let us give a rough complexity analysis of those $\text{SOS}(p, \text{Q})$ based on *interior point method*. Let $n = \#(\text{Q})$, the number of elements contained in Q. Then the size of matrix M in (9.5) is $n \times n$. Let m be the number of different elements occurring in QQ^T. It is easy to know that $n \le m \le n^2$. Suppose $m = O(n^c), c \in [1, 2]$ and we use *interior point method* in $\text{SOS}(p, \text{Q})$, which is a main method for solving SDP numerically. Then the algorithm will repeatedly solve *least squares* problems with m linear constraints and $\frac{(n+1)n}{2}$ unknowns. Suppose that the least squares procedure is called k times. Then, the total complexity is $O(kn^{2+2c})$. So, if n becomes $2n$, the time consumed will increase by at least 16 times. We demonstrate this fact by one polynomial of group $SQR(4, 5, 10, 3)$.

Example 9.10. $p = (-91w^4x^2yz^3 - 41k^4xy^2z^2 - 14kwx^3y^2z)^2 + (-40kx^7yz + 16w^4xy + 65w^2y^4)^2 + (11kx^2y^6z - 34k^5x^3z - 18kyz^5)^2 + (-26k^4w^3xyz - 35xy^6z^3 - 57kw^2x^2z^3)^2.$

In Example 9.10, QuickSOS divides p into four simpler polynomials p_1, p_2, p_3, p_4. For each simpler polynomial p_i, QuickSOS constructs a set Q_i whose size is 3 and $\text{SOSS}(p_i, \text{Q}_i)$ holds. YALMIP constructs one Q for p whose size is 97 and SOSTOOLS also constructs one Q for p whose size is 104. If the time consumed by constructing Q is short compared with the total time and assume these three tools use the same SDP solver, the ratio of total time of three tools is $4(3^{2+2c}) : 97^{2+2c} : 104^{2+2c}$ where $1 \le c \le 2$. In fact, in our experiments, the total time of these three tools on this example is 0.02 seconds, 23.91 seconds and 48.47 seconds, respectively.

For the performance of QuickSOS on more examples, see [Dai and Xia (2015)].

Remark 9.6. In general, if a polynomial can be written as a linear combination with positive coefficients of nonnegative polynomials, the polynomial is nonnegative. In addition to SOS decomposition, there are some other methods of this fashion for proving inequalities. See for example [Huang and Chen (2005); Chen and Huang (2006)].

---

[1]The meaning of "correct" is that the output is right with respect to a certain numerical error.

Chapter 10

Successive Difference Substitution

Successive difference substitution was first introduced in [Yang (2005, 2006)] as a heuristic algorithm for proving nonnegativity of polynomials when all the variables are nonnegative. The method is incomplete but very efficient for many examples and can solve some large examples which cannot be solved by other tools. The method has been developed into a complete algorithm with some interesting connections to other methods by a series of work, see for example [Yao (2009); Yang and Yao (2009); Yao (2010); Xu and Yao (2011, 2012); Hou and Shao (2011)]. Based on these references, we introduce in this chapter basic concepts, ideas and recent advances of the method.

10.1　Basic Idea

It is well known that when talking about nonnegativity of polynomials, we can only consider homogeneous polynomials, which are also called *forms*. So in this chapter, if not specified, all polynomials are homogeneous.

Definition 10.1. A form $f \in \mathbb{R}[\boldsymbol{x}]$ is *positive semi-definite* (*nonnegative*) on $\mathbb{R}^n_+$ if $f(\boldsymbol{x}) \geq 0$ for all $\boldsymbol{x} \in \mathbb{R}^n_+$. If $f(\boldsymbol{x}) > 0$ for all $\boldsymbol{x}(\neq \boldsymbol{0}) \in \mathbb{R}^n_+$, f is said to be *positive definite*. If there exist $\alpha, \beta \in \mathbb{R}^n_+$ such that $f(\alpha)f(\beta) < 0$, f is said to be *indefinite*. Denote by PSD (PD) the set of all positive semi-definite (positive definite) forms.

10.1.1　*An Example*

Let us begin with a simple example. Given a polynomial

$$p(x,y,z) = 3\,x^3 - 3\,x^2y - 3\,x^2z - 3\,xy^2 + 9\,xyz - 3\,xz^2 + 8\,y^3 - 8\,y^2z - 8\,yz^2 + 8\,z^3,$$

271

prove that $p(x, y, z) \geq 0$ if $x \geq 0, y \geq 0, z \geq 0$.

There are several effective methods to verify the nonnegativity of p. For example, we may use the methods presented in previous chapters or may express p as

$$p(x, y, z) = \frac{(-256\, z^2 + 86\, y\, z + 170\, y^2 + 47\, x\, z - 78\, x\, y + 31\, x^2)^2}{8192\, (x + y + z)}$$

$$+ \frac{(18318\, y\, z - 5989\, x\, z - 18318\, y^2 + 6603\, x^2 - 614\, x\, y)^2}{75030528\, (x + y + z)}$$

$$+ \frac{(-127\, z + 54\, y + 73\, x)^2\, x\, z}{2032\, (x + y + z)} + \frac{(-54\, y + 29\, z + 25\, x)^2\, x\, y}{464\, (x + y + z)}$$

$$+ \frac{43\, (y - z)^2\, y\, z}{8\, (x + y + z)} + \frac{767\, (x - y)^2\, x\, z}{2032\, (x + y + z)} + \frac{322878817\, x^2\, (x - y)^2}{161870848\, (x + y + z)}$$

$$+ \frac{767\, (x - y)^2\, x\, y}{464\, (x + y + z)} + \frac{x^2\, (-2529232\, z + 1854661\, y + 674571\, x)^2}{1482575096832\, (x + y + z)}.$$

However, getting this expression needs profound mathematics. We introduce a method which is based on the following simple observation and needs much less mathematics.

Split x, y, z into some nonnegative quantities, say t_1, t_2, t_3, and collect all the terms with respect to new variables t_1, t_2, t_3. If all the coefficients are nonnegative, p is nonnegative.

For example, let

$$x = t_1 + t_2 + t_3,$$

$$y = t_2 + t_3,$$

$$z = t_3.$$

That is

$$\begin{bmatrix} x \\ y \\ z \end{bmatrix} = \begin{bmatrix} 1 & 1 & 1 \\ 0 & 1 & 1 \\ 0 & 0 & 1 \end{bmatrix} \begin{bmatrix} t_1 \\ t_2 \\ t_3 \end{bmatrix}. \tag{10.1}$$

Obviously, $x \geq y \geq z \geq 0$ if and only if $t_1 \geq 0, t_2 \geq 0, t_3 \geq 0$.

Then $p(x, y, z)$ is transformed into a polynomial in t_1, t_2, t_3:

$$p_1(t_1, t_2, t_3) = 3\, t_1^3 + 6\, t_1^2 t_2 + 3\, t_1^2 t_3 + 3\, t_1 t_2 t_3 + 5\, t_2^3 + 13\, t_2^2 t_3,$$

whose coefficients are all nonnegative. Therefore, if $x \geq y \geq z$,

$$p(x, y, z) = p_1(t_1, t_2, t_3) \geq 0.$$

If $x \geq z \geq y$, we use the following transformation

$$\begin{bmatrix} x \\ y \\ z \end{bmatrix} = \begin{bmatrix} 1 & 1 & 1 \\ 0 & 0 & 1 \\ 0 & 1 & 1 \end{bmatrix} \begin{bmatrix} t_1 \\ t_2 \\ t_3 \end{bmatrix}.$$

Then we have $t_1 \geq 0, t_2 \geq 0, t_3 \geq 0$ and $p(x, y, z)$ is also transformed into

$$p_1(t_1, t_2, t_3) = 3\,t_1^3 + 6\,t_1^2 t_2 + 3\,t_1^2 t_3 + 3\,t_1 t_2 t_3 + 5\,t_2^3 + 13\,t_2^2 t_3.$$

Therefore, in this case, we still have $p(x, y, z) = p_1(t_1, t_2, t_3) \geq 0$.

Similarly, if $y \geq x \geq z$ or $z \geq x \geq y$, we use respectively the following transformations

$$\begin{bmatrix} x \\ y \\ z \end{bmatrix} = \begin{bmatrix} 0 & 1 & 1 \\ 1 & 1 & 1 \\ 0 & 0 & 1 \end{bmatrix} \begin{bmatrix} t_1 \\ t_2 \\ t_3 \end{bmatrix}, \qquad \begin{bmatrix} x \\ y \\ z \end{bmatrix} = \begin{bmatrix} 0 & 1 & 1 \\ 0 & 0 & 1 \\ 1 & 1 & 1 \end{bmatrix} \begin{bmatrix} t_1 \\ t_2 \\ t_3 \end{bmatrix}.$$

Then we have $t_1 \geq 0, t_2 \geq 0, t_3 \geq 0$ and $p(x, y, z)$ is transformed into

$$p_2(t_1, t_2, t_3) = 8\,t_1^3 + 21\,t_1^2 t_2 + 13\,t_1^2 t_3 + 15\,t_1 t_2^2 + 23\,t_1 t_2 t_3 + 5\,t_2^3 + 13\,t_2^2 t_3.$$

Therefore, in these two cases, we have $p(x, y, z) = p_2(t_1, t_2, t_3) \geq 0$.

Finally, if $y \geq z \geq x$ or $z \geq y \geq x$, we use

$$\begin{bmatrix} x \\ y \\ z \end{bmatrix} = \begin{bmatrix} 0 & 0 & 1 \\ 1 & 1 & 1 \\ 0 & 1 & 1 \end{bmatrix} \begin{bmatrix} t_1 \\ t_2 \\ t_3 \end{bmatrix}, \qquad \begin{bmatrix} x \\ y \\ z \end{bmatrix} = \begin{bmatrix} 0 & 0 & 1 \\ 0 & 1 & 1 \\ 1 & 1 & 1 \end{bmatrix} \begin{bmatrix} t_1 \\ t_2 \\ t_3 \end{bmatrix},$$

respectively. Then we have $t_1 \geq 0, t_2 \geq 0, t_3 \geq 0$ and $p(x, y, z)$ is transformed into

$$p_3(t_1, t_2, t_3) = 8\,t_1^3 + 16\,t_1^2 t_2 + 13\,t_1^2 t_3 + 3\,t_1 t_2 t_3 + 3\,t_2^2 t_3.$$

Therefore $p(x, y, z) = p_3(t_1, t_2, t_3) \geq 0$ in these two cases.

We summarize the above method as follows. There are 6 permutations for variables x, y, z. Each of the permutations corresponds to a linear transformation splitting x, y, z into smaller nonnegative quantities t_1, t_2, t_3 and transforming $p(x, y, z)$ into a polynomial in t_1, t_2, t_3. If the coefficients of the new polynomials are all nonnegative, p must be nonnegative.

10.1.2 *Difference Substitution*

Continue the example of last section. The polynomial set $\{p_1, p_2, p_3\}$ is called the *difference substitution* of p, denoted by $\mathrm{DS}(p)$. Since the inverse of

$$\begin{cases} x = t_1 + t_2 + t_3 \\ y = t_2 + t_3 \\ z = t_3 \end{cases} \quad \text{is} \quad \begin{cases} t_1 = x - y \\ t_2 = y - z \\ t_3 = z, \end{cases}$$

$\{t_1, t_2, t_3\}$ is a difference sequence of $\{x, y, z\}$. That is why *difference substitution* is named.

In general, $\mathsf{DS}(q)$ of a polynomial q in three variables is a set of at most 6 members. Let us see another example.

$$q(x, y, z) = 2\,x^4 - 3\,x^2y^2 - 6\,x^2yz + 9\,x^2z^2 + 2\,xy^3 - 6\,xyz^2$$
$$-4\,xz^3 + 2\,y^3z + 3\,y^2\,z^2 + z^4.$$

Compute $\mathsf{DS}\,(q) = \{q_1, q_2, q_3, q_4, q_5, q_6\}$, where

$$q_1 = 2\,t_1^4 + 8\,t_1^3t_2 + 8\,t_1^3t_3 + 9\,t_1^2t_2^2 + 12\,t_1^2t_2t_3 + 12\,t_1^2t_3^2 + 4\,t_1t_2^3 + t_2^4,$$

$$q_2 = 2\,t_1^4 + 8\,t_1^3t_2 + 8\,t_1^3t_3 + 21\,t_1^2t_2^2 + 36\,t_1^2t_2t_3 + 12\,t_1^2t_3^2 + 22\,t_1t_2^3$$
$$+ 48\,t_1t_2^2t_3 + 24\,t_1t_2t_3^2 + 8\,t_2^4 + 20\,t_2^3t_3 + 12\,t_2^2t_3^2,$$

$$q_3 = 2\,t_1^3t_2 + 4\,t_1^3t_3 + 3\,t_1^2t_2^2 + 12\,t_1^2t_2t_3 + 12\,t_1^2t_3^2 + t_2^4,$$

$$q_4 = 2\,t_1^3t_2 + 4\,t_1^3t_3 + 9\,t_1^2t_2^2 + 24\,t_1^2t_2t_3 + 12\,t_1^2t_3^2 + 12\,t_1t_2^3 + 36\,t_1t_2^2t_3$$
$$+ 24\,t_1t_2t_3^2 + 6\,t_2^4 + 16\,t_2^3t_3 + 12\,t_2^2t_3^2,$$

$$q_5 = t_1^4 + 3\,t_1^2t_2^2 + 10\,t_1t_2^3 + 12\,t_1t_2^2t_3 + 8\,t_2^4 + 20\,t_2^3t_3 + 12\,t_2^2t_3^2,$$

$$q_6 = t_1^4 + 4\,t_1^3t_2 + 9\,t_1^2t_2^2 + 12\,t_1t_2^3 + 12\,t_1t_2^2t_3 + 6\,t_2^4 + 16\,t_2^3t_3 + 12\,t_2^2t_3^2.$$

Note that all the coefficients of every polynomial of $\mathsf{DS}(q)$ are nonnegative. So, if x, y, z are nonnegative, $q(x, y, z) \geq 0$.

Generally speaking, for a polynomial in n variables $x_1, \ldots, x_n$, there are $n!$ permutations on the variables. Each permutation, *e.g.* $x_1 \geq x_2 \geq \cdots \geq x_n$, corresponds to a splitting:

$$\begin{cases} x_1 = t_1 + t_2 + \cdots + t_n \\ x_2 = t_2 + \cdots + t_n \\ \quad \cdots \\ x_n = t_n, \end{cases}$$

where $t_1, \ldots, t_n$ is just a difference sequence of $x_1, \ldots, x_n$.

Similarly, the difference substitution $\mathsf{DS}(f)$ of an n-variate polynomial $f(x_1, \ldots, x_n)$ is a set of at most $n!$ polynomials. It is easy to see that if f is symmetric, $\mathsf{DS}(f)$ has only one polynomial. If all the coefficients of all polynomials of $\mathsf{DS}(f)$ are nonnegative, it is clear that $f \geq 0$ when $x_1, \ldots, x_n$ are nonnegative, *i.e.* f is positive semi-definite on $\mathbb{R}_+^n$.

Definition 10.2. A polynomial is said to be *trivially nonnegative* if its coefficients are all nonnegative. A polynomial set is said to be *trivially nonnegative* if all polynomials of it are trivially nonnegative.

Although $\mathtt{DS}(f)$ being trivially nonnegative is only a sufficient condition for f to be positive semi-definite on $\mathbb{R}_+^n$, it is effective on many examples.

Example 10.1. Prove that

$$f = x^3 + y^3 + z^3 - x^2y - xy^2 - x^2z - xz^2 - y^2z - yz^2 + 3\,xyz$$

is positive semi-definite on $\mathbb{R}_+^3$. The polynomial is the well-known Robinson's polynomial [Reznick (2000)]. By computation, $\mathtt{DS}(f)$ has only one polynomial $t_1^3 + 2\,t_1^2t_2 + t_1^2t_3 + t_1t_2t_3 + t_2^2t_3$, whose coefficients are all nonnegative, *i.e.* $\mathtt{DS}(f)$ is trivially nonnegative. Thus f is positive semi-definite on $\mathbb{R}_+^3$.

Example 10.2. Prove that

$$\left(\frac{1}{2}(x^2 + y^2 + z^2)(x + y + z) - xyz\right)^2 \le \frac{1}{2}\left(x^2 + y^2 + z^2\right)^3$$

on $\mathbb{R}_+^3$. In other words, prove

$$\begin{aligned}
f = {}& x^6 - 2\,x^5y - 2\,x^5z + 3\,x^4y^2 + 2\,x^4yz + 3\,x^4z^2 - 4\,x^3y^3 - 4\,x^3z^3 \\
&+ 3\,x^2y^4 + 2\,x^2y^2z^2 + 3\,x^2z^4 - 2\,xy^5 + 2\,xy^4z + 2\,xyz^4 - 2\,xz^5 + y^6 \\
&- 2\,y^5z + 3\,y^4z^2 - 4\,y^3z^3 + 3\,y^2z^4 - 2\,yz^5 + z^6
\end{aligned}$$

is positive semi-definite on $\mathbb{R}_+^3$. By computation, $\mathtt{DS}(f)$ has only one polynomial

$$\begin{aligned}
& t_1^6 + 4\,t_1^5t_2 + 2\,t_1^5t_3 + 8\,t_1^4t_2^2 + 8\,t_1^4t_2t_3 + 3\,t_1^4t_3^2 + 8\,t_1^3t_2^3 + 12\,t_1^3t_2^2t_3 \\
&\quad + 12\,t_1^3t_2t_3^2 + 4\,t_1^3t_3^3 + 4\,t_1^2t_2^4 + 8\,t_1^2t_2^3t_3 + 20t_1^2t_2^2t_3^2 + 20\,t_1^2t_2t_3^3 + 7\,t_1^2t_3^4 \\
&\quad + 16\,t_1t_2^3t_3^2 + 36\,t_1t_2^2t_3^3 + 32\,t_1t_2t_3^4 + 10\,t_1t_3^5 + 8\,t_2^4t_3^2 + 24\,t_2^3t_3^3 + 32\,t_2^2t_3^4 \\
&\quad + 20\,t_2t_3^5 + 5\,t_3^6,
\end{aligned}$$

and is trivially nonnegative. Thus f is positive semi-definite on $\mathbb{R}_+^3$. This example is from `http:// www.mathlinks.ro/Forum/topic-54136.html`.

Example 10.3. Prove that

$$f = a^4b + b^4c + c^4d + d^4a - abcd(a + b + c + d)$$

is positive semi-definite on $\mathbb{R}_+^4$. This problem is from `http:// www.mathlinks.ro/Forum/topic-45218.html`. There are 24 permutations on a, b, c, d but $\mathtt{DS}(f)$ has only 6 members, whose coefficients are all nonnegative. Since $\mathtt{DS}(f)$ is trivially nonnegative, f is positive semi-definite on $\mathbb{R}_+^4$.

Example 10.4. Prove that

$$
\begin{aligned}
f = {} & 1056\, x_4 x_5^4 + 744 x_4^4 x_5 + 1120 x_3 x_5^4 + (672\, x_2 + 192\, x_5 + 352\, x_4 \\
& + 512\, x_3) x_1^4 + (-3360 x_5 x_4 + 912\, x_5^2 - 1440 x_2 x_3 + 752\, x_3^2 + 672\, x_2^2 \\
& - 2400 x_3 x_4 - 2400 x_5 x_2 + 832\, x_4^2 - 2880 x_5 x_3 - 1920 x_4 x_2) x_1^3 \\
& + 1224 x_3^4 x_4 + 1064 x_5 x_3^4 + (320 x_4^3 + 2016 x_2^2 x_3 - 96 x_2^3 - 3456 x_3 x_5 x_4 \\
& + 528 x_5^3 + 3312 x_5^2 x_4 + 112 x_3^3 + 2736 x_3^2 x_4 + 2016 x_2^2 x_5 + 3312 x_5^2 x_2 \\
& + 3312 x_5^2 x_3 + 2736 x_3^2 x_5 + 2592 x_2 x_4^2 - 3456 x_2 x_5 x_4 + 1872 x_2 x_3^2 \\
& + 2016 x_4 x_2^2 - 3456 x_2 x_3 x_4 + 3456 x_4^2 x_5 - 3456 x_3 x_5 x_2 + 2592 x_3 x_4^2) x_1^2 \\
& + 1200 x_2^4 x_3 + (2736 x_2^2 x_3^2 - 4992 x_5^3 x_2 - 3744 x_2^3 x_5 - 4800 x_4^3 x_2 \\
& - 2784 x_2^3 x_3 - 4992 x_3 x_4^3 - 3264 x_4 x_2^3 + 3456 x_4^2 x_2^2 - 4320 x_3^3 x_4 \\
& + 1152\, x_2^2 x_3 x_4 + 2304 x_2 x_3^2 x_4 + 1152\, x_2^2 x_4 x_5 + 2304 x_3 x_4^2 x_5 \\
& + 2304 x_2 x_3^2 x_5 + 1152\, x_2^2 x_3 x_5 + 2304 x_2 x_4^2 x_5 + 1152\, x_3^2 x_4 x_5 \\
& - 4608 x_3^3 x_2 + 336 x_2^4 + 1248 x_5^4 + 1448 x_4^4 + 1144 x_3^4 + 4752 x_5^2 x_4^2 \\
& + 3744 x_4^2 x_3^2 - 5184 x_5^3 x_3 + 4176 x_5^2 x_2^2 - 5376 x_4 x_5^3 - 4800 x_5 x_3^3 \\
& + 4464 x_5^2 x_3^2 - 5856 x_4^3 x_5) x_1 + 1184 x_5^4 x_2 + 528 x_2^3 x_3^2 + 384 x_3^2 x_4^3 \\
& - 4992 x_3^3 x_2 x_5 + 384 x_3^3 x_4^2 + 240 x_5^2 x_4^3 + 1320 x_4^4 x_3 + 144 x_3^3 x_2^2 \\
& + 1080 x_3^4 x_2 + 432 x_1^5 + 560 x_5^3 x_2^2 + 880 x_2^4 x_5 + 688 x_3^2 x_5^3 + 1152 x_5^5 \\
& - 5376 x_5^3 x_2 x_3 - 5568 x_5^3 x_2 x_4 + 3600 x_5^2 x_2^2 x_3 + 3024 x_2^2 x_3^2 x_5 - 5280 x_4 x_2^3 x_5 \\
& + 3744 x_4^2 x_2^2 x_5 + 3024 x_3^2 x_4 x_2^2 + 3744 x_3^2 x_4^2 x_2 - 5184 x_3 x_4^3 x_2 + 2880 x_3 x_4^2 x_2^2 \\
& - 4512 x_3^3 x_4 x_2 - 4320 x_2^3 x_3 x_4 + 3600 x_2^2 x_4 x_5^2 + 3888 x_3^2 x_4 x_5^2 + 4752 x_2 x_4^2 x_5^2 \\
& + 4464 x_2 x_3^2 x_5^2 - 6240 x_3 x_4^3 x_5 - 5760 x_3 x_4 x_5^3 + 4752 x_3 x_5^2 x_4^2 + 4032 x_3^2 x_4^2 x_5 \\
& - 7200 x_3^3 x_4 x_5 - 6048 x_4^3 x_2 x_5 - 4800 x_2^3 x_3 x_5 + 864 x_2^5 + 1224 x_4^5 + 1128 x_3^5 \\
& + 608 x_4^2 x_2^3 + 352\, x_4^3 x_2^2 + 1384 x_4^4 x_2 + 1040 x_2^4 x_4 + 624 x_4^2 x_5^3 + 464 x_5^2 x_3^3 \\
& + 592\, x_5^3 x_3^2 - 3456 x_3 x_2^2 x_4 x_5 + 1152\, x_2 x_3^2 x_4 x_5 + 2304 x_3 x_2 x_4^2 x_5
\end{aligned}
$$

is positive semi-definite on $\mathbb{R}_+^5$. The problem is from
`http://guestbook.nease.net/read.php?user=zgbdsyjxz&id=`
`1118121244&curpage=36&page=2`.

There are 120 permutations on x_1, x_2, x_3, x_4, x_5 and $\mathrm{DS}(f)$ has 120 members, whose coefficients are all nonnegative. Since $\mathrm{DS}(f)$ is trivially nonnegative, f is positive semi-definite on $\mathbb{R}_+^5$.

Example 10.5. Prove that

$$
\frac{a_1}{a_2 + a_3} + \frac{a_2}{a_3 + a_4} + \frac{a_3}{a_4 + a_5} + \frac{a_4}{a_5 + a_1} + \frac{a_5}{a_1 + a_2} \ge \frac{5}{2}
$$

holds on $\mathbb{R}_+^5$. In other words, prove that

$$
\begin{aligned}
f = {}& 2\,a_1^3 a_3 a_4 + 2\,a_1^3 a_3 a_5 + 2\,a_1^3 a_4^2 + 2\,a_1^3 a_4 a_5 + 2\,a_1^2 a_2^2 a_4 + 2\,a_1^2 a_2^2 a_5 \\
& + 2\,a_1^2 a_2 a_3^2 + a_1^2 a_2 a_3 a_4 - a_1^2 a_2 a_3 a_5 - 3a_1^2 a_2 a_4^2 - 3a_1^2 a_2 a_4 a_5 + 2\,a_1^2 a_3^3 \\
& - 3a_1^2 a_3^2 a_4 - 5a_1^2 a_3^2 a_5 - 5a_1^2 a_3 a_4^2 - 3a_1^2 a_3 a_4 a_5 + 2\,a_1^2 a_3 a_5^2 + 2\,a_1^2 a_4^2 a_5 \\
& + 2\,a_1^2 a_4 a_5^2 + 2\,a_1 a_2^3 a_4 + 2\,a_1 a_2^3 a_5 + 2\,a_1 a_2^2 a_3^2 - a_1 a_2^2 a_3 a_4 - 3a_1 a_2^2 a_3 a_5 \\
& - 5a_1 a_2^2 a_4^2 - 3a_1 a_2^2 a_4 a_5 + 2\,a_1 a_2^2 a_5^2 + 2\,a_1 a_2 a_3^3 - 3a_1 a_2 a_3^2 a_4 - 3a_1 a_2 a_3^2 a_5 \\
& - 3a_1 a_2 a_3 a_4^2 + a_1 a_2 a_3 a_5^2 + 2\,a_1 a_2 a_4^3 + a_1 a_2 a_4^2 a_5 - a_1 a_2 a_4 a_5^2 + 2\,a_1 a_3^3 a_5 \\
& + 2\,a_1 a_3^2 a_4^2 + a_1 a_3^2 a_4 a_5 - 3a_1 a_3^2 a_5^2 + 2\,a_1 a_3 a_4^3 - a_1 a_3 a_4^2 a_5 - 3a_1 a_3 a_4 a_5^2 \\
& + 2\,a_2^3 a_4 a_5 + 2\,a_2^3 a_5^2 + 2\,a_2^2 a_3^2 a_5 + 2\,a_2^2 a_3 a_4^2 + a_2^2 a_3 a_4 a_5 - 3a_2^2 a_3 a_5^2 \\
& + 2\,a_2^2 a_4^3 - 3a_2^2 a_4^2 a_5 - 5a_2^2 a_4 a_5^2 + 2\,a_2 a_3^3 a_5 + 2\,a_2 a_3^2 a_4^2 - a_2 a_3^2 a_4 a_5 \\
& - 5a_2 a_3^2 a_5^2 + 2\,a_2 a_3 a_4^3 - 3a_2 a_3 a_4^2 a_5 - 3a_2 a_3 a_4 a_5^2 + 2\,a_2 a_3 a_5^3 + 2\,a_2 a_4^2 a_5^2 \\
& + 2\,a_2 a_4 a_5^3 + 2\,a_3^2 a_4 a_5^2 + 2\,a_3^2 a_5^3 + 2\,a_3 a_4^2 a_5^2 + 2\,a_3 a_4 a_5^3
\end{aligned}
$$

is positive semi-definite on $\mathbb{R}_+^5$. The original inequality is the so-called *Shapiro inequality* [Bushell (1994); Bushell and McLeod (2002)] in 5 variables.

There are 120 permutations on a_1, a_2, a_3, a_4, a_5 and $\mathrm{DS}(f)$ has 24 members, whose coefficients are all nonnegative. Since $\mathrm{DS}(f)$ is trivially nonnegative, f is positive semi-definite on $\mathbb{R}_+^5$.

Example 10.6. Prove that

$$
f = \sum_{k=1}^{10} a_k^{10} - 10 \prod_{k=1}^{10} a_k \ge 0
$$

on $\mathbb{R}_+^{10}$. Since f is symmetric, $\mathrm{DS}(f)$ has only one member, whose coefficients are all nonnegative. Since $\mathrm{DS}(f)$ is trivially nonnegative, f is positive semi-definite on $\mathbb{R}_+^{10}$.

Example 10.7. Prove that

$$
\begin{aligned}
f = {}& (-x_3^2 - 2\,x_4 x_1 + 6x_1^2 + 6x_2^2 + 4x_2 x_1 - x_4^2 - 2\,x_2 x_3 - 2\,x_3 x_1 - 2\,x_4 x_2) \cdot \\
& (x_1 - x_2)^{1000} + (-2\,x_4 x_1 - 2\,x_3 x_1 - 2\,x_4 x_2 + 4x_3 x_4 + 6x_4^2 - x_1^2 - x_2^2 \\
& - 2\,x_2 x_3 + 6x_3^2)(x_3 - x_4)^{1000} + (6x_4^2 - x_3^2 - 2\,x_2 x_3 + 6x_2^2 - 2\,x_2 x_1 \\
& - x_1^2 - 2\,x_3 x_4 + 4x_4 x_2 - 2\,x_4 x_1)(x_2 - x_4)^{1000} + (6x_3^2 - x_1^2 - x_4^2 \\
& + 6x_2^2 - 2\,x_3 x_1 - 2\,x_2 x_1 - 2\,x_3 x_4 - 2\,x_4 x_2 + 4x_2 x_3)(x_2 - x_3)^{1000} \\
& + (-x_2^2 + 6x_1^2 - 2\,x_3 x_1 - 2\,x_4 x_2 + 6x_4^2 - 2\,x_3 x_4 + 4x_4 x_1 - x_3^2 \\
& - 2\,x_2 x_1)(x_4 - x_1)^{1000} + (-2\,x_4 x_1 - 2\,x_2 x_1 - 2\,x_3 x_4 - x_4^2 + 6x_3^2 \\
& - 2\,x_2 x_3 + 6x_1^2 + 4x_3 x_1 - x_2^2)(x_3 - x_1)^{1000}
\end{aligned}
$$

is positive semi-definite on $\mathbb{R}_+^4$. The degree of f is 1002. Since f is symmetric, $\mathrm{DS}(f)$ has only one member, whose coefficients are all nonnegative. Thus f is positive semi-definite on $\mathbb{R}_+^4$. The problem is provided by Baoqian Liu, where the number 1000 can be replaced by some bigger even numbers.

10.1.3 *Successive Difference Substitution*

Given a form f, if $\mathrm{DS}(f)$ is not trivially nonnegative, what should we do next?

For example, to prove the Shapiro inequality in 4 variables

$$\frac{a_1}{a_2 + a_3} + \frac{a_2}{a_3 + a_4} + \frac{a_3}{a_4 + a_1} + \frac{a_4}{a_1 + a_2} \geq 2,$$

we need to prove that

$$\begin{aligned}
f = {}& a_1^3 a_3 + a_1^3 a_4 + a_1^2 a_2^2 - a_1^2 a_2 a_4 - 2\,a_1^2 a_3^2 - a_1^2 a_3 a_4 + a_1^2 a_4^2 + a_1 a_2^3 \\
& - a_1 a_2^2 a_3 - a_1 a_2^2 a_4 - a_1 a_2 a_3^2 + a_1 a_3^3 - a_1 a_3 a_4^2 + a_2^3 a_4 + a_2^2 a_3^2 - 2a_2^2 a_4^2 \\
& + a_2 a_3^3 - a_2 a_3^2 a_4 - a_2 a_3 a_4^2 + a_2 a_4^3 + a_3^2 a_4^2 + a_3 a_4^3
\end{aligned} \tag{10.2}$$

is positive semi-definite on $\mathbb{R}_+^4$. By computation, $\mathrm{DS}(f)$ has 6 members, one of which has negative coefficients:

$$\begin{aligned}
f_1 = {}& t_1^3 t_2 + t_1^3 t_3 + 2\,t_1^3 t_4 + t_1^2 t_2^2 + 2\,t_1^2 t_2 t_3 + 4t_1^2 t_2 t_4 + 2\,t_1^2 t_3^2 + 5\,t_1^2 t_3 t_4 \\
& + 4t_1^2 t_4^2 - t_1 t_2^2 t_3 - t_1 t_2 t_3^2 - 2t_1 t_2 t_3 t_4 + t_1 t_3^3 + t_1 t_3^2 t_4 + t_2^2 t_3^2 + 3t_2 t_3^3 \\
& + 4t_2 t_3^2 t_4 + 2t_3^4 + 6t_3^3 t_4 + 4t_3^2 t_4^2.
\end{aligned}$$

Thus $\mathrm{DS}(f)$ is not trivially nonnegative. A trick is to compute the difference substitution of $f_1(t_1, t_2, t_3, t_4)$ to prove that f_1 is positive semi-definite on $\mathbb{R}_+^4$. Fortunately, we find that all the coefficients of all the 24 polynomials of $\mathrm{DS}(f_1)$ are nonnegative. Therefore f_1 is positive semi-definite on $\mathbb{R}_+^4$ and so is f.

Generally, we have the following procedure:

1. For a given form f, compute $\mathrm{DS}(f)$.
2. Denote $\mathrm{DS}^{(0)}(f) = \{f\}$, $\mathrm{DS}^{(1)}(f) = \mathrm{DS}(f)$. Let $k = 1$.
3. If $\mathrm{DS}^{(k)}(f)$ is trivially nonnegative, quit.
4. Otherwise, denote the polynomials of $\mathrm{DS}^{(k)}(f)$ with negative coefficients by $f_{k,1}, f_{k,2}, \ldots, f_{k,l_k}$. Compute $\mathrm{DS}(f_{k,1})$, $\mathrm{DS}(f_{k,2})$, $\ldots$, $\mathrm{DS}(f_{k,l_k})$, respectively.
5. Let $\mathrm{DS}^{(k+1)}(f) = \displaystyle\bigcup_{i=1}^{l_k} \mathrm{DS}\,(f_{k,i})$, $k \leftarrow k + 1$ and goto Step 3.

Note that the above procedure may never terminate even if the given form f is positive semi-definite. The procedure is called *successive difference substitution* (SDS for short). We write a short Maple program SDS, which computes one $\mathrm{DS}^{(k)}(f)$ for each execution. For example, to verify the nonnegativity of (10.2), type in `sds(sds(f))`. The output is: *"The form is positive semi-definite"*. That means SDS has to be executed twice.

Example 10.8. To verify

$$h = x^4y^2 - 2\,x^4yz + x^4z^2 + 3x^3y^2z - 2\,x^3yz^2 - 2\,x^2y^4 - 2\,x^2y^3z$$
$$+ x^2y^2z^2 + 2\,xy^4z + y^6$$

is positive semi-definite on $\mathbb{R}^3_+$, we need to type in

```
> sds(sds(sds(sds(sds(h))))): or
> for i to 5 do sds(%) od:
```
That means SDS has to be executed 5 times.

Example 10.9. To verify

$$f = 8x^7 + (8z + 6y)x^6 + 2\,y(31y - 77z)x^5 - y(69y^2 - 2\,z^2 - 202\,yz)x^4$$
$$+ 2\,y(9y^3 + 57yz^2 - 85y^2z + 9z^3)x^3 + 2\,y^2z(-13z^2 - 62\,yz + 27y^2)x^2$$
$$+ 2\,y^3z^2(-11z + 27y)x + y^3z^3(z + 18y)$$

is positive semi-definite on $\mathbb{R}^3_+$, we need to call SDS 18 times. This problem is from "`http://guestbook.nease.net/read.php?user=zgbdsyjxz&id=1118234222&curpage=35`".

Example 10.10. The following polynomial

$$g = 2572755344x^4 - 20000000x^3y - 6426888360x^3z + 30000000x^2y^2$$
$$+ 5315682897x^2z^2 - 20000000xy^3 - 1621722090xz^3 + 170172209y^4$$
$$- 1301377672\,y^3z + 3553788598y^2z^2 - 3864133016yz^3$$
$$+ 1611722090z^4$$

has degree 4 and 12 terms. However, to verify its nonnegativity on $\mathbb{R}^3_+$, we have to call SDS 46 times.

Notation 10.1. Denote

$$A_n = \begin{bmatrix} 1 & 1 & \cdots & 1 \\ 0 & 1 & \ddots & \vdots \\ \vdots & \ddots & \ddots & 1 \\ 0 & \cdots & 0 & 1 \end{bmatrix}, \quad \text{then} \quad A_n^{-1} = \begin{bmatrix} 1 & -1 & \cdots & 0 \\ 0 & 1 & \ddots & \vdots \\ \vdots & \ddots & \ddots & -1 \\ 0 & \cdots & 0 & 1 \end{bmatrix}.$$

Notation 10.2. Assume σ is a permutation on $\{1, ..., n\}$ such that $\sigma(i) = k_i$ for $i = 1, ..., n$. Define an $n \times n$ matrix $P_\sigma = [a_{i,j}]$ where, for $i = 1, ..., n$, $a_{i,j} = 1$ if $j = k_i$ and $a_{i,j} = 0$ otherwise.

It is straightforward to verify the following

Proposition 10.1. *For any $\sigma \in SP_n$ (recall that SP_n is the symmetric group on $\{1, ..., n\}$ (see Notation 7.2)), $P_\sigma^{-1} = P_{\sigma^{-1}} = P_\sigma^{\mathrm{T}}$.*

Definition 10.3. Suppose σ is a permutation on $\{1, ..., n\}$, the matrix

$$B_\sigma = P_\sigma A_n$$

is called a *difference substitution matrix* (DS matrix for short) determined by σ. Denote by PA_n the set of all DS matrices.

Definition 10.4. For a form $f \in \mathbb{R}[\boldsymbol{x}]$ and $k = 1, 2, ...$, define

$$\mathrm{DS}^{(k)}(f) = \bigcup_{\sigma_k \in SP_n} \cdots \bigcup_{\sigma_1 \in SP_n} f(B_{\sigma_1} \cdots B_{\sigma_k} \boldsymbol{x}^{\mathrm{T}})$$

and call it the *kth difference substitution* of f.

Now, we may formally describe the above procedure as an incomplete algorithm as follows.

Algorithm 10.1 Successive Difference Substitution (SDS)

Input: A form $f \in \mathbb{Z}[\boldsymbol{x}]$
Output: "positive semi-definite" or "not positive semi-definite"
 1: $k \leftarrow 0$; $\mathrm{DS}^{(k)} \leftarrow \{f\}$;
 2: **do**
 3:     delete the trivially nonnegative members of $\mathrm{DS}^{(k)}$;
 4:     **if** $\mathrm{DS}^{(k)}$ is empty **then return** "positive semi-definite" **end if**;
 5:     **if** $\exists g \in \mathrm{DS}^{(k)}$ such that $g(\boldsymbol{e}_j) < 0$ for some j **then**
 6:         **return** "not positive semi-definite"
 7:     **end if**;
 8:     $\mathrm{DS}^{(k+1)} \leftarrow \bigcup_{g \in \mathrm{DS}^{(k)}} \bigcup_{\sigma \in SP_n} g(B_\sigma \boldsymbol{x}^{\mathrm{T}})$;
 9:     $k \leftarrow k + 1$;
10: **end do**

Definition 10.5. $\mathrm{SDS}(f)$ is said to be *positively terminating* if it outputs "positive semi-definite"; it is said to be *negatively terminating* if it outputs "not positive semi-definite".

Note that $\mathrm{SDS}(f)$ may not terminate.

Remark 10.1. It is not hard to see that we may modify Algorithm SDS a little to output a point α such that $f(\alpha) < 0$ if $\mathrm{SDS}(f)$ negatively terminates.

Lemma 10.1. *Suppose* $\boldsymbol{y} = (y_1, ..., y_n) \in \mathbb{Z}_+^n$ *and* $y_{k_1} \geq y_{k_2} \geq \cdots \geq y_{k_n}$. *Denote by* σ *the permutation such that* $\sigma(i) = k_i$ *for* $i = 1, ..., n$, *and denote*

$$\boldsymbol{y}'^{\mathrm{T}} = (y_1', ..., y_n')^{\mathrm{T}} = B_{\sigma^{-1}}^{-1} \boldsymbol{y}^{\mathrm{T}}.$$

Then

1. $\boldsymbol{y}' \in \mathbb{Z}_+^n$;
2. $y_1' + \cdots + y_n' \leq y_1 + \cdots + y_n$ *and the equality holds if and only if* $y_{\sigma(2)} = \cdots = y_{\sigma(n)} = 0$;
3. $\gcd(y_1', \ldots, y_n') = \gcd(y_1, \ldots, y_n)$.

Proof. By the definition of B_σ and Proposition 10.1, we have

$$\begin{aligned}
\boldsymbol{y}'^{\mathrm{T}} &= B_{\sigma^{-1}}^{-1} \boldsymbol{y}^{\mathrm{T}} \\
&= (P_{\sigma^{-1}} A_n)^{-1} \boldsymbol{y}^{\mathrm{T}} \\
&= A_n^{-1} P_{\sigma^{-1}}^{-1} \boldsymbol{y}^{\mathrm{T}} \\
&= A_n^{-1} P_{\sigma^{-1}}^{\mathrm{T}} \boldsymbol{y}^{\mathrm{T}} \\
&= A_n^{-1} (\boldsymbol{y} P_{\sigma^{-1}})^{\mathrm{T}} \\
&= A_n^{-1} (y_{\sigma(1)}, y_{\sigma(2)}, \ldots, y_{\sigma(n)})^{\mathrm{T}} \\
&= (y_{\sigma(1)} - y_{\sigma(2)}, y_{\sigma(2)} - y_{\sigma(3)}, \ldots, y_{\sigma(n)})^{\mathrm{T}}.
\end{aligned}$$

First, it is easy to see that $\boldsymbol{y}' \in \mathbb{Z}_+^n$ since $y_{\sigma(1)} \geq y_{\sigma(2)} \geq \cdots \geq y_{\sigma(n)}$. Second,

$$y_1' + \cdots + y_n' = y_{\sigma(1)} \leq y_1 + \cdots + y_n.$$

And $y_{\sigma(1)} = y_1 + \cdots + y_n$ if and only if $y_{\sigma(2)} = \cdots = y_{\sigma(n)} = 0$. Third, a number a is a common divisor of $y_1', ..., y_n'$ if and only if it is a common divisor of $y_1, ..., y_n$. That completes the proof. $\qquad\square$

Theorem 10.1. *Suppose* $\alpha = (a_1, ..., a_n) \in \mathbb{Z}_+^n$ *and* $\gcd(a_1, ..., a_n) = 1$. *Then there exist* $\sigma_1, ..., \sigma_m \in SP_n$ *and some* $j(1 \leq j \leq n)$ *such that*

$$\alpha^{\mathrm{T}} = B_{\sigma_1} \cdots B_{\sigma_m} \boldsymbol{e}_j^{\mathrm{T}}.$$

Proof. By Lemma 10.1, there exists $\sigma_1 \in SP_n$ such that

$$\alpha^{(1)} = B_{\sigma_1}^{-1} \alpha^{\mathrm{T}} = (a_1^{(1)}, ..., a_n^{(1)})$$

satisfies that

$$a_i^{(1)} \geq 0, \quad i = 1, ..., n,$$
$$a_1^{(1)} + \cdots + a_n^{(1)} \leq a_1 + \cdots + a_n,$$
$$\gcd(a_1^{(1)}, ..., a_n^{(1)}) = \gcd(a_1, ..., a_n) = 1.$$

By the same reasoning, there exists $\sigma_2 \in SP_n$ such that

$$\alpha^{(2)} = (a_1^{(2)}, ..., a_n^{(2)}) = B_{\sigma_2}^{-1} B_{\sigma_1}^{-1} \alpha^{\mathrm{T}}$$

satisfies similar conditions. Generally, we can compute $\alpha^{(k)}$ for $k = 1, 2, ...,$ such that

$$a_i^{(k+1)} \geq 0, \quad i = 1, ..., n,$$
$$a_1^{(k+1)} + \cdots + a_n^{(k+1)} \leq a_1^{(k)} + \cdots + a_n^{(k)},$$
$$\gcd(a_1^{(k+1)}, ..., a_n^{(k+1)}) = \gcd(a_1^{(k)}, ..., a_n^{(k)}) = 1.$$

Since $\sum_{i=1}^n a_i$ is fixed, there must exist m such that

$$a_1^{(m)} + \cdots + a_n^{(m)} = a_1^{(m-1)} + \cdots + a_n^{(m-1)}.$$

By Lemma 10.1,

$$\alpha^{(m)} = (a_1^{(m)}, \ldots, a_n^{(m)}) = e_j \quad \text{for some} \ \ j.$$

That is

$$\alpha^{\mathrm{T}} = B_{\sigma_1} \cdots B_{\sigma_m} e_j^{\mathrm{T}}. \qquad \qquad \square$$

Theorem 10.2. *[Yao (2009); Yang and Yao (2009); Yao (2010)]*

1. $\mathrm{SDS}(f)$ *is positively terminating* $\Longrightarrow f \in \mathrm{PSD}$.
2. $\mathrm{SDS}(f)$ *is negatively terminating* $\Longleftrightarrow f \notin \mathrm{PSD}$.
3. $\mathrm{SDS}(f)$ *does not terminate* $\Longrightarrow f \in \mathrm{PSD}$.

Proof.

1. $\mathrm{SDS}(f)$ is positively terminating means there exists m such that $\mathrm{DS}^{(m)}$ is trivially nonnegative. It is clear that all polynomials of $\mathrm{DS}^{(m-1)}$ are positive semi-definite. It is also easy to see that $\mathrm{DS}^{(i+1)} \subset \mathrm{PSD}$ $\Rightarrow \mathrm{DS}^{(i)} \subset \mathrm{PSD}$. The conclusion follows.
2. Necessity. Suppose there exists $g \in \mathrm{DS}^{(m)}$ such that $g(e_j) < 0$ for some j. Then by the definition of $\mathrm{DS}^{(m)}$, there exist $\sigma_1, ..., \sigma_m \in SP_n$ such that $g(e_j) = f(B_{\sigma_1} \cdots B_{\sigma_m} e_j^{\mathrm{T}}) < 0$. Thus $f \notin \mathrm{PSD}$.
   Sufficiency. Suppose $f \notin \mathrm{PSD}$. Then there exists a point $\alpha \in \mathbb{R}_+^n$ such that $f(\alpha) < 0$. By the continuity of f, we may assume that α is a rational point. Further, since f is homogeneous, we may assume all the

entries of α are pairwise coprime integers. By Theorem 10.1, there exist $\sigma_1, ..., \sigma_m \in SP_n$ and $j(1 \leq j \leq n)$ such that $B_{\sigma_1} \cdots B_{\sigma_m} e_j^{\mathrm{T}} = \alpha^{\mathrm{T}}$. So $f(B_{\sigma_1} \cdots B_{\sigma_m} e_j^{\mathrm{T}}) = f(\alpha) < 0$, *i.e.* $g(\boldsymbol{x}) = f(B_{\sigma_1} \cdots B_{\sigma_m} \boldsymbol{x}^{\mathrm{T}})$ satisfies $g(e_j) < 0$. Therefore $\mathrm{SDS}(f)$ is negatively terminating.

3. This is a direct corollary from the second conclusion. $\qquad\square$

Corollary 10.1. *Algorithm* SDS *is correct.*

10.2 Weighted Successive Difference Substitution

Let

$$
T_n = \begin{bmatrix} 1 & \frac{1}{2} & \cdots & \frac{1}{n} \\ 0 & \frac{1}{2} & \ddots & \vdots \\ \vdots & \ddots & \ddots & \frac{1}{n} \\ 0 & \cdots & 0 & \frac{1}{n} \end{bmatrix}, \quad \text{then} \quad T_n^{-1} = \begin{bmatrix} 1 & -1 & \cdots & 0 \\ 0 & 2 & \ddots & \vdots \\ \vdots & \ddots & \ddots & -(n-1) \\ 0 & \cdots & 0 & n \end{bmatrix}.
$$

Using T_n instead of A_n in difference substitution was first proposed in [Yao (2010)]. The new difference substitution based on T_n has geometric meaning and leads to a complete algorithm for deciding nonnegativity of forms.

10.2.1 *Concepts*

Let $A = [a_{ij}]$ be an $n \times n$ real matrix with nonnegative entries. If $\sum_{i=1}^{n} a_{ij} = 1$ for $j = 1, ..., n$, then A is called a *column stochastic matrix*. It is clear that T_n is a special column stochastic matrix. So is the matrix D_σ defined below.

Definition 10.6. Suppose σ is a permutation on $\{1, ..., n\}$, the matrix

$$
D_\sigma = P_\sigma T_n
$$

is called a *weighted difference substitution matrix* (WDS matrix for short) determined by σ. Denote by PT_n the set of all WDS matrices.

Definition 10.7. For a form $f \in \mathbb{R}[\boldsymbol{x}]$ and $k = 1, 2, ...$, define

$$
\mathrm{WDS}^{(k)}(f) = \bigcup_{\sigma_k \in SP_n} \cdots \bigcup_{\sigma_1 \in SP_n} f(D_{\sigma_1} \cdots D_{\sigma_k} \boldsymbol{x}^{\mathrm{T}})
$$

and call it the *kth weighted difference substitution* of f. Especially, $\mathrm{WDS}(f) = \mathrm{WDS}^{(1)}(f)$ is called the *weighted difference substitution* of f.

We may simply replace DS in Algorithm SDS with WDS to get a new algorithm. To be concrete, we describe the algorithm below.

Algorithm 10.2 Weighted Successive Difference Substitution (WSDS)

Input:　A form $f \in \mathbb{Z}[\boldsymbol{x}]$

Output:　"positive semi-definite" or "not positive semi-definite"

1: $k \leftarrow 0$; $\mathrm{WDS}^{(k)} \leftarrow \{f\}$;
2: **do**
3:　　delete the trivially nonnegative members of $\mathrm{WDS}^{(k)}$;
4:　　**if** $\mathrm{WDS}^{(k)}$ is empty **then return** "positive semi-definite" **end if**;
5:　　**if** $\exists g \in \mathrm{WDS}^{(k)}$ such that $g(\boldsymbol{e}_j) < 0$ for some j **then**
6:　　　　**return** "not positive semi-definite"
7:　　**end if**;
8:　　$\mathrm{WDS}^{(k+1)} \leftarrow \bigcup_{g \in \mathrm{WDS}^{(k)}} \bigcup_{\sigma \in SP_n} g(D_\sigma \boldsymbol{x}^{\mathrm{T}})$;
9:　　$k \leftarrow k + 1$;
10: **end do**

Definition 10.8. $\mathrm{WSDS}(f)$ is said to be *positively terminating* if it outputs "positive semi-definite"; it is said to be *negatively terminating* if it outputs "not positive semi-definite".

Remark 10.2. It is not hard to see that we may modify Algorithm WSDS a little to output a point α such that $f(\alpha) < 0$ if $\mathrm{WSDS}(f)$ negatively terminates. Like $\mathrm{SDS}(f)$, $\mathrm{WSDS}(f)$ may not terminate, either.

Example 10.11.
[http://www.mathlinks.ro/Forum/viewtopic.php?t=290780]
　　Determine the nonnegativity of the following polynomial,

$$f = (a^2 + b^2 + c^2)^2 - 4\,(a - b)(b - c)(c - a)(a + b + c).$$

$\mathrm{WSDS}(f)$ runs up to five thousand steps without outputs. This process spends CPU time 811 seconds and RAM 66.12 Mb on a PC (Intel Core i7-3770 CPU @ 3.40GHz, 8G memory, Windows 7 OS) with Maple 17.

10.2.2　*Geometric Meaning*

Definition 10.9. The *standard simplex* Δ_n in $\mathbb{R}^n_+$ is defined as

$$\Delta_n = \{(x_1, ..., x_n) \mid (x_1, ..., x_n) \in \mathbb{R}^n_+,\ \sum_{i=1}^{n} x_i = 1\}.$$

The following lemma is obvious.

Lemma 10.2. *A form f is positive semi-definite (positive definite / indefinite) on $\mathbb{R}^n_+$ if and only if it is positive semi-definite (positive definite / indefinite) on Δ_n.*

So in what follows, we may assume all forms are defined on Δ_n.

Lemma 10.3. *Suppose $B = [b_{ij}] = A_1 A_2 \cdots A_m$ where every $A_i(i = 1, ..., m)$ is a column stochastic matrix. Then we have*

1. B is also a column stochastic matrix, i.e. $\sum_{i=1}^{n} b_{ij} = 1$ for $j = 1, ..., n$.
2. If $(x_1, ..., x_n)^{\mathrm{T}} = B(t_1, ..., t_n)^{\mathrm{T}}$, then $\sum_{i=1}^{n} x_i = \sum_{i=1}^{n} t_i$.

Proof. 1. By induction on m. If $m = 1$, there is nothing to prove. Suppose the conclusion is true for $m - 1$. Denote $B = CA_m$ where

$$C = [c_{ij}] = A_1 \cdots A_{m-1}, \quad A_m = [a_{ij}].$$

Then

$$\sum_{i=1}^{n} b_{ij} = \sum_{i=1}^{n}\sum_{k=1}^{n} c_{ik}a_{kj} = \sum_{k=1}^{n}(\sum_{i=1}^{n} c_{ik})a_{kj} = \sum_{k=1}^{n} a_{kj} = 1.$$

2. $\sum_{i=1}^{n} x_i = \sum_{i=1}^{n}\sum_{j=1}^{n} b_{ij}t_j = \sum_{j=1}^{n}(\sum_{i=1}^{n} b_{ij})t_j = \sum_{j=1}^{n} t_j.$ $\qquad\square$

Corollary 10.2. *If $\alpha \in \Delta_n$, then $D_\sigma \alpha \in \Delta_n$ for any WDS matrix $D_\sigma \in PT_n$.*

Denote by $t_1, ..., t_n$ the column vectors of T_n, respectively. It is clear that the transformation

$$(x_1, ..., x_n)^{\mathrm{T}} = T_n(u_1, ..., u_n)^{\mathrm{T}}$$

sends $e_1, ..., e_n$ to $t_1, ..., t_n$, respectively, *i.e.*

$$(x_1, ..., x_n)^{\mathrm{T}} = \begin{cases} t_1 = (1, 0, ..., 0)^{\mathrm{T}}, \ (u_1, ..., u_n) = e_1 = (1, 0, ..., 0); \\ t_2 = (\frac{1}{2}, \frac{1}{2}, ..., 0)^{\mathrm{T}}, \ (u_1, ..., u_n) = e_2 = (0, 1, ..., 0); \\ \cdots \\ t_n = (\frac{1}{n}, \frac{1}{n}, ..., \frac{1}{n})^{\mathrm{T}}, \ (u_1, ..., u_n) = e_n = (0, ..., 0, 1). \end{cases}$$

Obviously, t_j is the barycenter of the $(j-1)$-dimensional proper face of Δ_n containing $t_1, ..., t_j$. By Lemma 10.3, $\sum_{i=1}^{n} x_i = \sum_{i=1}^{n} u_i = 1$. Therefore in the Δ_n simplex coordinate system, $t_1, ..., t_n$ is a subsimplex of the first barycentric subdivision of Δ_n. Similarly, it is easy to verify that all the $n!$ transformations

$$(x_1, ..., x_n)^{\mathrm{T}} = D_\sigma(u_1, ..., u_n)^{\mathrm{T}} = P_\sigma T_n(u_1, ..., u_n)^{\mathrm{T}}, \ \sigma \in SP_n$$

correspond one by one to the $n!$ subsimplexes of the first barycentric subdivision of Δ_n. In what follows, we may use

$$\bigcup_{\sigma \in SP_n} \Delta_n D_\sigma$$

to denote the first barycentric subdivision of Δ_n. So

$$\bigcup_{\sigma_1 \in SP_n} \cdots \bigcup_{\sigma_k \in SP_n} \Delta_n D_{\sigma_1} \cdots D_{\sigma_k}$$

is the kth barycentric subdivision of Δ_n.

Lemma 10.4. *For a form $f \in \mathbb{R}[\boldsymbol{x}]$ and any positive integer m, we have*

$$f \in \mathrm{PSD} \Longleftrightarrow \mathrm{WDS}^{(m)}(f) \subset \mathrm{PSD}.$$

Proof.

$$
\begin{aligned}
f \in \mathrm{PSD} \ &\Longleftrightarrow f(\alpha) \geq 0, \ \forall \alpha \in \Delta_n \\
&\Longleftrightarrow f(\alpha) \geq 0, \ \forall \alpha \in \bigcup_{\sigma_1 \in SP_n} \cdots \bigcup_{\sigma_m \in SP_n} \Delta_n D_{\sigma_1} \cdots D_{\sigma_m} \\
&\Longleftrightarrow f(D_{\sigma_1} \cdots D_{\sigma_m} \alpha) \geq 0, \ \forall \alpha \in \Delta_n, \ \forall \sigma_i \in SP_n \\
&\Longleftrightarrow g \in \mathrm{PSD}, \ \forall g \in \mathrm{WDS}^{(m)}(f) \\
&\Longleftrightarrow \mathrm{WDS}^{(m)}(f) \subset \mathrm{PSD}.
\end{aligned}
$$

$\square$

Corollary 10.3. *For a form $f \in \mathbb{R}[\boldsymbol{x}]$ and any positive integer m, we have*

$$f \notin \mathrm{PSD} \Longleftrightarrow \exists g \in \mathrm{WDS}^{(m)}(f), g \notin \mathrm{PSD}.$$

Denote by $\mathrm{d}(\delta)$ the diameter of a simplex δ, *i.e.* the maximal distance between two vertices of δ. The following lemma is well-known.

Lemma 10.5. *Let δ be an n-dimensional simplex and δ' a subsimplex of the barycentric subdivision of δ, then*

$$\mathrm{d}(\delta') \leq \frac{n}{n+1}\mathrm{d}(\delta).$$

Theorem 10.3. *[Yao (2009, 2010)]*

1. $\mathrm{WSDS}(f)$ *is positively terminating* $\Longrightarrow f \in \mathrm{PSD}$.
2. $\mathrm{WSDS}(f)$ *is negatively terminating* $\Longleftrightarrow f \notin \mathrm{PSD}$.

3. WSDS(f) *does not terminate* $\Longrightarrow f \in$ PSD.

Proof.

1. WSDS(f) being positively terminating implies there exists m such that all polynomials of WDS$^{(m)}(f)$ are trivially nonnegative. By Lemma 10.4, $f \in$ PSD.

2. Necessity is evident by Lemma 10.4.
   Sufficiency. If $f \notin$ PSD, then there exists $\alpha \in \Delta_n$ such that $f(\alpha) < 0$. By the continuity of f, there exists a ball neighborhood $O(\alpha, \epsilon) \subset \Delta_n$ such that $f(\beta) < 0$ for all $\beta \in O(\alpha, \epsilon)$. We may choose a sequence of subsimplexes
   $$\Delta_n D_{\sigma_1} \supset \Delta_n D_{\sigma_1} D_{\sigma_2} \supset \cdots \supset \Delta_n D_{\sigma_1} \cdots D_{\sigma_k} \supset \cdots$$
   such that α is contained in all of the subsimplexes. By Lemma 10.5, there exists m such that
   $$O(\alpha, \epsilon) \supset \Delta_n D_{\sigma_1} \cdots D_{\sigma_m} \supset \cdots .$$
   So,
   $$f(\beta) < 0, \forall \beta \in \Delta_n D_{\sigma_1} \cdots D_{\sigma_m}.$$
   Let $g(\boldsymbol{x}) = f(D_{\sigma_1} \cdots D_{\sigma_m} \boldsymbol{x}^{\mathrm{T}}) \in$ WDS$^{(m)}$. Then $g(\boldsymbol{e}_j) < 0$ (indeed for all $1 \leq j \leq n$). Therefore WSDS(f) is negatively terminating.

3. This is a direct corollary from the second conclusion. $\qquad\square$

Corollary 10.4. *Algorithm* WSDS *is correct.*

10.2.3 *Termination*

In this section, we introduce the main result of [Hou and Shao (2011)] without proof, which gives a bound on the executed times of the loop in Algorithm WSDS and therefore proves the termination of the algorithm. Other analogous bounds were given in [Xu and Yao (2012); Han (2013)].

Theorem 10.4. *[Hou and Shao (2011)] Suppose*
$$f(x_1, ..., x_n) = \sum_{i_1 + \cdots + i_n = d} c_{i_1 \cdots i_n} x_1^{i_1} \cdots x_n^{i_n} \in \mathbb{Z}[x_1, ..., x_n]$$
is a form of degree d and
$$M = \max\{|c_{i_1 \cdots i_n}|\}, \quad \lambda = \min_{\boldsymbol{x} \in \Delta_n} f(\boldsymbol{x}).$$
Then

1. $f(\boldsymbol{x}) > 0$ for all $\boldsymbol{x} \in \Delta_n$ if and only if there exists $k \leq C_p(\lambda, M, n, d)$ such that the coefficients of every form in $\mathtt{WDS}^{(k)}(f)$ are all positive, where

$$C_p(\lambda, M, n, d) = \frac{\ln(2d^{d(n+1)}n^{d(n+2)}M) - \ln \lambda}{\ln n - \ln(n-1)}.$$

2. there exists $\alpha \in \Delta_n$ such that $f(\alpha) < 0$ if and only if there exists $k \leq C_n(\lambda, M, n, d)$ such that the coefficients of some form in $\mathtt{WDS}^{(k)}(f)$ are all negative, where

$$C_n(\lambda, M, n, d) = \frac{\ln(8d^{d(n+1)}n^{d(n+2)}M) - \ln |\lambda|}{\ln n - \ln(n-1)}.$$

Denote

$$\tilde{\Delta}_n = \{(x_1, ..., x_n)^{\mathrm{T}} \mid (x_1, ..., x_n) \in \mathbb{R}_+^n, \sum_{i=1}^{n} x_i \leq 1\}.$$

Lemma 10.6. *[Jeronimo and Perrucci (2010)] Suppose $f \in \mathbb{Z}[x_1, ..., x_n]$ is positive on $\tilde{\Delta}_n$. If the degree of f is d and the magnitudes of its coefficients are bounded by M, then*

$$\min_{\tilde{\Delta}_n} f \geq (2M)^{-d^{n+1}} \, d^{-(n+1)d^{n+1}}.$$

Lemma 10.7. *[Jeronimo and Perrucci (2010)] Suppose the minimum of $f \in \mathbb{Z}[x_1, ..., x_n]$ on $\tilde{\Delta}_n$ is not zero. If the degree of f is d and the magnitudes of its coefficients are bounded by M, then*

$$\left| \min_{\tilde{\Delta}_n} f \right| \geq (2M)^{-d^{n+1}} \, d^{-(n+1)d^{n+1}}.$$

Lemma 10.8. *[Hou and Shao (2011)] Suppose the minimum of the form $f \in \mathbb{Z}[x_1, ..., x_n]$ on Δ_n is not zero. If the degree of f is d and the magnitudes of its coefficients are bounded by M, then*

$$\min_{\Delta_n} f \geq (2M)^{-d^n} \, n^{-d^{n+1}-d} \, d^{-nd^n}$$

or

$$\min_{\Delta_n} f \leq -(2M)^{-d^{n+1}} \, d^{-(n+1)d^{n+1}}.$$

By Theorem 10.4 and Lemma 10.8, we have

Theorem 10.5. *[Hou and Shao (2011)] Suppose*

$$f(x_1, ..., x_n) = \sum_{i_1 + \cdots + i_n = d} c_{i_1 \cdots i_n} x_1^{i_1} \cdots x_n^{i_n} \in \mathbb{Z}[x_1, ..., x_n]$$

is a form of degree d and

$$M = \max\{|c_{i_1 \cdots i_n}|\}.$$

Then

1. $f(\boldsymbol{x}) > 0$ *for all* $\boldsymbol{x} \in \Delta_n$ *if and only if there exists* $k \le C_p(M, n, d)$ *such that the coefficients of every form in* $\mathrm{WDS}^{(k)}(f)$ *are all positive, where*

$$C_p(M, n, d) = \frac{(d^n + 1)\ln(2M) + (d^{n+1} + nd + 3d)\ln n + (nd^n + nd + d)\ln d}{\ln n - \ln(n-1)}.$$

2. *there exists* $\alpha \in \Delta_n$ *such that* $f(\alpha) < 0$ *if and only if there exists* $k \le C_n(M, n, d)$ *such that the coefficients of some form in* $\mathrm{WDS}^{(k)}(f)$ *are all negative, where*

$$C_n(M, n, d) = \frac{(d^{n+1} + 1)\ln(2M) + (n + 2)d\ln n + (n + 1)(d^{n+1} + d)\ln d + \ln 4}{\ln n - \ln(n-1)}.$$

By Theorem 10.5, we may modify Algorithm WSDS to get a complete algorithm for detecting nonnegativity of forms on $\mathbb{R}_+^n$.

Algorithm 10.3 Complete Weighted Successive Difference Substitution (CWSDS)

Input: A form $f \in \mathbb{Z}[\boldsymbol{x}]$

Output: "positive semi-definite" or "not positive semi-definite"

1: $k \leftarrow 0$; $\mathrm{WDS}^{(k)} \leftarrow \{f\}$; $C \leftarrow C_n(M, n, d)$;

2: **do**

3:    **if** $k \ge C$ **then return** "positive semi-definite" **end if**;

4:    delete the trivially nonnegative members of $\mathrm{WDS}^{(k)}$;

5:    **if** $\mathrm{WDS}^{(k)}$ is empty **then return** "positive semi-definite" **end if**;

6:    **if** $\exists g \in \mathrm{WDS}^{(k)}$ such that $g(\boldsymbol{e}_j) < 0$ for some j **then**

7:       **return** "not positive semi-definite"

8:    **end if**;

9:    $\mathrm{WDS}^{(k+1)} \leftarrow \bigcup_{g \in \mathrm{WDS}^{(k)}} \bigcup_{\sigma \in SP_n} g(D_\sigma \boldsymbol{x}^{\mathrm{T}})$;

10:    $k \leftarrow k + 1$;

11: **end do**

Theorem 10.6. *Algorithm* CWSDS *terminates correctly.*

Proof.   Termination is obvious. Correctness is guaranteed by Theorem 10.5. $\qquad\square$

Example 10.12. (Example 10.11 continued) Determine the nonnegativity of the following polynomial,

$$f = (a^2 + b^2 + c^2)^2 - 4\,(a - b)(b - c)(c - a)(a + b + c).$$

The steps of Algorithm CWSDS are bounded by $C_n(M, n, d)$. In the case of f, we have $M = 4$, $d = 4$, $n = 3$ and

$$C_n(4, 3, 4) = 4931.425937\cdots.$$

Since we have known that WSDS(f) runs up to five thousand steps without outputs, CWSDS(f) returns at Line 3, *i.e.* the computational steps exceed the bound C_n and f is positive semi-definite.

Note that, by Theorem 10.5, we may modify Algorithm CWSDS a little so that it can further decide whether a form is positive definite or positive semi-definite. The computational steps of the modified algorithm are bounded by $C(M, n, d) = \max(C_n(M, n, d), C_p(M, n, d))$. That is, after $C(M, n, d)$ iterations,

1. If the coefficients of every form in WDS$^{(C(M,n,d))}(f)$ are all positive, then $f \in$ PD;
2. If there exists a form in WDS$^{(C(M,n,d))}(f)$ such that its coefficients are all negative, then $f \notin$ PSD;
3. Otherwise, $f \in$ PSD\PD.

10.3   Examples

A prototype tool, called "tsds5", implementing Algorithms SDS and WSDS using Maple can be downloaded at "http://pan.baidu.com/s/1o68GDaM". We list some more examples solved automatically by the tool. In the following examples, all variables are in $\mathbb{R}_+$ if not specified. The first four were regarded as "open inequalities" in the book [Cîrtoaje *et al.* (2009)].

Example 10.13 (Problem 7.2 of [Cîrtoaje *et al.* (2009)]).

$$\frac{a}{b} + \frac{b}{c} + \frac{c}{a} - \frac{6(a-c)^2}{(a+b+c)^2} - 3 \geq 0.$$

Example 10.14 (Problem 7.3 of [Cîrtoaje *et al.* (2009)]).

$$\frac{a^2}{b^2} + \frac{b^2}{c^2} + \frac{c^2}{a^2} + \frac{10(ab+bc+ca)}{a^2+b^2+c^2} - 13 \geq 0.$$

Example 10.15 (Problem 7.14 of [Cîrtoaje *et al.* (2009)]).

$$\frac{a-b}{b+c} + \frac{b-c}{c+d} + \frac{c-d}{d+e} + \frac{d-e}{e+a} + \frac{e-a}{a+b} \geq 0.$$

Example 10.16 (Problem 7.15 of [Cîrtoaje *et al.* (2009)]).

$$\frac{a-b}{a+2b+c} + \frac{b-c}{b+2c+d} + \frac{c-d}{c+2d+e} + \frac{d-e}{d+2e+a} + \frac{e-a}{e+2a+b} \geq 0.$$

The inequality does not hold. The modified Algorithm WSDS outputs a counterexample as follows:

$$a = \frac{6680627611}{12960000}, b = \frac{1}{625}, c = \frac{965041}{12960000}, d = \frac{5576431}{12960000}, e = \frac{2101}{160000}.$$

Example 10.17. An example with radical coefficients.

$$2(x^3 y^3 + y^3 z^3 + z^3 x^3) - 3\sqrt[3]{2}\, xyz(x^2 y + y^2 z + z^2 x) - (6 - 9\sqrt[3]{2}\,)x^2 y^2 z^2 \geq 0.$$

SDS positively terminates on the input in 77.86 seconds, running 1011 steps, while WSDS positively terminates in 101.43 seconds, running 276 steps. The computation was executed on a PC (Intel Core i7-3770 CPU @ 3.40GHz, 8G memory, Windows 7 OS) with Maple 17.

Example 10.18 (Motzkin).

$$z^6 + x^4 y^2 + y^4 x^2 - 3x^2 y^2 z^2 \geq 0.$$

Example 10.19 (Vasc's Conjecture with 6 variables).

$$\frac{a_1 - a_2}{a_2 + a_3} + \frac{a_2 - a_3}{a_3 + a_4} + \frac{a_3 - a_4}{a_4 + a_5} + \frac{a_4 - a_5}{a_5 + a_6} + \frac{a_5 - a_6}{a_6 + a_1} + \frac{a_6 - a_1}{a_1 + a_2} \geq 0$$

where $a_i > 0$, $(i = 1, \ldots, 6)$. It is easy to see that Example 10.15 is a special case of Vasc's Conjecture when the number of variables is 5.

The inequality does not hold. An output counterexample is

$$a_1 = 84,\ a_2 = 7,\ a_3 = 79,\ a_4 = 5,\ a_5 = 76,\ a_6 = 1.$$

Let n be the number of variables in Vasc's Conjecture. By SDS method, we have proven that the inequality holds for $n = 3, 4, 5, 7$.

Example 10.20 (Vasc's Conjecture with 7 variables).

$$\frac{a_1 - a_2}{a_2 + a_3} + \frac{a_2 - a_3}{a_3 + a_4} + \frac{a_3 - a_4}{a_4 + a_5} + \frac{a_4 - a_5}{a_5 + a_6} + \frac{a_5 - a_6}{a_6 + a_7} + \frac{a_6 - a_7}{a_7 + a_1} + \frac{a_7 - a_1}{a_1 + a_2} \geq 0$$

where $a_i > 0$ for $i = 1, \ldots, 7$.

Chen (2008) made use of the *Parallel Successive Difference Substitution* to verify this inequality. It runs on 18 nodes of a set of HP Proliant DL360, Intel xeon 2.8GHz, 2GB memory, Windows 2003 Server, with Maple 10, and proved that the inequality holds in CPU time 9128.92 seconds.

Example 10.21. [Hou *et al.* (2010)] Suppose $x \geq 0, y \geq 0, z \geq 0$ and at least two of x, y, z are nonzero. What is the least m such that

$$\frac{2}{3}\left(\frac{x^2}{y+z} + \frac{y^2}{z+x} + \frac{z^2}{x+y}\right) - \left(\frac{x^m + y^m + z^m}{3}\right)^{\frac{1}{m}} \geq 0 \qquad (10.3)$$

does not hold?

By SDS method, we know that (10.3) holds for $m = 1, ..., 5$ and does not hold for $m = 6$. An output counterexample is

$$x = \frac{2159}{5832}, y = \frac{3685}{11664}, z = \frac{3661}{11664}.$$

Example 10.22. Let

$$\mathrm{LL}_n = \sum_{i=1}^{n} \prod_{j \neq i}(x_i - x_j).$$

For $n = 5$, Lax and Lax (1978) proved that LL_5 is nonnegative but not SOS. Safey El Din (2008) compared three methods (Algo, SOS, CAD) on determining the nonnegativity of $\mathrm{LL}_5, \mathrm{LL}_7$ and LL_9. Algo and CAD are symbolic methods and SOS is numerical.

According to [Safey El Din (2008)], CAD could not get output on any of the three examples within a week. SOS worked out LL_5 and LL_7 in several seconds. The timings of Algo on LL_5 and LL_7 are 67 seconds and 10 hours, respectively. Algo and SOS both failed to get output for LL_9 within a week.

By Algorithm SDS, LL_7 is proved to be not nonnegative on a PC (Intel Core i7-3770 CPU @ 3.40GHz, 8G memory, Windows 7 OS) with Maple 17 in 2.46 seconds with a counterexample

$$x_1 = 132, x_2 = 128, x_3 = 125, x_4 = 64, x_5 = 4, x_6 = 2, x_7 = 1.$$

LL_9 is also proved to be not nonnegative in 38162 seconds with a counterexample

$$x_1 = 804, x_2 = 797, x_3 = 791, x_4 = 786,$$
$$x_5 = 782, x_6 = 7, x_7 = 4, x_8 = 2, x_9 = 1.$$

The program of (complete weighted) successive difference substitute can also be used as an interactive tool to prove some complicated inequalities or inequalities involving negative variables. See for example [Yang (2005); Yang and Xia (2008); Huang (2016)].

10.4   Pólya's Theorem

Suppose $f \in \mathbb{R}[x_1, ..., x_n]$ is a form of degree d. We transform f by the following simultaneous replacement on the variables:

$$x_i \leftarrow (x_1 + \cdots + x_n)x_i, \text{ for } i = 1, ..., n.$$

It is clear that, after the transformation, the resulting polynomial is

$$(x_1 + \cdots + x_n)^d f(x_1, ..., x_n).$$

Let us call the replacement on the variables *p-substitution*. Now, if we apply p-substitution again to the resulting polynomial, we will get

$$(x_1 + \cdots + x_n)^{3d} f(x_1, ..., x_n).$$

Generally, if p-substitution is iterated k times (the process may be called *successive p-substitution*), the resulting polynomial will be

$$(x_1 + \cdots + x_n)^m f(x_1, ..., x_n),$$

where $m = (2^k - 1)d$.

Difference substitution is a kind of linear replacements on variables while p-substitution is non-linear. By Pólya's Theorem, p-substitution can also be used for detecting nonnegativity of forms.

Theorem 10.7. *[Pólya (1928)][Theorem 56 of [Hardy et al. (1952)]] If the form $f \in \mathbb{R}[x_1, ..., x_n]$ is positive definite on Δ_n, then there exists a positive integer N such that all the coefficients of*

$$(x_1 + \cdots + x_n)^N f$$

are strictly positive.

Naturally, one may use Pólya's theorem to design an incomplete algorithm verifying whether a form f is positive, *i.e.* multiply f by

$$(x_1 + \cdots + x_n)^i \text{ for } i = 1, 2, ...,$$

and check whether all the coefficients of the resulting polynomials are strictly positive. In some sense, this method based on Pólya's theorem and the SDS method are somehow similar, *e.g.* it keeps transforming the polynomial and checking the signs of all the coefficients.

The above naive method based on Pólya's theorem has been greatly improved, see for example [Powers and Reznick (2001); Castle *et al.* (2009, 2011)], and the improvements have been applied to other topics, see for example [Schweighofer (2002)]. In this section, we list some results in the above references without proof.

An explicit bound for the exponent N in terms of the degree, the size of the coefficients, and the minimum of f on the simplex was given by Theorem 1 of [Powers and Reznick (2001)]. That means it can be determined by Pólya's theorem within a finite number of steps whether a given form is positive definite on Δ_n.

How to characterize positive semi-definite forms which can be proved by Pólya's theorem? Such a characterization and a bound for the exponent N was given in [Castle *et al.* (2011)].

For $I \subseteq \{1, ..., n\}$, denote by $F(I)$ the *face* of Δ_n given by

$$F(I) = \{(x_1, ..., x_n)^{\mathrm{T}} \in \Delta_n \mid x_i = 0, i \in I\}.$$

The *relative interior* of the face $F(I)$ is the set

$$\{(x_1, ..., x_n)^{\mathrm{T}} \in F(I) \mid x_j > 0, j \in \{1, ..., n\} \setminus I\}.$$

Definition 10.10. [Castle *et al.* (2011)] Suppose $\alpha = (a_1, ..., a_n), \beta = (b_1, ..., b_n) \in \mathbb{N}^n$.

1. We write $\alpha \preceq \beta$ if $a_i \geq b_i$ for all i, and $\alpha \prec \beta$ if $\alpha \preceq \beta$ and $\alpha \neq \beta$.
2. For $\alpha = (a_1, ..., a_n) \in \mathbb{N}^n$ and a face $F = F(I)$ of Δ_n, let α_F denote $(\tilde{a}_1, ...\tilde{a}_n)$ where $\tilde{a}_i = a_i$ for $i \in I$ and $\tilde{a}_i = 0$ for $i \notin I$.
3. Assume form $f = \sum_{\alpha \in \mathbb{N}^n} c_\alpha x^\alpha$. Recall that the support of f is $\mathsf{S}(f) = \{\alpha \mid c_\alpha \neq 0\}$. Define

$$\mathsf{S}^+(f) = \{\alpha \in \mathsf{S}(f) \mid c_\alpha > 0\}, \quad \mathsf{S}^-(f) = \{\alpha \in \mathsf{S}(f) \mid c_\alpha < 0\}.$$

4. For a form f and a face F of Δ_n, we say that $\alpha \in \mathsf{S}^+(f)$ is minimal with respect to F if there is no $\beta \in \mathsf{S}^+(f)$ such that $\beta_F \prec \alpha_F$.
5. For $\Gamma \subseteq \mathsf{S}(f)$, denote $f(\Gamma) = \sum_{\gamma \in \Gamma} c_\gamma x^\gamma$. For $\alpha \in \mathsf{S}(f)$ and a face F of Δ_n, define

$$f(\alpha, F) = f(\{\gamma \in \mathsf{S}(f) \mid \gamma_F = \alpha_F\})/x^{\alpha_F}.$$

Theorem 10.8. *[Castle* et al. *(2011)] Suppose f is a nonzero form of degree d such that $f \geq 0$ on Δ_n and $\mathsf{V}_{\mathbb{R}}(f) \cap \Delta_n$ is a union of faces. Then there exists a positive integer N such that all the coefficients of $(x_1 + \cdots + x_n)^N f$ are nonnegative if and only if for every face $F \subseteq \mathsf{V}_{\mathbb{R}}(f)$ the following two conditions hold:*

1. *For every $\beta \in \mathsf{S}^-(f)$, there is $\alpha \in \mathsf{S}^+(f)$ such that $\alpha_F \preceq \beta_F$.*
2. *For every $\alpha \in \mathsf{S}^+(f)$ which is minimal with respect to F, the form $f(\alpha, F)$ is strictly positive on the relative interior of F.*

A bound for the exponent N was given by Theorem 3 of [Castle *et al.* (2011)]. Theorem 10.8 means that (1) Pólya's theorem is also applicable to a class of positive semi-definite forms; (2) Pólya's theorem is not a complete method for determining nonnegativity of polynomials. Because Theorem 10.8 gives the necessary and sufficient condition, it is not hard to find some non-strict inequalities which cannot be proved by Pólya's theorem.

Example 10.23 (Robinson).

$$p = x^6 + y^6 + z^6 - x^4y^2 - y^4x^2 - x^4z^2 - z^4x^2 - z^4y^2 - y^4z^2 + 3x^2y^2z^2 \geq 0.$$

It is easy to verify that p has a real zero, $(\frac{1}{3}, \frac{1}{3}, \frac{1}{3})$, in the interior of Δ_3. So, by Theorem 10.8, there does not exist a positive integer N such that all the coefficients of $(x_1 + \cdots + x_n)^N p$ are nonnegative.

For many geometric (non-strict) inequalities on triangles, the condition for the equalities hold is that the triangle is equilateral. So the corresponding forms all have a real zero, $(\frac{1}{3}, \frac{1}{3}, \frac{1}{3})$, in the interior of Δ_3. Therefore this kind of inequalities cannot be proved by Pólya's theorem.

Chapter 11

Proving Inequalities Beyond the Tarski Model

Propositions of elementary algebra and geometry (Tarski's model) are decidable by Tarski's work. For any given proposition inside Tarski's model, all the functions involved are polynomials in a fixed number of variables. In this chapter, we consider inequality proving problems where the number of variables is a variable or the functions are not polynomials, *i.e.* propositions beyond the Tarski model.

To have an idea of what we will talk about in this chapter, let us first see two concrete examples below.

Example 11.1. Given a polynomial in n variables

$$
f = -\left(\sum_{k=1}^{n} x_k^5\right) - 6\left(\sum_{k=1}^{n} x_k^4\right)\left(\sum_{k=1}^{n} x_k\right) + 2\left(\sum_{k=1}^{n} x_k^3\right)\left(\sum_{k=1}^{n} x_k^2\right) + 8\left(\sum_{k=1}^{n} x_k^3\right)
$$

$$
\cdot \left(\sum_{k=1}^{n} x_k\right)^2 + 3\left(\sum_{k=1}^{n} x_k^2\right)^2\left(\sum_{k=1}^{n} x_k\right) - 6\left(\sum_{k=1}^{n} x_k^2\right)\left(\sum_{k=1}^{n} x_k\right)^3 + \left(\sum_{k=1}^{n} x_k\right)^5,
$$

(11.1)

where $x_i \in \mathbb{R}_+$, $i = 1, \ldots, n$, and $\mathbb{R}_+$ stands for nonnegative real numbers. Decide whether or not $f \geq 0$ for all positive integers n.

Example 11.2. Does the following integral inequality (11.2) hold for any function $g(s)$ which is integrable on $[0, 1]$?

$$
G = \int_0^1 |g(s)|^7 ds \int_0^1 |g(s)| ds - 3 \int_0^1 g^6(s) ds \int_0^1 g^2(s) ds
$$

$$
+ 3 \int_0^1 |g(s)|^5 ds \int_0^1 |g(s)|^3 ds - \left(\int_0^1 g^4(s) ds\right)^2 \geq 0.
$$

(11.2)

By the definition of the Riemann integral, the problem in Example 11.2

297

can be transformed into proving the following inequality.

$$\sum_{i=1}^{n} \frac{x_i^7}{n} \sum_{i=1}^{n} \frac{x_i}{n} - 3 \sum_{i=1}^{n} \frac{x_i^6}{n} \sum_{i=1}^{n} \frac{x_i^2}{n} + 3 \sum_{i=1}^{n} \frac{x_i^5}{n} \sum_{i=1}^{n} \frac{x_i^3}{n} - \left(\sum_{i=1}^{n} \frac{x_i^4}{n}\right)^2 \geq 0. \quad (11.3)$$

A special feature of these two examples is that the number of variables is not fixed. So, the problem cannot be characterized inside Tarski's model. In this chapter, we prove that (1) the nonnegativity of symmetric forms of degree less than 5 (*e.g.* Example 11.1) is decidable; and (2) the nonnegativity of a class of symmetric forms of any degrees (*e.g.* Example 11.2) is also decidable. The second result can be applied to proving a class of integral inequalities.

The main content of this chapter is from [Yang *et al.* (2007, 2010a)].

11.1 Symmetric Forms of Degrees Less Than Five

11.1.1 *Problem*

Problem Is the following problem decidable?

For any given symmetric homogeneous polynomial $f(x_1, \ldots, x_n)$ with rational coefficients, whether or not $f(x_1, \ldots, x_n) \geq 0$ holds on $\mathbb{R}_+^n$ for all positive integers n?

We first introduce some basic concepts and known results.

Definition 11.1. Homogeneous polynomial $f(x_1, \ldots, x_n)$ is said to be *symmetric*, if

$$f(x_1, \ldots, x_n) = f(\sigma(x_1, \ldots, x_n))$$

for all $\sigma \in SP_n$ where SP_n is the symmetric group in n symbols (see also Notation 7.2).

We denote by $S_{n,m}$ the set of symmetric forms of degree m in n variables with real coefficients. Under addition and scalar multiplication, $S_{n,m}$ is a vector space whose dimension is denoted by $\dim(S_{n,m})$.

Definition 11.2. For a positive integer k and $x = (x_1, \ldots, x_n) \in \mathbb{R}^n$, define

$$P_{(n,k)}(x) = \sum_{j=1}^{n} x_j^k.$$

Lemma 11.1 (Basic Theorem of Symmetric Forms). *Any symmetric form $f(x_1, \ldots, x_n) \in S_{n,m}$ can be expressed uniquely as a polynomial in $P_{(n,1)}, \ldots, P_{(n,d)}$ where $d = \min(n, m)$. Moreover, $P_{(n,1)}, \ldots, P_{(n,d)}$ are algebraic independent, i.e. there does not exist a nonzero polynomial g such that $g(P_{(n,1)}, \ldots, P_{(n,d)}) = 0$.*

Denote by Ω the set of nonnegative integer solutions to the Diophantine equation

$$y_1 + 2y_2 + \cdots + dy_d = m.$$

Lemma 11.2. *The set $B_{n,m} = \{P_{(n,1)}^{\lambda_1} P_{(n,2)}^{\lambda_2} \cdots P_{(n,d)}^{\lambda_d} | (\lambda_1, \lambda_2, \ldots, \lambda_d) \in \Omega\}$ where $d = \min(n, m)$ is a base of vector space $S_{n,m}$ and the dimension of $S_{n,m}$ equals the number of elements in Ω.*

By Lemma 11.2, any symmetric form of degree 5 can be expressed as

$$
\begin{aligned}
g &= aP_{(n,5)} + bP_{(n,4)}P_{(n,1)} + cP_{(n,3)}P_{(n,2)} + dP_{(n,3)}P_{(n,1)}^2 + eP_{(n,2)}^2 P_{(n,1)} \\
&\quad + \alpha P_{(n,2)} P_{(n,1)}^3 + \beta P_{(n,1)}^5 && (11.4) \\
&= [a, b, c, d, e, \alpha, \beta]_P. && (11.5)
\end{aligned}
$$

For example, the form in Example 11.1 can be denoted simply by

$$f = [-1, -6, 2, 8, 3, -6, 1]_P.$$

Notation 11.1.

- $1_k = \underbrace{(1, 1, \ldots, 1)}_{k}; \quad 0_k = \underbrace{(0, 0, \ldots, 0)}_{k};$
- $\lfloor d \rfloor$ stands for the biggest integer smaller than or equal to d;
- For any $x = (x_1, \ldots, x_n) \in \mathbb{R}^n$,

$$v(x) = |\{x_j | j = 1, \ldots, n\}|,$$

$$v(x)^* = |\{x_j | x_j \neq 0, j = 1, \ldots, n\}|,$$

where $|A|$ denotes the number of elements in A, i.e. $v(x)$ is the number of distinct coordinates of x and $v(x)^*$ is the number of distinct nonzero coordinates of x.

Lemma 11.3. *[Timofte (2003, 2005)]*
(1) An inequality of a symmetric form of degree d in n variables is valid on $\mathbb{R}_+^n$ if and only if it is valid on the set $\{x | x \in \mathbb{R}_+^n, v(x)^* \leq \max(\lfloor \frac{d}{2} \rfloor, 1)\}$.
(2) An inequality of a symmetric form of degree d in n variables is valid on $\mathbb{R}^n$ if and only if it is valid on the set $\{x | x \in \mathbb{R}^n, v(x) \leq \max(\lfloor \frac{d}{2} \rfloor, 2)\}$.

Lemma 11.3 is important since it means that a problem concerning a multivariate polynomial can be reduced to some similar problems with less variables. A different method for reducing the number of variables can be found in [Wen *et al.* (2003); Gao and Wen (2013)].

Let $d = 3$, a corollary of Lemma 11.3 is the following famous result:

Lemma 11.4. *[Choi et al. (1987)] For any $f \in S_{n,3}$, $f \geq 0$ is valid on $\mathbb{R}^n_+$ if and only if $f(1_k, 0_{n-k}) \geq 0$ holds for every $k = 1, \ldots, n$.*

So, to prove an inequality of a symmetric form of degree 3, we only need to verify the inequality at the following n points

$$(1_k, 0_{n-k}), \ k = 1, \ldots, n.$$

Lemma 11.5. *For any symmetric form $f \in S_{n,p}$, where $p \in \{4, 5\}$,*

$$\forall x \in \mathbb{R}^n_+ (f(x) \geq 0) \Longleftrightarrow \forall t \in \mathbb{R}_+ \forall (r, s) \in N_n (f(t \cdot 1_r, 1_s, 0_{n-r-s}) \geq 0),$$

where $N_n = \{(r, s) \mid r, s \text{ are positive integers and } r + s \leq n\}$.

Proof. By Lemma 11.3,

$$f(x) \geq 0, x \in \mathbb{R}^n_+ \Longleftrightarrow f(y) \geq 0, \forall y \in \mathbb{R}^n_+ \text{ and } v^*(y) \leq 2.$$

Since f is symmetric, we may let $y = (t_1 \cdot 1_r, t_2 \cdot 1_s, 0_{n-r-s}), t_1, t_2 \in \mathbb{R}_+$. Note that f is a homogeneous polynomial, we have

$$f(y) \geq 0, \forall y \in \mathbb{R}^n_+ \text{ and } v^*(y) \leq 2$$
$$\Longleftrightarrow (t_2)^p f(\frac{t_1}{t_2} \cdot 1_r, 1_s, 0_{n-r-s}) \geq 0, \forall t_1, t_2 \in \mathbb{R}_+ \text{ and } t_2 \neq 0, \forall (r, s) \in N_n$$
$$\Longleftrightarrow f(\frac{t_1}{t_2} \cdot 1_r, 1_s, 0_{n-r-s}) \geq 0, \forall t_1, t_2 \in \mathbb{R}_+ \text{ and } t_2 \neq 0, \forall (r, s) \in N_n$$
$$\Longleftrightarrow f(t \cdot 1_r, 1_s, 0_{n-r-s}) \geq 0, \forall t \in \mathbb{R}_+, \forall (r, s) \in N_n.$$

That completes the proof. $\qquad\square$

Set

$$\begin{aligned}
f_{r,s}(t) = &\, a(rt^5 + s) + b(rt^4 + s)(rt + s) + c(rt^3 + s)(rt^2 + s) \\
&+ d(rt^3 + s)(rt + s)^2 + e(rt^2 + s)^2(rt + s) \\
&+ \alpha(rt^2 + s)(rt + s)^3 + \beta(rt + s)^5.
\end{aligned}$$

Expand $f_{r,s}(t)$ and collect it as a polynomial in t:

$$f_{r,s}(t) = A_{r,s}t^5 + B_{r,s}t^4 + C_{r,s}t^3 + D_{r,s}t^2 + E_{r,s}t + H_{r,s}, \qquad (11.6)$$

where

$$\begin{cases} A_{r,s} = r(\beta r^4 + \alpha r^3 + (d+e)r^2 + (b+c)r + a), \\ B_{r,s} = sr(5\beta r^3 + 3\alpha r^2 + (2d+e)r + b), \\ C_{r,s} = sr((\alpha + 10s\beta)r^2 + (3\alpha s + 2e)r + ds + c), \\ D_{r,s} = sr((\alpha + 10r\beta)s^2 + (3\alpha r + 2e)s + dr + c), \\ E_{r,s} = sr(5\beta s^3 + 3\alpha s^2 + (2d+e)s + b), \\ H_{r,s} = s(\beta s^4 + \alpha s^3 + (d+e)s^2 + (b+c)s + a). \end{cases} \tag{11.7}$$

By Eq. (11.4), $f_{r,s}(t)$ is the value of symmetric form $f(x_1, \ldots, x_n)$ of degree 5 at a point whose r coordinates are t, s coordinates are 1, and the others are 0.

By Lemma 11.5, we have

Lemma 11.6. *Suppose $f(x_1, \ldots, x_n) \in S_{n,5}$ and $(x_1, \ldots, x_n) \in \mathbb{R}^n_+$. Then $f(x_1, \ldots, x_n) \geq 0$ holds for all positive integers n if and only if $f_{r,s}(t) \geq 0$ holds for all positive integers r, s and all $t \in \mathbb{R}_+$.*

11.1.2   *Algorithm*

Theorem 11.1. *The following problem is decidable:*

For any given symmetric form $g(x_1, \ldots, x_n)$ of degree no bigger than 5 with rational coefficients, does $g(x_1, \ldots, x_n) \geq 0$ hold on $\mathbb{R}^n_+$ for all positive integers n?

We prove the theorem by directly giving an algorithm solving the problem and then proving the correctness and termination of the algorithm. Note that Lemma 11.6 cannot be used directly to design an algorithm for proving Theorem 11.1 since it needs to check infinitely many pairs of (r, s). Our basic idea is: reduce the problem to one inside the Tarski model.

We only prove the theorem for the case that degree is 5. A symmetric form of degree $i < 5$ can be transformed to a symmetric form of degree 5 by multiplying a factor $(x_1 + \cdots + x_n)^{5-i}$.

Lemma 11.7. *Assume x_0 is any real root of a real polynomial*

$$g(x) = x^n + a_{n-1}x^{n-1} + \cdots + a_0,$$

then $x_0 \leq 1 + \max\{|a_0|, \ldots, |a_{n-1}|\}$.

An obvious corollary is that, for any $x > 1 + \max\{|a_0|, \ldots, |a_{n-1}|\}$, $g(x) > 0$.

Lemma 11.8. *Suppose* $f(x_1, \ldots, x_n) = [a, b, c, d, e, \alpha, \beta]_P \in S_{n,5}$. *If* $f(x_1, \ldots, x_n) \geq 0$ *on* $\mathbb{R}^n_+$ *for all positive integers* n, *then*

$$\beta > 0 \text{ or}$$
$$\beta = 0, \ \alpha > 0 \text{ or}$$
$$\beta = \alpha = 0, \ d + e > 0, \ d \geq 0 \text{ or}$$
$$\beta = \alpha = d + e = 0, \ d > 0, \ b + c > 0 \text{ or}$$
$$\beta = \alpha = d = e = 0, \ b + c > 0, \ b \geq 0 \text{ or}$$
$$\beta = \alpha = d = e = b + c = 0, \ b > 0 \text{ or}$$
$$\beta = \alpha = d = e = b = c = 0, \ a \geq 0.$$

Proof. By Lemma 11.6, $f_{r,s}(t) \geq 0$ for any positive integers r, s and any $t \in \mathbb{R}_+$. Especially, for any positive integer s, we have $f_{r,s}(0) \geq 0$, *i.e.* $H_{r,s} \geq 0$. If s is large enough, the sign of $H_{r,s}$ is determined by the leading coefficient of $H_{r,s}$, *i.e.* $H_{r,s}$ and β have the same sign if s is large enough. Thus $\beta \geq 0$.

If $\beta = 0$, we may obtain $\alpha \geq 0$ similarly.

If $\beta = \alpha = 0$, we may obtain $d + e \geq 0$ similarly. We claim that $d \geq 0$ holds, too. Set $r = 1, s = t^2$ in Eq. (11.7), then

$$f_{r,s}(t) = (t + 1)t^2(dt^4 + (2d + b + 4e)t^3 + (a + 2c + d)t^2 + (b - a)t + a).$$

It is clear that $f_{r,s}(t)$ and d have the same sign if t is large enough. So $d \geq 0$.

The other cases can be discussed similarly. $\qquad\square$

Lemma 11.9. *Suppose* $f(x_1, \ldots, x_n) = [a, b, c, d, e, \alpha, \beta]_P \in S_{n,5}$. *If* $f(x_1, \ldots, x_n) \geq 0$ *holds on* $\mathbb{R}^n_+$ *for all positive integers* n, *then there exist real numbers* r_0, s_0 *such that* $f_{r,s}(t) \geq 0$ *for any real numbers* $r > r_0, \ s > s_0$ *and* $t \in \mathbb{R}_+$.

Proof. (Case I) $[\beta > 0]$ or $[\beta = 0, \ \alpha > 0]$.

Since $[\beta > 0]$ or $[\beta = 0, \ \alpha > 0]$, by Lemma 11.7, there exist positive real numbers $r_1, \ s_1$ such that $A_{r,s} \geq 0, \ B_{r,s} \geq 0, \ E_{r,s} \geq 0, \ H_{r,s} \geq 0$ whenever $r > r_1, \ s > s_1$.

From the expression of $C_{r,s}, D_{r,s}$ we see that r and s can be exchanged. So, without loss of generality, we suppose $r \geq s$. Set $r = s + y \ (y \in R_+)$ and substitute it in $C_{r,s}, D_{r,s}$, we have

$$C_{r,s} = (10\beta s + \alpha)y^2 + (20\beta s^2 + 5\alpha s + 2e)y + 10\beta s^3 + 4\alpha s^2 + (d + 2e)s + c,$$

$$D_{r,s} = (10\beta s^2 + 3\alpha s + d)y + 10\beta s^3 + 4\alpha s^2 + (d + 2e)s + c.$$

By Lemma 11.7 again, there exists a positive real number s_2 such that, for any $s > s_2$,

$$\begin{cases} 10\beta s + \alpha \geq 0, \\ 20\beta s^2 + 5\alpha s + 2e \geq 0, \\ 10\beta s^3 + 4\alpha s^2 + (d + 2e)s + c \geq 0, \\ 10\beta s^2 + 3\alpha s + d \geq 0, \\ 10\beta s^3 + 4\alpha s^2 + (d + 2e)s + c \geq 0. \end{cases}$$

Because $y \in R_+$, for any $r > s_2, s > s_2$, we have $C_{r,s} \geq 0$ and $D_{r,s} \geq 0$.

Let $r_0 = s_0 = \max\{r_1,\ s_1,\ s_2\}$, for $r > r_0, s > s_0$, we have

$$A_{r,s} \geq 0, B_{r,s} \geq 0, C_{r,s} \geq 0, D_{r,s} \geq 0, E_{r,s} \geq 0, H_{r,s} \geq 0.$$

Note that $t \in R_+$, so $f_{r,s}(t) \geq 0$.

(Case II) $[\beta = \alpha = 0,\ d + e > 0,\ d \geq 0]$.

Since $d + e > 0$, we assume $e = h - d$ where $h > 0$. Substitute it in $f_{r,s}(t)$:

$$f_{r,s}(t) = dsrt(rt + s)(t - 1)^2 + A_{r,s}^{(1)}t^5 + B_{r,s}^{(1)}t^4 + C_{r,s}^{(1)}t^3 + D_{r,s}^{(1)}t^2$$
$$+ E_{r,s}^{(1)}t + H_{r,s}^{(1)},$$
$$A_{r,s}^{(1)} = r(hr^2 + (b + c)r + a),\ \ H_{r,s}^{(1)} = s(hs^2 + (b + c)s + a),$$
$$B_{r,s}^{(1)} = rs(hr + b),\ \ E_{r,s}^{(1)} = rs(hs + c),$$
$$C_{r,s}^{(1)} = rs(2hr + c),\ \ D_{r,s}^{(1)} = rs(2hs + b).$$

By Lemma 11.7, there exist positive real numbers r_0, s_0 such that, for any $r > r_0$ and $s > s_0$,

$$A_{r,s}^{(1)} \geq 0,\ B_{r,s}^{(1)} \geq 0,\ C_{r,s}^{(1)} \geq 0,\ D_{r,s}^{(1)} \geq 0,\ E_{r,s}^{(1)} \geq 0,\ H_{r,s}^{(1)} \geq 0.$$

Since $dsrt(rt + s)(t - 1)^2 \geq 0$, for any $r > r_0, s > s_0$, we have

$$f_{r,s}(t) \geq 0.$$

(Case III) $[\beta = \alpha = d + e = 0,\ d > 0,\ b + c > 0]$.

Assume $h = b + c > 0$, i.e. $b = h - c$, and $e = -d$. Substitute in $f_{r,s}(t)$:

$$f_{r,s}(t) = rst(t - 1)^2[(dr - c)t + ds - c] + (hr^2 + ar)t^5 + hrst^4$$
$$+ hrst + (hs^2 + as).$$

By Lemma 11.7, there exist positive real numbers r_0, s_0, for any $r > r_0$ and $s > s_0$,

$$(dr - c) \geq 0,\ (ds - c) \geq 0,\ (hr^2 + ar) \geq 0,\ hrs \geq 0,\ (hs^2 + as) \geq 0.$$

Thus, for any $r > r_0, s > s_0$, we have $f_{r,s}(t) \geq 0$.

(Case IV) For the last three cases of Lemma 11.8, we may use similar discussion to complete the proof. $\qquad\square$

Lemma 11.10. *If $f(x, u) \geq 0$ for all positive integers u and all $x \in \mathbb{R}_+$, then there exists positive real number u_0 such that, for any real number $u \geq u_0$, $f(x, u) \geq 0$ for all $x \in \mathbb{R}_+$.*

Proof. We regard $f(x, u)$ as a univariate polynomial in x with parameter u. Then

$$f(x, u) = p_m(u)x^m + p_{m-1}(u)x^{m-1} + \cdots + p_0(u),$$

where $p_m(u)$ is not a zero polynomial.

Since $p_m(u)$ is not a zero polynomial, either $p_m(u) > 0$ or $p_m(u) < 0$ for all sufficiently large u. By the premise of the lemma, $f(x, u) \geq 0$ for all $x \in \mathbb{R}_+$ and all positive integers u. So the leading coefficient of $f(x, u)$ must be nonnegative. Therefore $p_m(u) > 0$. That means there exists a real number u_1 such that, for any real number $u > u_1$, $p_m(u) > 0$.

Set $\Delta = \mathrm{discrim}(f(x^2, u), x)$ be the discriminant of $f(x^2, u)$ with respect to x. Obviously Δ is a univariate polynomial in u (may be zero polynomial). If u is large enough, Δ and its leading coefficient have the same sign. That is, there exists a positive real number u_2 such that, for any real number $u \geq u_2$, the sign of Δ is invariant.

Let $u_0 = \max\{u_1, u_2\}$. By Lemma 3.1, the number and multiplicities of real roots of $f(x^2, u)$, viewed as a univariate polynomial in x, are invariant in the interval $(u_0, +\infty)$. On the other hand, we know by the premise that

$$f(x^2, \lfloor u_0 \rfloor + 1) \geq 0.$$

So the equation $f(x^2, \lfloor u_0 \rfloor + 1) = 0$ has no real roots with odd multiplicities. Combining the above facts, we have that the equation $f(x^2, u) = 0$ has no real roots with odd multiplicities when $u \geq u_0$. Because $p_m(u) > 0$ for $u \geq u_0$,

$$f(x^2, u) \geq 0 \ \text{for } u \geq u_0,$$

i.e. $f(x, u) \geq 0$ for $u \geq u_0$, $x \in \mathbb{R}_+$. $\qquad\qquad\qquad\square$

Corollary 11.1. *Suppose $f(x_1, \ldots, x_n) \in S_{n,5}$. If $f(x_1, \ldots, x_n) \geq 0$ holds on $\mathbb{R}_+^n$ for all positive integers n, for any given nonnegative integer $\tilde{r}$, there exists a real number $s_0(\tilde{r})$ such that, for any real number $s > s_0(\tilde{r})$ and $t \in \mathbb{R}_+$, $f_{\tilde{r},s}(t) \geq 0$.*

Proof. For any given nonnegative integer $\tilde{r}$, $f_{\tilde{r},s}(t)$ can be viewed as a polynomial in t, s (see Eq. (11.7)). By Lemma 11.6, for any positive integer s and real number $t \in \mathbb{R}_+$, $f_{\tilde{r},s}(t) \geq 0$. Thus the premise of Lemma 11.10 is verified. That completes the proof. $\qquad\qquad\square$

Algorithm 11.1 nprove

Input: A symmetric form $f(x_1, \ldots, x_n) = [a, b, c, d, e, \alpha, \beta]_P$ of degree 5 with rational coefficients

Output: `true`, if $f(x_1, \ldots, x_n) \geq 0$ holds on $\mathbb{R}_+^n$ for all positive integers n; `false`, otherwise.

1: Compute the expression (11.6) of $f_{r,s}(t)$;

2: Compute corresponding r_0, s_0 in Lemma 11.9 and denote
$$D = \{(r, s) \mid (r, s) \in \mathbb{R}_+^2, \ r \geq 1 + \lfloor r_0 \rfloor, \ s \geq 1 + \lfloor s_0 \rfloor\};$$

3: **if** $\forall t \in \mathbb{R}_+ \ (f_{r,s}(t) \geq 0)$ does not hold on D **then**

4:     **return false**

5: **end if**

6: Compute $s_0(\tilde{r})$ in Corollary 11.1 for $\tilde{r} = 1, \ldots, \lfloor r_0 \rfloor$ one by one and denote
$$D\tilde{r} = \{s \mid s \in \mathbb{R}_+, s \geq 1 + \lfloor s_0(\tilde{r}) \rfloor\};$$

7: **if** $\forall t \in \mathbb{R}_+ \ (f_{\tilde{r},s}(t) \geq 0)$ does not hold on some $D\tilde{r}$ **then**

8:     **return false**

9: **end if**

10: Denote by L the finite set $\bigcup\limits_{\tilde{r}=1}^{\lfloor r_0 \rfloor} Lset[\tilde{r}]$, where
$$Lset[\tilde{r}] = \{(\tilde{r}, 1), \ldots, (\tilde{r}, s_0(\tilde{r}))\};$$

11: **if** there exist $(r, s) \in L$ such that $\forall t \in \mathbb{R}_+ \ (f_{r,s}(t) \geq 0)$ does not hold **then**

12:     **return false**

13: **else**

14:     **return true**

15: **end if**

Note that $s_0(\tilde{r})$ in the above corollary can be estimated effectively by Lemma 11.7. We are now ready to give the algorithm.

Theorem 11.2. *Algorithm 11.1 terminates correctly.*

Proof. Note that the conditions of the three "if" statements are propositions inside the Tarski model and thus decidable. Then the termination is obvious and the correctness is guaranteed by Lemma 11.9, Corollary 11.1 and Lemma 11.6. $\qquad\square$

11.1.3 *Examples and Discussion*

Algorithm 11.1 has been implemented as a function **nprove** in our program BOTTEMA (see Chapter 8). We report several examples solved by **nprove** and raise some questions in this section.

Example 11.3 (Example 11.1 continued). Decide whether or not $f \geq 0$ holds on $\mathbb{R}^n_+$ for all positive integers n where f is defined by (11.1) in Example 11.1, *i.e.*

$$f = -(\sum_{k=1}^{n} x_k^5) - 6(\sum_{k=1}^{n} x_k^4)(\sum_{k=1}^{n} x_k) + 2(\sum_{k=1}^{n} x_k^3)(\sum_{k=1}^{n} x_k^2) + 8(\sum_{k=1}^{n} x_k^3)$$
$$\cdot(\sum_{k=1}^{n} x_k)^2 + 3(\sum_{k=1}^{n} x_k^2)^2(\sum_{k=1}^{n} x_k) - 6(\sum_{k=1}^{n} x_k^2)(\sum_{k=1}^{n} x_k)^3 + (\sum_{k=1}^{n} x_k)^5.$$

After loading BOTTEMA, we type in

$$\texttt{nprove}([-1, -6, 2, 8, 3, -6, 1]);$$

the output is "true". The time consumed is 39.4 seconds on a PC (Intel Core i5-3470 CPU @ 3.20GHz, 8G memory, Windows 7 OS) with Maple 17.

Example 11.4.

$$f = -7(\sum_{k=1}^{n} x_k^4) + 8(\sum_{k=1}^{n} x_k^3)(\sum_{k=1}^{n} x_k) + 4(\sum_{k=1}^{n} x_k^2)^2 - 6(\sum_{k=1}^{n} x_k^2)(\sum_{k=1}^{n} x_k)^2$$
$$+(\sum_{k=1}^{n} x_k)^4.$$

It is proved, by typing in

$$\texttt{nprove}([-7, 8, 4, -6, 1]);$$

that $f \geq 0$ holds on $\mathbb{R}^n_+$ for all positive integers n. The time consumed is 1.7 seconds on the same machine.

Example 11.5.

$$f = -(\sum_{k=1}^{n} x_k^5) - 6(\sum_{k=1}^{n} x_k^4)(\sum_{k=1}^{n} x_k) + 195(\sum_{k=1}^{n} x_k^3)(\sum_{k=1}^{n} x_k^2) - 11(\sum_{k=1}^{n} x_k^3)$$
$$\cdot(\sum_{k=1}^{n} x_k)^2 - 2(\sum_{k=1}^{n} x_k^2)^2(\sum_{k=1}^{n} x_k) - 9(\sum_{k=1}^{n} x_k^2)(\sum_{k=1}^{n} x_k)^3 + (\sum_{k=1}^{n} x_k)^5.$$

By typing in

$$\texttt{nprove}([-1, -6, 195, -11, -2, -9, 1]);$$

we find that $f \geq 0$ does not hold on $\mathbb{R}^n_+$ for all positive integers n. The time consumed is 276.3 seconds on the same machine. In fact, the inequality holds for $n \leq 6$ and does not hold for $n \geq 7$.

Theorem 11.1 partly answers the problem presented at the beginning of Section 11.1.1, *i.e.* nonnegativity of symmetric form of degree less than 5 is decidable. We conjecture that the whole problem should be decidable, *i.e.* nonnegativity of symmetric form of any degree is decidable. The difficulty of the proof is that similar result as Lemma 11.10 does not hold in general for multivariate polynomials. In addition, we think the following two questions are also interesting.

(1) If the coefficients of symmetric form $g(x_1, \ldots, x_n)$ are functions in n, what is the answer to the problem?

Example 11.6.

$$f = 2n(n-1)(\sum_{k=1}^{n} x_k^5) - (n+6)(n-1)(\sum_{k=1}^{n} x_k^4)(\sum_{k=1}^{n} x_k)$$

$$-(2n^2 - 4n + 4)(\sum_{k=1}^{n} x_k^3)(\sum_{k=1}^{n} x_k^2) + (6n - 4)(\sum_{k=1}^{n} x_k^3)(\sum_{k=1}^{n} x_k)^2$$

$$+(n^2 - n + 3)(\sum_{k=1}^{n} x_k^2)^2(\sum_{k=1}^{n} x_k) - (2n + 2)(\sum_{k=1}^{n} x_k^2)(\sum_{k=1}^{n} x_k)^3$$

$$+(\sum_{k=1}^{n} x_k)^5.$$

Decide whether or not $f \geq 0$ holds on $\mathbb{R}^n_+$ for all positive integers n.

(2) If the variables of symmetric form $g(x_1, \ldots, x_n)$ take values in $\mathbb{R}^n$ instead of $\mathbb{R}^n_+$, what is the answer to the problem?

11.2 A Class of Symmetric Forms of Any Degrees

11.2.1 *Problem*

Recall that Ω is the set of nonnegative integer solutions to the Diophantine equation $y_1 + 2y_2 + \cdots + dy_d = m$. By Lemma 11.2, any symmetric form

$f \in S_{n,m}$ can be expressed as

$$f = a_1 P_{(n,m)} + a_2 P_{(n,m-1)} P_{(n,1)} + a_3 P_{(n,m-2)} P_{(n,2)} + a_4 P_{(n,m-2)} P_{(n,1)}^2$$

$$+ \cdots + a_{|\Omega|} P_{(n,1)}^m,$$

where $|\Omega|$ stands for the cardinal number of Ω. The expression is simply denoted by

$$f = [a_1, a_2, \ldots, a_{|\Omega|}]_P.$$

Definition 11.3. For $x = (x_1, \ldots, x_n) \in \mathbb{R}^n$ and any positive integer k, define

$$A_k(x) = \frac{\sum_{i=1}^n x_i^k}{n}.$$

Moreover, define

$$S_{n,m}^0 = \{f \in S_{n,m} | f \text{ is a polynomial in } A_1, \ldots, A_{\min(n,m)}\}.$$

For example, any symmetric form of degree 5 in n variables $f^0 \in S_{n,5}^0$ can be expressed as

$$f^0 = [a_1, a_2, \ldots, a_7]_A$$
$$= a_1 A_5 + a_2 A_4 A_1 + a_3 A_3 A_2 + a_4 A_3 A_1^2 + a_5 A_2^2 A_1 + a_6 A_2 A_1^3 + a_7 A_1^5$$

and any symmetric form of degree 6 in n variables $f^0 \in S_{n,6}^0$ can be expressed as

$$f^0 = [a_1, a_2, \ldots, a_{11}]_A$$
$$= a_1 A_6 + a_2 A_5 A_1 + a_3 A_4 A_2 + a_4 A_4 A_1^2 + a_5 A_3^2 + a_6 A_3 A_2 A_1 + a_7 A_3 A_1^3$$
$$+ a_8 A_2^3 + a_9 A_2^2 A_1^2 + a_{10} A_2 A_1^4 + a_{11} A_1^6.$$

Generally, any symmetric homogeneous polynomial of degree m in n variables $f^0 \in S_{n,m}^0$ can be expressed as

$$f^0 = [a_1, a_2, \ldots, a_{|\Omega|}]_A$$
$$= a_1 A_m + a_2 A_{m-1} A_1 + a_3 A_{m-2} A_2 + a_4 A_{m-2} A_1^2 + \cdots + a_{|\Omega|} A_1^m.$$

Now we describe the problem discussed in this section.

Problem *Suppose $f^0(x_1, \ldots, x_n) \in S_{n,m}^0$. Decide whether or not $f^0(x_1, \ldots, x_n) \geq 0$ holds on $\mathbb{R}_+^n$ for all positive integers n.*

Obviously, the nonnegativity of

$$f^0 = [a_1, a_2, a_3, a_4, a_5, a_6, a_7]_A \in S_{n,5}^0$$

is equivalent to the nonnegativity of

$$f = [n^4 a_1, n^3 a_2, n^3 a_3, n^2 a_4, n^3 a_5, n a_6, a_7]_P \in S_{n,5}.$$

The nonnegativity of

$$f^0 = [a_1, a_2, a_3, a_4, a_5, a_6, a_7, a_8, a_9, a_{10}, a_{11}]_A \in S^0_{n,6}$$

is equivalent to the nonnegativity of

$$f = [n^5 a_1, n^4 a_2, n^4 a_3, n^3 a_4, n^4 a_5, n^3 a_6, n^2 a_7, n^3 a_8, n^2 a_9, n a_{10}, a_{11}]_P \in S_{n,6}.$$

Generally speaking, the nonnegativity of

$$f^0 = [a_1, a_2, a_3, a_4, \ldots, a_{|\Omega|}]_A \in S^0_{n,m}$$

is equivalent to the nonnegativity of

$$f = [n^{m-1} a_1, n^{m-2} a_2, n^{m-2} a_3, n^{m-3} a_4, \ldots, a_{|\Omega|}]_P \in S_{n,m}.$$

11.2.2 *Algorithm*

In this section, we denote $\ell = \max\left(\lfloor \frac{m}{2} \rfloor, 1\right)$ where m is the degree of given symmetric form and

$$N_{\ell,n} = \{(r_1, \ldots, r_\ell) | r_1, \ldots, r_\ell \text{ are positive integers and } r_1 + \cdots + r_\ell \leq n\}.$$

We can easily obtain an analogue of Lemma 11.5 for symmetric forms of degree m.

Lemma 11.11. *For any* $f \in S_{n,m}$, $\forall x \in \mathbb{R}^n_+ (f(x) \geq 0)$ *if and only if* $\forall (t_1, \ldots, t_\ell) \in \mathbb{R}^\ell_+ \forall (r_1, \ldots, r_\ell) \in N_{\ell,n}(f(t_1 \cdot 1_{r_1}, \ldots, t_\ell \cdot 1_{r_\ell}, 0_{n-r_1-\cdots-r_\ell}) \geq 0)$.

Proof. From Lemma 11.3, it follows that

$$f(x) \geq 0, x \in \mathbb{R}^n_+ \iff f(y) \geq 0, \forall y \in \mathbb{R}^n_+, v^*(y) \leq \ell = \max\left(\lfloor \frac{m}{2} \rfloor, 1\right).$$

The conclusion follows immediately because f is a symmetric form. $\qquad\square$

For any $f^0 \in S^0_{n,m}$, any $t = (t_1, \ldots, t_\ell)$ and any $r = (r_1, \ldots, r_\ell)$, denote

$$q_k = \sum_{i=1}^{\ell} r_i t_i^k, \quad k = 1, \ldots, m,$$

$$\bar{q}_k = \frac{\sum_{i=1}^{\ell} r_i t_i^k}{n}, \quad k = 1, \ldots, m,$$

$$\bar{f}_r^0(t) = a_1 \bar{q}_m + a_2 \bar{q}_{m-1} \bar{q}_1 + a_3 \bar{q}_{m-2} \bar{q}_2 + a_4 \bar{q}_{m-2} \bar{q}_1^2 + \cdots + a_{|\Omega|} \bar{q}_1^m$$
$$= [a_1, a_2, \ldots, a_{|\Omega|}]_{\bar{q}(r,t)},$$

$$f_r^0(t) = a_1 q_m + a_2 q_{m-1} q_1 + a_3 q_{m-2} q_2 + a_4 q_{m-2} q_1^2 + \cdots + a_{|\Omega|} q_1^m$$
$$= [a_1, a_2, \ldots, a_{|\Omega|}]_{q(r,t)}.$$

That is

$$
\begin{aligned}
\bar{f}_r^0(t) = a_1 &\frac{\left(r_1 t_1^m + \cdots + r_\ell t_\ell^m\right)}{n} \\
+ a_2 &\frac{\left(r_1 t_1^{m-1} + \cdots + r_\ell t_\ell^{m-1}\right)}{n} \frac{\left(r_1 t_1 + \cdots + r_\ell t_\ell\right)}{n} \\
+ a_3 &\frac{\left(r_1 t_1^{m-2} + \cdots + r_\ell t_\ell^{m-2}\right)}{n} \frac{\left(r_1 t_1^2 + \cdots + r_\ell t_\ell^2\right)}{n} \\
+ a_4 &\frac{\left(r_1 t_1^{m-3} + \cdots + r_\ell t_\ell^4\right)}{n} \frac{\left(r_1 t_1 + \cdots + r_\ell t_\ell\right)^2}{n^2} \\
+ \cdots &+ a_{|\Omega|} \frac{\left(r_1 t_1 + \cdots + r_\ell t_\ell\right)^m}{n^m}
\end{aligned}
$$

and

$$
\begin{aligned}
f_r^0(t) = a_1 &\left(r_1 t_1^m + \cdots + r_\ell t_\ell^m\right) \\
+ a_2 &\left(r_1 t_1^{m-1} + \cdots + \bar{r}_\ell t_\ell^{m-1}\right)\left(r_1 t_1 + \cdots + r_\ell t_\ell\right) \\
+ a_3 &\left(r_1 t_1^{m-2} + \cdots + r_\ell t_\ell^{m-2}\right)\left(r_1 t_1^2 + \cdots + r_\ell t_\ell^2\right) \\
+ a_4 &\left(r_1 t_1^{m-2} + \cdots + r_\ell t_\ell^{m-2}\right)\left(r_1 t_1 + \cdots + r_\ell t_\ell\right)^2 \\
+ \cdots &+ a_{|\Omega|}\left(r_1 t_1 + \cdots + r_\ell t_\ell\right)^m .
\end{aligned}
$$

According to Lemma 11.11, it follows that

Lemma 11.12. *Suppose that $f^0(x_1, \ldots, x_n) \in S_{n,m}^0$ and $(x_1, \ldots, x_n) \in \mathbb{R}_+^n$. Then*

$$f^0 \geq 0, \ (\forall n \in \mathbb{Z}_+) \tag{11.8}$$

$$\Leftrightarrow \bar{f}_r^0(t) \geq 0, \ (\forall n \in \mathbb{Z}_+, \forall (r_1, \ldots, r_\ell) \in \mathbb{Z}_+^\ell, r_1 + \cdots + r_\ell \leq n, \forall t \in \mathbb{R}_+^\ell) \tag{11.9}$$

$$\Leftrightarrow f_r^0(t) \geq 0, \ (\forall (r_1, \ldots, r_\ell) \in \mathbb{Q}_+^\ell, r_1 + \cdots + r_\ell \leq 1, \forall t \in \mathbb{R}_+^\ell) \tag{11.10}$$

$$\Leftrightarrow f_r^0(t) \geq 0, \ (\forall (r_1, \ldots, r_\ell) \in \mathbb{R}_+^\ell, r_1 + \cdots + r_\ell \leq 1, \forall t \in \mathbb{R}_+^\ell). \tag{11.11}$$

Proof. Note that $A_k = \bar{q}_k$ at any $(t_1 \cdot 1_{r_1}, \ldots, t_\ell \cdot 1_{r_\ell}, 0_{n - r_1 - \cdots - r_\ell})$. Then (11.8)$\Leftrightarrow$(11.9) is just Lemma 11.11.

Note that r_i and n in (11.9) are arbitrary. Then (11.9)$\Leftrightarrow$(11.10).

To prove (11.10)$\Leftrightarrow$(11.11), assume that there exist nonnegative real numbers $r_1, \ldots, r_\ell, r_1 + \cdots + r_\ell \leq 1$ and $t \in \mathbb{R}_+^\ell$ such that $f_r^0(t) < 0$. Because of the continuity of function $f_r^0(t)$, there must exist nonnegative rational numbers $r_1', \ldots, r_\ell', r_1' + \cdots + r_\ell' \leq 1$ and $t' \in \mathbb{R}_+^\ell$ such that $f_{r'}^0(t') < 0$. This contradicts with (11.10). That completes the proof. $\qquad\square$

Note that formula (11.11) of Lemma 11.12 is inside the Tarski model. So, we know that the problem presented in the last subsection is decidable as stated in the following main theorem.

Theorem 11.3. *Given a symmetric form $f^0(x_1, \ldots, x_n) \in S^0_{n,m}$ with rational coefficients, the following problem is decidable: Decide whether or not $f^0(x_1, \ldots, x_n) \geq 0$ holds on $\mathbb{R}^n_+$ for all positive integers n.*

Theoretically speaking, by Theorem 11.3 and Lemma 11.12, the problem of deciding whether or not $f^0(x_1, \ldots, x_n) \geq 0$ holds on $\mathbb{R}^n_+$ for all positive integers n can be solved by any quantifier elimination algorithms. However, from formula (11.11) we know the QE problem contains many variables if m is big, which in practice makes general QE algorithms hardly work. We reduce the problem further so that the difference substitution method (see Chapter 10) can be applied. This provides more possibility to work out more examples.

Introduce a new variable $r_{\ell+1}$ and define

$$F^0_r(t) = (\sum_{i=1}^{\ell+1} r_i)^{m-1} a_1 \left(r_1 t_1^m + \cdots + r_\ell t_\ell^m\right)$$

$$+ (\sum_{i=1}^{\ell+1} r_i)^{m-2} a_2 \left(r_1 t_1^{m-1} + \cdots + r_\ell t_\ell^{m-1}\right) \left(r_1 t_1 + \cdots + r_\ell t_\ell\right)$$

$$+ (\sum_{i=1}^{\ell+1} r_i)^{m-2} a_3 \left(r_1 t_1^{m-2} + \cdots + r_\ell t_\ell^{m-2}\right) \left(r_1 t_1^2 + \cdots + r_\ell t_\ell^2\right)$$

$$+ (\sum_{i=1}^{\ell+1} r_i)^{m-3} a_4 \left(r_1 t_1^{m-2} + \cdots + r_\ell t_\ell^{m-2}\right) \left(r_1 t_1 + \cdots + r_\ell t_\ell\right)^2$$

$$+ \cdots + a_{|\Omega|} \left(r_1 t_1 + \cdots + r_\ell t_\ell\right)^m$$

$$= [(\sum_{i=1}^{\ell+1} r_i)^{m-1} a_1, (\sum_{i=1}^{\ell+1} r_i)^{m-2} a_2, \ldots, a_{|\Omega|}]_{q(r,t)}.$$

Theorem 11.4. *Suppose that $f^0(x_1, \ldots, x_n) \in S^0_{n,m}$ and $(x_1, \ldots, x_n) \in \mathbb{R}^n_+$. Then $f^0(x_1, \ldots, x_n) \geq 0$ holds for any positive integer n if and only if $F^0_r(t) \geq 0$ holds for any $r = (r_1, \ldots, r_{\ell+1}) \in \mathbb{R}^{\ell+1}_+$ and any $t = (t_1, \ldots, t_\ell) \in \mathbb{R}^\ell_+$.*

Proof. Sufficiency. For any $(r_1, \ldots, r_\ell) \in \mathbb{R}^\ell_+, r_1 + \cdots + r_\ell \leq 1$, let $r_{\ell+1} = 1 - \sum_{i=1}^\ell r_i$. Obviously, $f^0_r(t) = F^0_r(t) \geq 0$.

Necessity. For any $r = (r_1, \ldots, r_{\ell+1}) \in \mathbb{R}_+^{\ell+1}$, let $s = \sum_{i=1}^{\ell+1} r_i$ and $r' = (r_1', \ldots, r_{\ell+1}') = (\frac{r_1}{s}, \ldots, \frac{r_{\ell+1}}{s})$. Then $\frac{F_r^0(t)}{s^m} = f_{r'}^0(t) \geq 0$. That completes the proof. $\qquad\qquad\square$

The inequality $F_r^0(t) \geq 0$ in Theorem 11.4 is of the type to which difference substitution method is applicable. By Theorem 11.3 and Theorem 11.4, to determine whether $f^0(x_1, \ldots, x_n) \geq 0$, we may perform the following two steps: First, compute ℓ and $f_r^0(t)$ (or $F_r^0(t)$); Second, decide whether or not $f_r^0(t) \geq 0$ (or $F_r^0(t) \geq 0$). We report some examples in the next section.

11.2.3   *Examples*

Example 11.7 (Example 11.2 continued). Does the following integral inequality hold for any function $g(s)$ which is integrable on $[0, 1]$?

$$G = \int_0^1 |g(s)|^7 ds \int_0^1 |g(s)| ds - 3 \int_0^1 g^6(s) ds \int_0^1 g^2(s) ds \qquad (11.12)$$
$$+ \, 3 \int_0^1 |g(s)|^5 ds \int_0^1 |g(s)|^3 ds - \left(\int_0^1 g^4(s) ds \right)^2 \geq 0.$$

By the definition of the Riemann integral, we divide the interval $[0, 1]$ into n parts

$$0 = s_0 < s_1 < \cdots < s_n = 1.$$

Obviously,

$$\int_0^1 g(s) ds = \lim_{n \to \infty} \sum_{i=1}^n \frac{1}{n} g\left(\frac{i}{n}\right).$$

It follows that

$$G = \lim_{n \to \infty} \left(\left(\sum_{i=1}^n \frac{1}{n} |g(\tfrac{i}{n})|^7 \right) \left(\sum_{i=1}^n \frac{1}{n} |g(\tfrac{i}{n})| \right) - 3 \left(\sum_{i=1}^n \frac{1}{n} g^6(\tfrac{i}{n}) \right) \left(\sum_{i=1}^n \frac{1}{n} g^2(\tfrac{i}{n}) \right) \right.$$
$$\left. + \, 3 \left(\sum_{i=1}^n \frac{1}{n} |g(\tfrac{i}{n})|^5 \right) \left(\sum_{i=1}^n \frac{1}{n} |g(\tfrac{i}{n})|^3 \right) - \left(\sum_{i=1}^n \frac{1}{n} g^4(\tfrac{i}{n}) \right)^2 \right).$$

It is not hard to see that proving $G \geq 0$ is equivalent to verifying that the following symmetric form of degree 8 in n variables holds for all positive integers n and arbitrary nonnegative variables x_i.

$$f^0 = \sum_{i=1}^n \frac{x_i^7}{n} \sum_{i=1}^n \frac{x_i}{n} - 3 \sum_{i=1}^n \frac{x_i^6}{n} \sum_{i=1}^n \frac{x_i^2}{n} + 3 \sum_{i=1}^n \frac{x_i^5}{n} \sum_{i=1}^n \frac{x_i^3}{n} - \left(\sum_{i=1}^n \frac{x_i^4}{n} \right)^2 \geq 0.$$

According to Theorem 11.3, we need to prove

$$f_r^0(t) = \sum_{i=1}^4 r_i t_i^7 \sum_{i=1}^4 r_i t_i - 3 \sum_{i=1}^4 r_i t_i^6 \sum_{i=1}^4 r_i t_i^2 + 3 \sum_{i=1}^4 r_i t_i^5 \sum_{i=1}^4 r_i t_i^3 - (\sum_{i=1}^4 r_i t_i^4)^2 \geq 0$$

for all $t = (t_1, t_2, t_3, t_4) \in \mathbb{R}_+^4$ and all $r_1, r_2, r_3, r_4 \in \mathbb{R}_+$ subject to $r_1 + r_2 + r_3 + r_4 \leq 1$. It is not so easy to prove this inequality since it contains 8 variables. Alternatively, according to Theorem 11.4, we prove

$$F_r^0(t) = (\sum_{i=1}^5 r_i)^6 \sum_{i=1}^4 r_i t_i^7 \sum_{i=1}^4 r_i t_i - 3(\sum_{i=1}^5 r_i)^6 \sum_{i=1}^4 r_i t_i^6 \sum_{i=1}^4 r_i t_i^2$$

$$+ 3(\sum_{i=1}^5 r_i)^6 \sum_{i=1}^4 r_i t_i^5 \sum_{i=1}^4 r_i t_i^3 - (\sum_{i=1}^5 r_i)^6 (\sum_{i=1}^4 r_i t_i^4)^2 \geq 0$$

for any $t = (t_1, t_2, t_3, t_4) \in \mathbb{R}_+^4$ and any $r_1, r_2, r_3, r_4, r_5 \in \mathbb{R}_+$.

Removing the positive factor $(\sum_{i=1}^5 r_i)^6$ of F_r^0 and calling difference substitution program `tsds`, we know that the inequality holds.

Example 11.8. Does the inequality $G \geq 0$ hold for any $g(s)$ which is integrable on $[0, 1]$?

$$G = 4 \int_0^1 |g(s)|^5 ds - \int_0^1 g^4(s) ds \int_0^1 |g(s)| ds - 4 \int_0^1 |g(s)|^3 ds \int_0^1 g^2(s) ds$$

$$+ 2 \int_0^1 |g(s)|^3 ds (\int_0^1 |g(s)| ds)^2 - 3(\int_0^1 g^2(s) ds)^2 \int_0^1 |g(s)| ds$$

$$+ 3 \int_0^1 g^2(s) ds (\int_0^1 |g(s)| ds)^3 - (\int_0^1 |g(s)| ds)^5 \geq 0.$$

Equivalently, we consider the following polynomial inequality

$$f^0 = 4 \sum_{i=1}^n \frac{x_i^5}{n} - \sum_{i=1}^n \frac{x_i^4}{n} \sum_{i=1}^n \frac{x_i}{n} - 4 \sum_{i=1}^n \frac{x_i^3}{n} \sum_{i=1}^n \frac{x_i^2}{n} + 2 \sum_{i=1}^n \frac{x_i^3}{n} (\sum_{i=1}^n \frac{x_i}{n})^2$$

$$- 3(\sum_{i=1}^n \frac{x_i^2}{n})^2 \sum_{i=1}^n \frac{x_i}{n} + 3 \sum_{i=1}^n \frac{x_i^2}{n} (\sum_{i=1}^n \frac{x_i}{n})^3 - (\sum_{i=1}^n \frac{x_i}{n})^5 \geq 0$$

for all positive integers n and all $x_i \in \mathbb{R}_+, i = 1, \ldots, n$.

According to Theorem 11.4, we need to prove

$$F_r^0(t) = [4s^4, -s^3, -4s^3, 2s^2, -3s^2, 3s, -1]_{q(r,t)} \geq 0$$

for any $t_1, t_2, r_1, r_2, r_3 \in \mathbb{R}_+$, where $s = r_1 + r_2 + r_3$.

Calling successive difference substitution program `tsds`, we immediately know that the inequality holds.

Example 11.9. Prove that

$$G = \int_0^1 g^8(s)ds - (\int_0^1 |g(s)|ds)^8 \geq 0$$

for all $g(s)$ integrable on interval $[0,1]$.

Equivalently, consider the following polynomial inequality

$$f^0 = \sum_{l=1}^n \frac{x_i^8}{n} - (\sum_{i=1}^n \frac{x_i}{n})^8 \geq 0$$

for all positive integers n and all $x_i \in \mathbb{R}_+, i = 1, \ldots, n$.

According to Theorem 11.4, we verify that

$$F_r^0(t) = (\sum_{i=1}^5 r_i)^7 \sum_{i=1}^4 r_i t_i^8 - (\sum_{i=1}^4 r_i t_i)^8 \geq 0$$

for all $t_1, t_2, t_3, t_4, r_1, r_2, r_3, r_4, r_5 \in \mathbb{R}_+$ by difference substitution method.

Example 11.10. Prove that

$$- 20(\int_0^1 g^4(s)ds)^2 + 3 \int_0^1 |g(s)|^3 ds \int_0^1 |g(s)|^5 ds$$

$$+ 13 \int_0^1 g^2(s)ds \int_0^1 g^6(s)ds + \int_0^1 g^8(s)ds + 7 \int_0^1 |g(s)|^7 ds \int_0^1 |g(s)|ds \geq 0.$$

for all $g(s)$ integrable on interval $[0,1]$.

Equivalently, we need to prove

$$f^0 = - 20(\sum_{i=1}^n \frac{x_i^4}{n})^2 + 3 \sum_{i=1}^n \frac{x_i^3}{n} \sum_{i=1}^n \frac{x_i^5}{n} + 13 \sum_{i=1}^n \frac{x_i^2}{n} \sum_{i=1}^n \frac{x_i^6}{n} + \sum_{i=1}^n \frac{x_i^8}{n}$$

$$+ 7 \sum_{i=1}^n \frac{x_i^7}{n} \sum_{i=1}^n \frac{x_i}{n} \geq 0$$

for all positive integers n and all $x_i \in \mathbb{R}_+, i = 1, \ldots, n$.

According to Theorem 11.4, we verify that

$$F_r^0(t) = - 20(\sum_{i=1}^5 r_i)^6(\sum_{i=1}^4 r_i t_i^4)^2 + 3(\sum_{i=1}^5 r_i)^6 \sum_{i=1}^4 r_i t_i^3 \sum_{i=1}^4 r_i t_i^5$$

$$+ 13(\sum_{i=1}^5 r_i)^6 \sum_{i=1}^4 r_i t_i^2 \sum_{i=1}^4 r_i t_i^6 + (\sum_{i=1}^5 r_i)^7 \sum_{i=1}^4 r_i t_i^8$$

$$+ 7(\sum_{i=1}^5 r_i)^6 \sum_{i=1}^4 r_i t_i^7 \sum_{i=1}^4 r_i t_i \geq 0$$

for all $t_1, t_2, t_3, t_4, r_1, r_2, r_3, r_4, r_5 \in \mathbb{R}_+$ by successive difference substitution method.

Bibliography

Achatz, M., McCallum, S. and Weispfenning, V. (2008). Deciding polynomial-exponential problems, in D. Jeffrey (ed.), *Proc. ISSAC'2008* (ACM), pp. 215–222.

Akritas, A. G., Bocharov, A. V. and Strzeboński, A. W. (1994). Implementation of real root isolation algorithms in mathematica, in *Abstracts of the International Conference on Interval and Computer-Algebraic Methods in Science and Engineering*, Interval'94, pp. 23–27.

Akritas, A. G. and Strzeboński, A. W. (2005). A comparative study of two real root isolation methods, *Nonlinear Analysis: Modelling and Control* **10**, pp. 297–304.

Alefeld, G. and Herzberger, J. (1983). *Introduction to Interval Computations* (Academic Press).

Angeli, D., Ferrell, J. E. J. and Sontag, E. D. (2004). Detection of multistability, bifurcations, and hysteresis in a large class of biological positive-feedback systems, *Proc. Nat. Acad. Sci. USA* **101**, pp. 1822–1827.

Artin, E. (1927). über die zerlegung definiter funktionen in quadrate, *Hamb. Abh.* **5**, pp. 100–115.

Basu, S., Pollack, R. and Roy, M.-F. (1998). A new algorithm to find a point in every cell defined by a family of polynomials, in *Quantifier Elimination and Cylindrical Algebraic Decomposition* (Springer), pp. 341–350.

Basu, S., Pollack, R. and Roy, M.-F. (2003). *Algorithms in Real Algebraic Geometry* (Springer-Verlag).

Ben-Or, M., Kozen, D. and Reif, J. (1986). The complexity of elementary algebra and geometry, *J. Computer and System Sciences* **32**, pp. 251–264.

Besson, F., Jensen, T. and Talpin, J.-P. (1999). Polyhedral analysis of synchronous languages, in *LNCS 1694*, SAS'99 (Springer-Verlag), pp. 51–69.

Blekherman, G. (2006). There are significantly more nonegative polynomials than sums of squares, *Israel Journal of Mathematics* **153**, pp. 355–380.

Bochnak, J., Coste, M. and Roy, M.-F. (1998). *Real Algebraic Geometry* (Springer).

Bottema, O., Dordevic, R. Z., Janic, R. R., Mitrinovic, D. S. and Vasic, P. M. (1969). *Geometric Inequalities* (Wolters-Noordhoff Publishing).

Boulier, F., Chen, C., Lemaire, F. and Moreno Maza, M. (2009). Real root isolation of regular chains, in *Proc. ASCM'2009*, pp. 15–29.

Brown, C. W. (2001). Improved projection for cylindrical algebraic decomposition, *J. Symbolic Computation* **32**, pp. 447–465.

Brown, C. W. and McCallum, S. (2005). On using bi-equational constraints in cad construction, in M. Kauers (ed.), *Proc. ISSAC'2005* (ACM Press), pp. 76–83.

Bushell, P. (1994). Shapiro's cyclic sum, *Bulletin of the London Mathematical Society* **26**, pp. 564–574.

Bushell, P. and McLeod, J. (2002). Shapiro's cyclic inequality for even, *Journal of Inequalities and Applications* **2002**, p. 509463.

Castle, M., Powers, V. and Reznick, B. (2009). A quantitative pólya's theorem with zeros, *Journal of Symbolic Computation* **44**, pp. 1285–1290.

Castle, M., Powers, V. and Reznick, B. (2011). Pólya's theorem with zeros, *Journal of Symbolic Computation* **46**, pp. 1039–1048.

Chen, C., Davenport, J. H., Lemaire, F., Maza, M. M., Xia, B., Xiao, R. and Xie, Y. (2012a). Computing the real solutions of polynomial systems with the regularchains library in maple, *ACM Communications in Computer Algebra* **45**, pp. 166–168.

Chen, C., Davenport, J. H., May, J. P., Maza, M. M., Xia, B. and Xiao, R. (2012b). Triangular decomposition of semi-algebraic systems, *Journal of Symbolic Computation* **49**, pp. 3–26.

Chen, C., Davenport, J. H., Maza, M. M., Xia, B. and Xiao, R. (2013). Computing with semi-algebraic sets: Relaxation techniques and effective boundaries, *Journal of Symbolic Computation* **52**, pp. 72–96.

Chen, C., Golubitsky, O., Lemaire, F., Moreno Maza, M. and Pan, W. (2007). Comprehensive triangular decomposition, in V. G. Ganzha, E. W. Mayr and E. V. Vorozhtsov (eds.), *Proc. CASC'2007*, LNCS 4770, pp. 73–101.

Chen, L. (2008). *Study on Several Parallel Algorithms of Symbolic Computation*, Ph.D. dissertation, East China Normal University, Shanghai, China, (in Chinese).

Chen, S. and Huang, F. (2006). Schur decomposition for symmetric ternary forms and readable proof to inequalities, *Acta Mathematica Sinica, Chinese Series* **49**, pp. 491–502, (in Chinese).

Chen, Z., Tang, X. and Xia, B. (2014). Hierarchical comprehensive triangular decomposition, in *Mathematical Software–ICMS 2014* (Springer), pp. 434–441.

Chen, Z., Tang, X. and Xia, B. (2015). Generic regular decompositions for parametric polynomial systems, *Journal of Systems Science and Complexity* **28**, pp. 1194–1211.

Cheng, J., Gao, X. and Guo, L. (2012). Root isolation of zero-dimensional polynomial systems with linear univariate representation, *Journal of Symbolic Computation* **47**, pp. 843–858.

Cheng, J.-S., Gao, X.-S. and Li, J. (2009). Root isolation for bivariate polynomial systems with local generic position method, in *Proc. ISSAC'2009*, pp. 103–110.

Cheng, J.-S., Gao, X.-S. and Yap, C.-K. (2007). Complete numerical isolation of real zeros in zero-dimensional triangular systems, in *Proc. ISSAC'2007*, pp. 92–99.

Choi, M. and Lam, T. (1977). Extremal positive semidefinite forms, *Mathematische Annalen* **231**, pp. 1–18.

Choi, M. D., Lam, T. Y. and Reznick, B. (1987). Even symmetric sextics, *Math. Z.* **195**, pp. 559–580.

Choi, M.-D., Lam, T. Y. and Reznick, B. (1995). Sums of squares of real polynomials, in *Proceedings of Symposia in Pure Mathematics*, Vol. 58 (American Mathematical Society), pp. 103–126.

Chou, S. C., Gao, X.-S. and Arnon, D. S. (1992). On the mechanical proof of geometry theorems involving inequalities, *Advances in Computing Research* **6**, pp. 139–181.

Cîrtoaje, V., Can, V. Q. B. and Anh, T. Q. (2009). *Inequalities with Beautiful Solutions* (GIL, Romania).

Collins, G. (1975). Quantifier elimination for real closed fields by cylindrical algebraic decompostion, in *Automata Theory and Formal Languages 2nd GI Conference Kaiserslautern, May 20–23, 1975, LNCS*, Vol. 33 (Springer), pp. 134–183.

Collins, G. E. and Akritas, A. G. (1976). Polynomial real root isolation using Descarte's rule of signs, in *Proceedings of the Third ACM Symposium on Symbolic and Algebraic Computation* (ACM), pp. 272–275.

Collins, G. E. and Hong, H. (1991). Partial cylindrical algebraic decomposition for quantifier elimination, *J. Symb. Comput.* **12**, pp. 299–328.

Collins, G. E. and Johnson, J. R. (1989). Quantifier elimination and the sign variation method for real root isolation, in *Proc. ISSAC'1989* (ACM Press), pp. 264–271.

Collins, G. E. and Loos, R. (1982). Real zeros of polynomials, in B. Buchberger, G. E. Collins and R. Loos (eds.), *Computer Algebra: Symbolic and Algebraic Computation* (Springer), pp. 83–94.

Cousot, P. (2001). Abstract interpretation based formal methods and future challenges, in R. Wilhelm (ed.), *Informatics, 10 Years Back - 10 Years Ahead*, LNCS 2000 (Springer), pp. 138–156.

Cousot, P. and Halbwachs, N. (1978). Automatic discovery of linear restraints among the variables of a program, in *Proc. POPL'78* (ACM Press), pp. 84–97.

Cox, D., Little, J. and O'Shea, D. (1992). *Ideas, Varieties and Algorithms: An Introduction to Computational Algebraic Geometry* (Springer).

Cox, D., Little, J. and O'Shea, D. (1998). *Using Algebraic Geometry* (Springer).

Dai, L., Han, J., Hong, H. and Xia, B. (2015). Open weak CAD and its applications, *arXiv:1507.03834*, accepted by J. Symbolic Computation.

Dai, L. and Xia, B. (2015). Smaller SDP for SOS decomposition, *Journal of Global Optimization* **63**, pp. 343–361.

Dayton, B., Li, T. and Zeng, Z. (2011). Multiple zeros of nonlinear systems, *Math. Comp.* **80**, pp. 2143–2168.

Dayton, B. H. and Zeng, Z. G. (2005). Computing the multiplicity structure in solving polynomial systems, in *Proc. ISSAC'2005* (ACM Press), pp. 116–123.

Dolzman, A. and Sturm, T. (1997). Redlog: Computer algebra meets computer logic, *ACM SIGSAM Bulletin* **31**, pp. 2–9.

Eigenwillig, A., Sharma, V. and Yap, C. K. (2006). Almost tight recursion tree bounds for the Descartes method, in *Proc. ISSAC'2006* (ACM), pp. 71–78.

Emiris, I. Z., Galligo, A. and Tsigaridas, E. P. (2010a). Random polynomials and expected complexity of bisection methods for real solving, in *Proc. ISSAC'2010* (ACM), pp. 235–242.

Emiris, I. Z., Mourrain, B. and Tsigaridas, E. P. (2010b). The dmm bound: multivariate (aggregate) separation bounds, in *Proc. ISSAC'2010*, pp. 243–250.

Ferrell, J. E. J. and Machleder, E. M. (1998). The biochemical basis of an all-or-none cell fate switch in xenopus oocytes, *Science* **280**, pp. 895–898.

Folke, E. (1994). Which triangles are plane sections of regular tetrahedra?, *Amer. Math. Monthly* **101**, pp. 788–789.

Gan, T., Chen, M., Dai, L., Xia, B. and Zhan, N. (2015). Decidability of the reachability for a family of linear vector fields, in *Automated Technology for Verification and Analysis*, LNCS 9364 (Springer), pp. 482–499.

Gan, T., Chen, M., Li, Y., Xia, B. and Zhan, N. (2016). Computing reachable sets of linear vector fields revisited, in *Proc. ECC'2016*.

Gantmacher, F. R. (1955). *The theory of matrices (Chinese version)* (Higher Education Press, Beijing), translated from Russian edition by Z. Ke.

Gao, C. and Wen, J. (2013). A dimensionality reduction principle on the optimization of function, *Journal of Mathematical Inequalities* **7**, pp. 357–375.

Gao, X.-S., Hou, X., Tang, J. and Chen, H. (2003). Complete solution classification for the perspective-three-point problem, *IEEE Trans. on PAMI* **25**, pp. 930–943.

Garcia, M. E. A. and Galligo, A. (2012). A root isolation algorithm for sparse univariate polynomials, in *Proc. ISSAC'2012* (ACM), pp. 35–42.

Gatermann, K. and Huber, B. (2002). A family of sparse polynomial systems arising in chemical reaction systems, *J. Symb. Comput.* **33**, pp. 275–305.

Gatermann, K. and Xia, B. (2003). Existence of 3 positive solutions of systems from chemistry, Tech. Rep. 108, Institute of Mathematics, Peking University.

Gonzalez, L., Lombardi, H., Recio, T. and Roy, M.-F. (1989). Sturm-habicht sequence, in *Proc. ISSAC'1989* (ACM), pp. 136–146.

Guergueb, A., Mainguené, J. and Roy, M.-F. (1994). Examples of automatic theorem proving in real geometry, in *Proc. ISSAC'94*, pp. 20–24.

Habicht, W. (1940). über die zerlegung strikte definiter formen in quadrate, *Comm. Math. Helv.* **12**, pp. 317–322.

Habicht, W. (1948). Einer verallgemeinerung des sturmschen wurzelzÄahlverfahrens, *Comm. Math. Helvetici* **21**, pp. 99–116.

Halbwachs, N., Proy, Y. E. and Roumanoff, P. (1997). Verification of real-time systems using linear relation analysis, *Formal Methods in System Design* **11**, pp. 157–185.

Han, J. (2011). *An Introduction to the Proving of Elementary Inequalities (in Chinese)*, Vol. 221 (Harbin Institute of Technology Press, Harbin).

Han, J. (2013). A complete method based on successive difference substitution method for deciding positive semi-definiteness of polynomials, *Acta Scientiarum Naturalium Universitatis Pekinensis* **49**, pp. 545–551, (in Chinese).

Han, J. (2016). *Some Topics on Automated Inequality Proving* (in preparation), (in Chinese).

Han, J., Dai, L. and Xia, B. (2014). Constructing fewer open cells by GCD computation in CAD projection, in *Proc. ISSAC 2014* (ACM Press), pp. 240–247.

Han, J., Jin, Z. and Xia, B. (2016). Proving inequalities and solving global optimization problems via simplified CAD projection, *J. Symbolic Computation* **72**, pp. 206–230.

Hardy, G. H., Littlewood, J. E. and Pólya, G. (1952). *Inequalities*, 2nd edn. (Cambridge University Press).

Hauenstein, J. D. and Sottile, F. (2012). Algorithm 921: alphacertified: Certifying solutions to polynomial systems, *ACM Trans. Math. Softw.* **38**, p. 28.

Henzinger, T. A. and Ho, P.-H. (1995). Algorithmic analysis of nonlinear hybrid systems, in *LNCS 939*, CAV'95, pp. 225–238.

Henzinger, T. A., Kopke, P. W., Puri, A. and Varaiya, P. (1998). What's decidable about hybrid automata?, *Journal of Computer and System Sciences* **57**, pp. 94–124.

Hilbert, D. (1888). über die darstellung definiter formen als summe von formenquadraten, *Math. Ann.* **32**, pp. 342–350.

Hilbert, D. (1901). Mathematische probleme, *Arch. Math. Phys.* **3**, pp. 44–63, 213–237, (English translation by M. W. Newson in *Bull. Amer. Math. Soc.*, **8**: 437–445, 478–479, 1902).

Ho, C.-J. and Yap, C. K. (1996). The Habicht approach to subresultants, *J. Symb. Comput.* **21**, pp. 1–14.

Hong, H. (1990). An improvement of the projection operator in cylindrical algebraic decomposition, in *Proceedings of ISSAC 1990*, pp. 261–264.

Hong, H. and Safey El Din, M. (2012). Variant quantifier elimination, *J. Symb. Comput.* **47**, pp. 883–901.

Hong, H., Tang, X. and Xia, B. (2015). Special algorithm for stability analysis of multistable biological regulatory systems, *Journal of Symbolic Computation* **70**, pp. 112–135.

Hou, X. and Shao, J. (2011). Bounds on the number of steps of WDS required for checking the positivity of integral forms, *Applied Mathematics and Computation* **217**, pp. 9978–9984.

Hou, X., Xu, S. and Shao, J. (2010). The weighted difference substitutions and nonnegativity decision of forms, *Acta Mathematica Sinica, Chinese Series* **53**, pp. 1171–1180, (in Chinese).

Huang, F. (2016). Proofs to two inequality conjectures for a point on the plane of a triangle, *Journal of Inequalities and Applications* **2016**, pp. 1–12.

Huang, F. and Chen, S. (2005). Schur partition for symmetric ternary forms and readable proof of inequalities, in *Proc. ISSAC'05* (ACM Press), pp. 185–192.

Janous, W. (1986). Problem 1137, *Crux Math.* **12**, pp. 79, 177.

Jeronimo, G. and Perrucci, D. (2010). On the minimum of a positive polynomial over the standard simplex, *J. Symbolic Computation* **45**, pp. 434–442.

Johnson, J. R. and Krandick, W. (1997). Polynomial real root isolation using approximate arithmetic, in *Proc. ISSAC'97* (ACM Press), pp. 225–232.

Kalkbrener, M. (1993). A generalized Euclidean algorithm for computing triangular representationa of algebraic varieties, *J. Symb. Comput.* **15**, pp. 143–167.

Kaltofen, E., Li, B., Yang, Z. and Zhi, L. (2008). Exact certification of global optimality of approximate factorizations via rationalizing sums-of-squares with floating point scalars, in *Proc. ISSAC'2008* (ACM Press), pp. 155–164.

Kaltofen, E., Yang, Z. and Zhi, L. (2009). A proof of the monotone column permanent (mcp) conjecture for dimension 4 via sums-of-squares of rational functions, in *Proc. SNC'2009* (ACM), pp. 65–70.

Kim, S., Kojima, M. and Waki, H. (2005). Generalized lagrangian duals and sums of squares relaxations of sparse polynomial optimization problems, *SIAM Journal on Optimization* **15**, pp. 697–719.

Krivine, J.-L. (1964). Anneaux préordonnés, *Journal d'analyse mathématique* **12**, pp. 307–326.

Kuang, J. C. (2010). *Applied Inequalities (in Chinese)*, 4th edn. (Shandong Science and Technology Press).

Lafferrierre, G., Pappas, G. J. and Yovine, S. (2001). Symbolic reachability computaion for families of linear vector fields, *J. Symb. Comput.* **11**, pp. 1–23.

Lasserre, J. (2001). Global optimization with polynomials and the problem of moments, *SIAM Journal on Optimization* **11**, pp. 796–817.

Lax, A. and Lax, P. D. (1978). On sums of squares, *Linear Algebra and Its Applications* **20**, pp. 71–75.

Lenstra, A. K., Lenstra, H. W. and Lovász, L. (1982). Factoring polynomial with rational coefficients, *Mathematische Annalen* **261**, pp. 515–534.

Li, B.-H. (2003). A method to solve algebraic equations up to multiplicities via Ritt-Wu's characteristic sets, *Acta Analysis Functionalis Applicata* **5**, pp. 98–109.

Li, X., Mou, C., Niu, W. and Wang, D. (2011). Stability analysis for discrete biological models using algebraic methods, *Mathematics in Computer Science* **5**, pp. 247–262.

Li, Y., Xia, B. and Zhang, Z. (2010). Zero decomposition with multiplicity of zero-dimensional polynomial systems, *System Sicences and Mathematics* **30**, pp. 1491–1500, (in Chinese, English version: arXiv:1011.1634v1).

Liang, S. and Zhang, J. (1999). A complete discrimination system for polynomials with complex coefficients and its automatic generation, *Science in China Series E: Technological Sciences* **42**, 2.

Liu, B. Q. (2003). *BOTTEMA, What we see (in Chinese)* (Tibet People's Publishing House, Lhasa).

Lofberg, J. (2004). Yalmip: A toolbox for modeling and optimization in matlab, in

2004 IEEE International Symposium on Computer Aided Control Systems Design, pp. 284–289.

Loos, R. (1983). Generalized polynomial remainder sequences, in B. Buchberger, G. E. Collins and R. Loos (eds.), *Computer Algebra: Symbolic and Algebraic Computation*, 2nd edn. (Springer), pp. 115–137.

Lu, Z., He, B. and Luo, Y. (2004). *An Algorithm of Real Root Isolation for Polynomial Systems with Applications* (Science Press), (in Chinese).

Mantzaflaris, A., Mourrain, B. and Tsigaridas, E. P. (2011). On continued fraction expansion of real roots of polynomial systems, complexity and condition numbers, *Theor. Comput. Sci.* **412**, pp. 2312–2330.

Marshall, M. (2008). *Positive Polynomials and Sums of Squares* (American Mathematical Soc.).

McCallum, S. (1988). An improved projection operation for cylindrical algebraic decomposition of three-dimensional space, *J. Symbolic Computation* **5**, pp. 141–161.

McCallum, S. (1998). An improved projection operation for cylindrical algebraic decomposition, in B. Caviness and J. Johnson (eds.), *Quantifier Elimination and Cylindrical Algebraic Decomposition*, Texts and Monographs in Symbolic Computation (Springer-Verlag), pp. 242–268.

Mehlhorn, K. and Sagraloff, M. (2009). Isolating real roots of real polynomials, in *Proc. ISSAC'2009* (ACM), pp. 247–254.

Mignotte, M. (1992). *Mathematics for Computer Algebra* (Springer-Verlag).

Miller, R. K. and Michel, A. N. (1982). *Ordinary Differential Equations* (Academic Press).

Mishra, B. (1993). *Algorithmic Algebra* (Springer).

Mitrinović, D. S., Pecaric, J. E. and Volenec, V. (1989). *Recent Advances in Geometric Inequalities* (Kluwer Academic Publishers).

Moore, R. E., Kearfott, R. B. and Cloud, M. J. (2009). *Introduction to Interval Analysis* (Society for Industrial and Applied Mathematics, Philadelphia).

Motzkin, T. S. (1967). The arithmetic-geometric inequality, in O. Shisha (ed.), *Inequalities* (Academic Press), pp. 205–224.

Niu, W. and Wang, D. (2008). Algebraic approaches to stability analysis of biological systems, *Mathematics in Computer Science* **1**, pp. 507–539.

Novák, B. and Tyson, J. J. (1993). Numerical analysis of a comprehensive model of m-phase control in xenopus oocyte extracts and intact embryos, *J. Cell Sci.* **106**, pp. 1153–1168.

Ojika, T. (1987). Modified deflation algorithm for the solution of singular problems, *J. Math. Anal. Appl.* **123**, pp. 199–221.

Papachristodoulou, A., Anderson, J., Valmorbida, G., Prajna, S., Seiler, P. and Parrilo, P. (2013). SOSTOOLS: Sum of squares optimization toolbox for MATLAB, *arXiv:1310.4716*.

Parrilo, P. A. (2000). *Structured semidefinite programs and semialgebraic geometry methods in robustness and optimization*, Ph.D. dissertation, California Institute of Technology.

Parrilo, P. A. (2003). Semidefinite programming relaxations for semialgebraic problems, *Mathematical Programming* **96**, pp. 293–320.

Parrilo, P. A. and Sturmfels, B. (2003). Minimizing polynomial functions, *Algorithmic and quantitative real algebraic geometry, DIMACS Series in Discrete Mathematics and Theoretical Computer Science* **60**, pp. 83–99.

Pólya, G. (1928). über positive darstellung von polynomen, *Vierteljschr. Natruforsch. Ges.* **73**, pp. 141–145.

Pomerening, J. R., Sontag, E. D. and Ferrell, J. E. J. (2003). Building a cell cycle oscillator: Hysteresis and bistability in the activation of cdc2, *Nature Cell Biol.* **5**, pp. 346–351.

Powers, V. and Reznick, B. (2001). A new bound for pólya's theorem with applications to polynomials positive on polyhedra, *Journal of Pure and Applied Algebra* **164**, pp. 221–229.

Powers, V. and Wörmann, T. (1998). An algorithm for sums of squares of real polynomials, *Journal of Pure and Applied Algebra* **127**, pp. 99–104.

Qian, X. (2013). *Improvements on a simplified CAD projection operator with application to global optimization (in Chinese)*, Master thesis, Peking University, Beijing.

Reznick, B. (1978). Extremal PSD forms with few terms, *Duke Mathematical Journal* **45**, pp. 363–374.

Reznick, B. (1989). Forms derived from the arithmetic-geometric inequality, *Mathematische Annalen* **283**, pp. 431–464.

Reznick, B. (2000). Some concrete aspects of Hilbert's 17th problem, *Comtemporary Mathematics* **253**, pp. 251–272.

Rouillier, F. and Zimmermann, P. (2004). Efficient isolation of polynomial's real roots, *J. Comput. Appl. Math.* **162**, pp. 33–50.

Safey El Din, M. (2008). Computing the global optimum of a multivariate polynomial over the reals, in *Proc. ISSAC'2008* (ACM), pp. 71–78.

Safey El Din, M. and Schost, É. (2003). Polar varieties and computation of one point in each connected component of a smooth real algebraic set, in *Proc. ISSAC'2003* (ACM), pp. 224–231.

Sagraloff, M. (2012). When Newton meets Descartes: A simple and fast algorithm to isolate the real roots of a polynomial, in *Proc. ISSAC'2012* (ACM), pp. 297–304.

Schweighofer, M. (2002). An algorithmic approach to Schmüdgen's Positivstellensatz, *Journal of Pure and Applied Algebra* **166**, pp. 307–319.

Schweighofer, M. (2005). Optimization of polynomials on compact semialgebraic sets, *SIAM Journal on Optimization* **15**, pp. 805–825.

Seiler, P. (2013). Sosopt: A toolbox for polynomial optimization, *arXiv: 1308.1889*.

Seiler, P., Zheng, Q. and Balas, G. (2013). Simplification methods for sum-of-squares programs, *arXiv:1303.0714*.

Shan, Z. (1996). *Geometric Inequality in China (in Chinese)* (Jiangsu Educational Publishing House, China).

Sharma, V. (2007). Complexity of real root isolation using continued fractions, in *Proc. ISSAC'2007* (ACM), pp. 339–346.

Sharma, V. and Batra, P. (2015). Near optimal subdivision algorithms for real root isolation, in *Proc. ISSAC'2015* (ACM), pp. 331–338.

Sharma, V. and Yap, C. K. (2012). Near optimal tree size bounds on a simple real root isolation algorithm, in *Proc. ISSAC'2012* (ACM), pp. 319–326.

Shen, F. (2012). *The Real Roots Isolation of Polynomial System Based on Hybrid Computation*, Master's thesis, Peking University, Beijing.

Shen, F., Wu, W. and Xia, B. (2014). Real root isolation of polynomial equations based on hybrid computation, in *Computer Mathematics - Proc. ASCM2012* (Springer), pp. 375–396.

Stengle, G. (1974). A nullstellensatz and a positivstellensatz in semialgebraic geometry, *Math. Ann.* **207**, pp. 87–97.

Strzeboński, A. (2000). Solving systems of strict polynomial inequalities, *J. Symbolic Computation* **29**, pp. 471–480.

Strzeboński, A. (2008). Real root isolation for exp-log functions, in *Proc. ISSAC'2008* (ACM), pp. 303–314.

Strzeboński, A. (2011). Cylindrical decomposition for systems transcendental in the first variable, *Journal of Symbolic Computation* **46**, pp. 1284–1290.

Strzeboński, A. W. and Tsigaridas, E. P. (2012). Univariate real root isolation in multiple extension fields, in *Proc. ISSAC'2012*, pp. 343–350.

Sturmfels, B. (1998). Polynomial equations and convex polytopes, *American Mathematical Monthly* **105**, pp. 907–922.

Tang, X., Chen, Z. and Xia, B. (2014). Generic regular decompositions for generic zero-dimensional systems, *Science China: Information Sciences* **57**, pp. 1–14.

Tarski, A. (1951). *A Decision Method for Elementary Algebra and Geometry*, 2nd edn. (University of California Press).

Timofte, V. (2003). On the positivity of symmetric polynomial functions. Part I: General results, *J. Math. Anal. Appl.* **284**, pp. 174–190.

Timofte, V. (2005). On the positivity of symmetric polynomial functions. Part II: Lattice general results and positivity criteria for degrees 4 and 5, *J. Math. Anal. Appl.* **304**, pp. 652–667.

Tiwari, A. (2004). Termination of linear programs, in *Proc. CAV'04*, LNCS 3114, pp. 70–82.

Vandenberghe, L. and Boyd, S. (1996). Semidefinite programming, *SIAM Review* **38**, pp. 49–95.

Wang, D. M. (2000). Computing triangular systems and regular systems, *J. Symb. Comput.* **30**, pp. 221–236.

Wang, D. M. (2001). *Elimination Methods* (Springer).

Wang, D. M. (2002). *Elimination Methods and Their Applications (in Chinese)* (Science Press).

Wang, D. M. and Xia, B. (2004). *Computer Algebra (in Chinese)* (Tsinghua University Press).

Wang, D. M. and Xia, B. (2005a). Algebraic analysis of stability for some biological systems, in H. Anai and K. Horimoto (eds.), *Algebraic Biology 2005 - Computer Algebra in Biology* (Universal Academy Press, Inc., Tokyo), pp. 75–83.

Wang, D. M. and Xia, B. (2005b). Stability analysis of biological systems with real solution classification, in *Proc. ISSAC'2005* (ACM Press), pp. 354–361.

Wang, W.-L. (2011). *Approaches to Prove Inequalities* (Harbin Institute of Technology Press), (in Chinese).

Wang, Z. H. and Hu, H. Y. (1999). Delay-independent stability of retarded dynamic systems of multiple degrees of freedom, *J. Sound and Vibration* **226**, pp. 57–81.

Wang, Z. H. and Hu, H. Y. (2000). Stability of time-delayed dynamic systems with unknown parameters, *J. Sound and Vibration* **233**, pp. 215–233.

Weispfenning, V. (1998). A new approach to quantifier elimination for real algebra, in B. F. Caviness and J. R. Johnson (eds.), *Quantifier Elimination and Cylindrical Algebraic Decomposition* (Springer), pp. 376–392.

Weiss, E. (1963). *Algebraic Number Theory* (McGrawHill).

Wen, J., Wang, W., Lu, Y. and Zhang, Y. (2003). The method of descending dimension for establishing inequalities (i), *Journal of Southwest University for Nationalities, Natural Science Edition* **29**, pp. 527–532.

Wu, W.-T. (1978). On the decision problem and the mechanization of theorem-proving in elementary geometry, *Sci. Sin.* **21**, pp. 159–172.

Wu, W.-T. (1994a). *Mechanical theorem proving in geometries: basic principles* (Springer), translated from the Chinese edition — published in 1984 by Science Press — by Jin, X. and Wang, D.

Wu, W.-T. (1994b). On a finiteness theorem about problem involving inequalities, *J. Sys. Sci. Math.* **7**, pp. 193–200.

Wu, W.-T. (1998). On global-optimization problems, in *Proc. ASCM '98* (Lanzhou University Press, Lanzhou, China), pp. 135–138.

Xia, B. (1998). *Automated discovering and proving of geometric inequalities*, Ph.D. thesis, Sichuan University, Chengdu, China.

Xia, B. (2000). Discoverer: A tool for solving problems involving polynomial inequalities, in W.-C. Yang (ed.), *Proc. ATCM'2000* (ATCM Inc.), pp. 472–481.

Xia, B. (2003). A note on the subresultant chain theorem, Preprint 172, Institute of Mathematics, Peking University.

Xia, B. (2007). Discoverer: a tool for solving semi-algebraic systems, *Software Demo at ISSAC 2007, ACM Commun. Comput. Algebra* **41**, pp. 102–103.

Xia, B. and Hou, X. (2002). A complete algorithm for counting real solutions of polynomial systems of equations and inequalities, *Computers & Mathematics with Applications* **44**, pp. 633–642.

Xia, B., Xiao, R. and Yang, L. (2005). Solving parametric semi-algebraic systems, in S. Pae and H. Park (eds.), *Proc. ASCM'2005*, pp. 153–156.

Xia, B. and Yang, L. (2002). An algorithm for isolating the real solutions of semi-algebraic systems, *J. Symb. Comput.* **34**, pp. 461–477.

Xia, B. and Yang, L. (2003). Properties of discrimination matrices of polynomials and their applications, *Acta Mathematicae Applicatae Sinica* **26**, pp. 652–663, (in Chinese).

Xia, B., Yang, L., Zhan, N. and Zhang, Z. (2011). Symbolic decision procedure for termination of linear programs, *Formal Aspects of Computing* **23**, pp. 171–190.

Xia, B. and Zhang, T. (2006). Real solution isolation using interval arithmetic,

Computers and Mathematics with Applications **52**, pp. 853–860.

Xiao, R. (2009). *Parametric Polynomial Systems Solving*, Ph.D. thesis, Peking University, Beijing, (in Chinese).

Xu, J. and Yao, Y. (2011). Completion of difference substitution method based on stochastic matrix, *Acta Mathematica Sinica, Chinese Series* **54**, pp. 219–226, (in Chinese).

Xu, J. and Yao, Y. (2012). Pólya method and the successive difference substitution method, *Science China Mathematics* **42**, pp. 203–213, (in Chinese).

Xu, M., Li, Z.-B. and Yang, L. (2015). Quantifier elimination for a class of exponential polynomial formulas, *Journal of Symbolic Computation* **68**, pp. 146–168.

Yang, L. (1998). A simplified algorithm for solution classification of the perspective-three-point problem, Tech. Rep. 17, MM Preprints, MMRC, Beijing.

Yang, L. (1999). Recent advances in automated theorem proving on inequalities, *J. Comput. Sci. Technol.* **14**, pp. 434–446.

Yang, L. (2001). Symbolic algorithm for global optimization and principle of finite kernel, in D. Lin (ed.), *Mathematics and Mathematical Mechanization* (Shandong Education Press), pp. 210–220, (in Chinese).

Yang, L. (2005). Solving harder problems with lesser mathematics, in *Proc. ATCM'2005* (ATCM Inc., Blacksburg), pp. 37–46.

Yang, L. (2006). Difference substitution and automated inequality proving, *J. Guangzhou University (Natural Science edition)* **5**, pp. 1–7, (in Chinese).

Yang, L., Hou, X. and Xia, B. (1999). Automated discovering and proving for geometric inequalities, in X.-S. Gao, D. Wang and L. Yang (eds.), *Automated Deduction in Geometry*, LNAI 1669 (Springer-Verlag), pp. 30–46.

Yang, L., Hou, X. and Xia, B. (2001). A complete algorithm for automated discovering of a class of inequality-type theorems, *Sci. China* **F 44**, pp. 33–49.

Yang, L., Hou, X. and Zeng, Z. (1996a). A complete discrimination system for polynomials, *Sci. China* **E 39**, pp. 628–646.

Yang, L. and Xia, B. (2004). Automated deduction in geometry, in F. Chen and D. Wang (eds.), *Geometric Computation* (World Scientific), pp. 248–298.

Yang, L. and Xia, B. (2005). Real solution classifications of parametric semialgebraic systems, in A. Dolzmann, A. Seidl and T. Sturm (eds.), *Algorithmic Algebra and Logic — Proc. A3L'2005* (Herstellung und Verlag, Norderstedt), pp. 281–289.

Yang, L. and Xia, B. (2006). Quantifier elimination for quartics, in T. Ida, J. Calmet and D. Wang (eds.), *LNAI 4120* (Springer), pp. 131–145.

Yang, L. and Xia, B. (2008). *Automated Proving and Discovering on Inequalities* (Science Press, Beijing), (in Chinese).

Yang, L. and Xia, S. (2000). An inequality-proving program applied to global optimization, in W.-C. Yang (ed.), *Proc. ATCM'2000* (ATCM Inc.), pp. 40–51.

Yang, L. and Yao, Y. (2009). Difference substitution matrices and the decision of nonnegativity of polynomials, *J. Sys. Sci. & Math. Scis.* **29**, pp. 1169–1177, (in Chinese).

Yang, L., Yao, Y. and Feng, Y. (2007). A kind of decision problems beyond Tarski's model, *Sci. China* **A 37**, pp. 513–522, (in Chinese).

Yang, L., Yu, W. and Yuan, R. (2010a). Mechanical decision for a class of integral inequalities, *Science China Information Sciences* **53**, pp. 1800–1815.

Yang, L., Zhan, N., Xia, B. and Zhou, C. (2005). Program verification by using discoverer, in *Proc. VSTTE'2005, LNCS 4171*, pp. 528–538.

Yang, L. and Zhang, J. (1991). Searching dependency between algebraic equations: an algorithm applied to automated reasoning, Technical Report ICTP/91/6, International Center for Theoretical Physics, Trieste, Italy.

Yang, L. and Zhang, J. (1994). Searching dependency between algebraic equations: an algorithm applied to automated reasoning, in *Artificial Intelligence in Mathematics* (Oxford University Press), pp. 147–156.

Yang, L. and Zhang, J. (2001). A practical program of automated proving for a class of geometric inequalities, in *Automated Deduction in Geometry*, LNAI 2061 (Springer-Verlag), pp. 41–57.

Yang, L., Zhang, J. and Hou, X. (1992). A criterion of dependency between algebraic equations and its applications, in W.-T. Wu and M.-D. Cheng (eds.), *Proc. International Workshop on Mathematics Mechanization 1992* (International Academic Publishers), pp. 110–134.

Yang, L., Zhang, J. and Hou, X. (1995). An efficient decomposition algorithm for geometry theorem proving without factorization, in *Proc. ASCM'1995* (Scientists Inc.), pp. 33–41.

Yang, L., Zhang, J. and Hou, X. (1996b). *Nonlinear Algebraic Equation System and Automated Theorem Proving* (Shanghai Scientific and Technological Education Publishing House), (in Chinese).

Yang, L., Zhou, C., Zhan, N. and Xia, B. (2010b). Recent advances in program verification through computer algebra, *Frontiers of Computer Science in China* **4**, pp. 1–16.

Yao, Y. (2009). Termination of the sequence of sds sets and machine decision for positive semi-definite forms, *arXiv:0904.4030v2*.

Yao, Y. (2010). Infinite product convergence of column stochastic mean matrix and machine decision for positive semi-definite forms, *Science China Mathematics* **40**, pp. 251–264, (in Chinese).

Zhang, T. (2004). *Isolating Real Roots of Nonlinear Polynomials*, Master's thesis, Peking University, Beijing.

Zhang, T., Xiao, R. and Xia, B. (2005). Real solution isolation based on interval krawczyk operator, in S.-I. Pae and H. Park (eds.), *Proc. ASCM'2005*, pp. 235–237.

Zhang, Z., Fang, T. and Xia, B. (2011). Real solution isolation with multiplicity of zero-dimensional systems, *Science China Information Science* **54**, pp. 60–69.

Index